Lecture Notes in Physics

Springer
Berlin
Heidelberg
New York
Barcelona
Hong Kong
London
Milan
Paris
Singapore
Tokyo

Physics and Astronomy ONLINE LIBRARY

http://www.springer.de/phys/

Editorial Policy

The series *Lecture Notes in Physics* (LNP), founded in 1969, reports new developments in physics research and teaching -- quickly, informally but with a high quality. Manuscripts to be considered for publication are topical volumes consisting of a limited number of contributions, carefully edited and closely related to each other. Each contribution should contain at least partly original and previously unpublished material, be written in a clear, pedagogical style and aimed at a broader readership, especially graduate students and nonspecialist researchers wishing to familiarize themselves with the topic concerned. For this reason, traditional proceedings cannot be considered for this series though volumes to appear in this series are often based on material presented at conferences, workshops and schools (in exceptional cases the original papers and/or those not included in the printed book may be added on an accompanying CD ROM, together with the abstracts of posters and other material suitable for publication, e.g. large tables, colour pictures, program codes, etc.).

Acceptance

A project can only be accepted tentatively for publication, by both the editorial board and the publisher, following thorough examination of the material submitted. The book proposal sent to the publisher should consist at least of a preliminary table of contents outlining the structure of the book together with abstracts of all contributions to be included.
Final acceptance is issued by the series editor in charge, in consultation with the publisher, only after receiving the complete manuscript. Final acceptance, possibly requiring minor corrections, usually follows the tentative acceptance unless the final manuscript differs significantly from expectations (project outline). In particular, the series editors are entitled to reject individual contributions if they do not meet the high quality standards of this series. The final manuscript must be camera-ready, and should include both an informative introduction and a sufficiently detailed subject index.

Contractual Aspects

Publication in LNP is free of charge. There is no formal contract, no royalties are paid, and no bulk orders are required, although special discounts are offered in this case. The volume editors receive jointly 30 free copies for their personal use and are entitled, as are the contributing authors, to purchase Springer books at a reduced rate. The publisher secures the copyright for each volume. As a rule, no reprints of individual contributions can be supplied.

Manuscript Submission

The manuscript in its final and approved version must be submitted in camera-ready form. The corresponding electronic source files are also required for the production process, in particular the online version. Technical assistance in compiling the final manuscript can be provided by the publisher's production editor(s), especially with regard to the publisher's own Latex macro package which has been specially designed for this series.

Online Version/ LNP Homepage

LNP homepage (list of available titles, aims and scope, editorial contacts etc.):
http://www.springer.de/phys/books/lnpp/

LNP online (abstracts, full-texts, subscriptions etc.):
http://link.springer.de/series/lnpp/

J. Bricmont D. Dürr M.C. Galavotti
G. Ghirardi F. Petruccione N. Zanghì (Eds.)

Chance in Physics

Foundations and Perspectives

Springer

Editors

Jean Bricmont
Fyma: Unité de physique theorique
et de physique mathematique
Université catholique de Louvain
Chemin du Cyclotron 2
1348 Louvain la Neuve, Belgium

Giancarlo Ghirardi
Department of Theoretical Physics
International Centre
for Theoretical Physics
Strada Costiera 11
34100 Trieste, Italy

Detlef Dürr
Mathematisches Institut
Universität München
Theresienstrasse 39
80333 München, Germany

Francesco Petruccione
Fakultät für Physik
Albert-Ludwigs-Universität
Hermann-Herder-Strasse 3
79104 Freiburg, Germany

Maria Carla Galavotti
Dipartimento di Filosofia
Via Zamboni 38
40126 Bologna, Italy

Nino Zanghì
Dipartimento di Fisica
Via Dodecanesco 33
16146 Genova, Italy

Cover-figure by D. Dürr, courtesy of WIR (www.webdesign-regensburg.de)

Library of Congress Cataloging-in-Publication Data applied for.

Die Deutsche Bibliothek - CIP-Einheitsaufnahme

Chance in physics : foundations and perspectives / J. Bricmont ... (ed.). - Berlin ; Heidelberg ; New York ; Barcelona ; Hong Kong ; London ; Milan ; Paris ; Singapore ; Tokyo : Springer, 2001
(Lecture notes in physics ; 574)
(Physics and astronomy online library)
ISBN 3-540-42056-8

ISSN 0075-8450
ISBN 3-540-42056-8 Springer-Verlag Berlin Heidelberg New York

Typesetting: Camera-ready by the authors/editors
Camera-data conversion by Steingraeber Satztechnik GmbH Heidelberg
Cover design: *design & production*, Heidelberg

Printed on acid-free paper
SPIN: 10837158 57/3141/du - 5 4 3 2 1 0

Preface

The conference *Chance in Physics: Foundations and Perspectives* was held from 29th November to 3rd December 1999 in Ischia, Italy. It was sponsored by the *Istituto Italiano Per Gli Studi Filosofici* in Naples, by the *Deutsche Forschungsgemeinschaft* (DFG), and by the *Società Italiana Di Fondamenti Della Fisica.* Sponsoring by the *International School for Advanced Studies* (ISAS) of Trieste, Italy, made the compilation of this volume possible; the funding by the *Istituto Italiano Per Gli Studi Filosofici* was crucial for the conference and is very gratefully acknowledged. The Istituto managed to provide a unique atmosphere for an interdisciplinary meeting, and these proceedings reflect indeed the very friendly but nevertheless intense and neverending discussions on one of the most debated issues of science: probability, and in particular probability in physics. We gratefully acknowlegde the organisational work as well as the editorial work done by our secretary of the meeting PhD student Roderich Tumulka.

The meeting was intended to stimulate renewed reflection on the fundamental and practical aspects of probability in physics, in particular the foundations of statistical sechanics, the probability in the foundations of quantum mechanics, the algebraic view of probability and the philosophy of probability in its interrelation with physics.

Questions like what probability is, or what it is about, or how probability enters physics are of a subtle kind. They are difficult in various ways, often mixed up with the enormous complexity and the inescapable lack of mathematical rigor in the physical application, or with the foundational problems of quantum mechanics, where the probabilistic ignorance concerning the values of certain physical quantities has even been elevated to a matter of principle. At present, the understanding of probability in physics is almost as personal as the understanding of quantum theory.

The aim of the conference was thus to focus on ideas about probability in physics, its meaning and its philosophical implications, by reviewing the different facets of probability in physics in its modern settings and by taking into account modern quantum theories without observers, where the origin of probability is not mystified by dogmatism.

The reviews were given in one-hour talks, and the discussions were held in the form of roundtables, where shorter contributions were also given.

The speakers were asked not to dilute the main themes of the conference with technicalities and to focus sharply on the issue of probability. This was taken to heart by all speakers and the meeting thus proved very successful. The contributions in this volume consequently focus on conceptual issues, and they make worthwhile reading for specialists in the field of foundations as well as for nonspecialists, because extensive technical prior knowledge is not required. The contributions have been left in the order they were discussed in the meeting, which proved to be a very natural one:

1. *Classical Statistical Mechanics*, where Boltzmann's understanding of statistical mechnanics as arising from kinetic gas theory is reviewed and put into modern perspectives, with an outlook on relativistic statistical mechanics. The relative lack of emphasis on the effect of chaotic behaviour on the foundations of probability is noteworthy.
2. *Quantum Mechanics*, where we review those *ontological* quantum theories, that have been most seriously discussed in the recent years. Among these are a deterministic theory (Bohmian mechanics) and both the intrinsically random theories of wavepacket reduction and the operator-based consistent (decoherent) histories. It starts with the "orthodox" view, again with emphasis on the probabilistic aspects of these theories.
3. *Chaotic systems*, where the dynamical aspects for the foundations of probability in physics are adressed.
4. *Philosophy of Probability*, where the issues of the earlier sections are further scrutinized on philosophical grounds. These contributions have no abstracts.

The book starts with an introductory paper, in which almost all the topics which are discussed by the later contributions are critically presented.

München,
January 2001

J. Bricmont, D. Dürr, M.C. Galavotti, G.C. Ghirardi, F. Petruccione, N. Zanghì

Table of Contents

List of Participants

Adami, Riccardo, Roma, Italy
adami@mat.uniroma1.it

Adams, Stefan, München, Germany
adams@rz.mathematik.uni-muenchen.de

Adler, Stephen, IAS Princeton, USA
adler@sns.ias.edu

Agnese, Angelo, Genova, Italy
agnese@ge.infn.it

Albert, David Z., New York, USA
da5@columbia.edu

Allori, Valia, Genova, Italy,
allori@mi.infn.it

Arntzenius, Frank, New Brunswick, USA
arntzeni@rci.rutgers.edu

Bacciagaluppi, Guido, Cambridge, UK
guido.bacciagaluppi@philosophy.oxford.ac.uk

Bassani, Franco, Pisa, Italy
bassani@sns.it

Bassi, Angelo, Trieste, Italy
bassi@ts.infn.it

Beltrametti, Enrico, Genova, Italy
beltrametti@ge.infn.it

Benci, Vieri, Pisa, Italy,
benci@dm.unipi.it

Ben-Menahem, Yemima, Jerusalem, Israel
msbenhy@mscc.huji.ac.il

Bergia, Silvio, Bologna, Italy
silvio.bergia@bo.infn.it

Bernardini, Carlo, Roma, Italy
carlo.bernardini@roma1.infn.it

Blanchard, Philipp, Bielefeld, Germany
blanchard@physik.uni-bielefeld.de

Booss-Bavnbek, Bernhelm, Roskilde, Denmark
booss@mmf.ruc.dk

Breuer, Heinz-Peter, Freiburg, Germany
breuer@physik.uni-freiburg.de

Bricmont, Jean, Louvain-la-Neuve, Belgium
bricmont@fyma.ucl.ac.be

Butterfield, Jeremy N., Cambridge, UK
jb56@cus.cam.ac.uk

Casati, Giulio, Como, Italy
casati@fis.unico.it

Castellani, Elena, Firenze, Italy
castella@philos.unifi.it

Cercignani, Carlo, Milano, Italy
carcer@mate.polimi.it

Clark, Peter, St. Andrews, UK
pjc@st-andrews.ac.uk

Corregi, Michele, Pisa, Italy
micorre@libero.it

Costantini, Domenico, Bologna, Italy
costanti@stat.unibo.it

Cufaro Petroni, Nicola, Bari, Italy
nicola.cufaro@ba.infn.it

Dalla Chiara, Marisa, Firenze, Italy
dachiara@risc.idg.fi.cnr.it

Dell'Antonio, Gianfausto, Roma, Italy
dellantonio@mat.uniroma1.it

Dewdney, Chris, Portsmouth, UK
chris.dewdney@port.ac.uk

Dinges, Herrmann, Frankfurt, Germany
dinges@mi.informatik.uni-frankfurt.de

Dorner, Uwe, Freiburg, Germany
dorner@physik.uni-freiburg.de

Dürr, Detlef, München, Germany
duerr@rz.mathematik.uni-muenchen.de

Festa, Roberto, Genova, Italy
festa@fisica.unige.it

Fronzoni, Leone, Pisa, Italy
fronzoni@mailbox.difi.unipi.it

Fuchs, Christopher, Los Alamos, USA
cfuchs@lanl.gov

Galavotti, Maria Carla, Bologna, Italy
galavott@philo.unibo.it

Galgani, Luigi, Milano, Italy
galgani@mi.infn.it

Garuccio, Augusto, Bari, Italy
garuccio@bari.infn.it

Gentili, Fausto, Bologna, Italy
gentili@bo.infn.it

Ghirardi, Gian Carlo, Trieste, Italy
ghirardi@trieste.infn.it

Giannerini, Simone, Bologna, Italy
giannerini@posta.statfac.unibo.it

Goldstein, Sheldon, New Brunswick, USA
oldstein@math.rutgers.edu

González Jiménez, Edgar, Bogotá, Colombia
edgar.gonzalez@jol.net.co

Gray, Lawrence, Minneapolis, USA
gray@math.umn.edu

Grigolini, Paolo, Denton, USA
gpaolo@unt.edu

Guerra, Francesco, Roma, Italy
guerra@romagtc.roma1.infn.it

Guttmann, Yair, Stanford, USA
guttmann@csli.stanford.edu

Gwirceman, Orit, New Brunswick, USA
orit@spacelab.net

Hemmo, Meir, Haifa, Israel
meir@research.haifa.ac.il

Horton, George, Portsmouth, UK
george.horton@port.ac.uk

Huggett, Nicholas, Chicago, USA
huggett@uic.edu

Jona-Lasinio, Giovanni, Roma, Italy
jona@roma1.infn.it

Kastner, Ruth, Maryland, USA
rkastner@wam.umd.edu

Kent, Adrian, Cambridge, UK
a.p.a.kent@damtp.cam.ac.uk

Khrennikov, Andrei, Växjö, Sweden
andrei.khrennikov@masda.vxu.se

Kiessling, Michael, New Brunswick, USA
miki@math.rutgers.edu

Knauf, Andreas, Erlangen, Germany
knauf@mi.uni-erlangen.de

Loewer, Barry, New Brunswick, USA
loewer@rci.rutgers.edu

Marinatto, Luca, Trieste, Italy
marinat@ts.infn.it

Maudlin, Tim, New Brunswick, USA
maudlin@rci.rutgers.edu

Morato, Laura, Verona, Italy
morato@sci.univr.it

Nelson, Edward, Princeton, USA
nelson@math.princeton.edu

Noja, Diego, Milano, Italy
noja@berlioz.mat.unimi.it

Omnès, Roland, Paris, France
roland.omnes@th.u-psud.fr

Panati, Gianluca, Trieste, Italy
panati@sissa.it

Penrose, Oliver, Edinburgh, UK
o.penrose@ma.hw.ac.uk

Peruzzi, Giulio, Padova, Italy
peruzzi@pd.infn.it

Petruccione, Francesco, Freiburg, Germany
petruccione@physik.uni-freiburg.de

Pizzo, Alessandro, Trieste, Italy
pizzo@sissa.it

Pohlmeyer, Klaus, Freiburg, Germany
trunk@physik.uni-freiburg.de

Poschadel, Norbert, Saarbrücken, Germany
norbert@math.uni-sb.de

Posilicano, Andrea, Como, Italy
posilicano@mat.unimi.it

Prestipino, Santi, Trieste, Italy
prestip@sissa.it

Redei, Miklos, Budapest, Hungary
redei@hps.elte.hu

Ridderbos, Katinka, Cambridge, UK
tmr23@cus.cam.ac.uk

Rimini, Alberto, Pavia, Italy
rimini@pv.infn.it

Robotti, Nadia, Genova, Italy
robotti@fisica.unige.it

Rosa, Rodolfo, Bologna, Italy
rosa@stat.unibo.it

Ruffo, Stefano, Firenze, Italy
ruffo@avanzi.de.unifi.it

Saunders, Simon, Oxford, UK
simon.saunders@linacre.oxford.ac.uk

Satzger, Helmut, München, Germany
helmut.satzger@physik.uni-muenchen.de

Shenker, Orly, Jerusalem, Israel
oshenker@cc.huji.ac.il

Sklar, Lawrence, Michigan, USA
lsklar@umich.edu

Spohn, Herbert, München, Germany
spohn@mathematik.tu-muenchen.de

Stokley, Martin, Portsmouth, UK
martin.stokley@port.ac.uk

Storck, Tim, München, Germany
tim.storck@physik.uni-muenchen.de

Suppes, Patrick, Stanford, USA
suppes@csli.stanford.edu

Teta, Alessandro, Roma, Italy
teta@mercurio.mat.uniroma1.it

Teufel, Stefan, München, Germany
teufel@mathematik.tu-muenchen.de

Tumulka, Roderich, München, Germany
tumulka@rz.mathematik.uni-muenchen.de

Uffink, Jos, Utrecht, The Netherlands
j.b.m.uffink@phys.uu.nl

Ugolini, Stefania, Verona, Italy
ugolini@sci.univr.it

Ullmann, Philipp, Frankfurt, Germany
ullmann@math.uni-frankfurt.de

Valentini, Antony, London, UK
antonyvalentini@hotmail.com

van Lith, Janneke, Utrecht, The Netherlands
j.h.vanlith@phys.uu.nl

Vulpiani, Angelo, Roma, Italy
angelo.vulpiani@roma1.infn.it

Yorke, James A., Maryland, USA
yorke@ipst.umd.edu

Zanghì, Nino, Genova, Italy
zanghi@ge.infn.it

Part I

Introduction

Bayes, Boltzmann and Bohm: Probabilities in Physics

Jean Bricmont

Physique Théorique, UCL, 2, chemin du Cyclotron, B-1348, Louvain-la-Neuve, Belgium

Abstract. In this introductory essay I shall make some remarks on the role of probabilities in physics, and discuss some concrete examples illustrating Boltzmann's explanation of approach to equilibrium.

1 Introduction

Boltzmann's explanation of approach to equilibrium, is very clearly discussed, in a general framework, by Sheldon Goldstein in his contribution to this volume [7]; so, I will not repeat it here. I will rather make some general comments about the notion of probability and its role in physics, thus trying to answer questions that are left open in the papers of Sheldon Goldstein and of Detlef Dürr [5] (see Sections 2 and 3). In particular, I will argue that a Bayesian view on probabilities is quite natural, and I will try to dispel various misconsceptions concerning that view.

I will then discuss simple models (in Sections 4 and 5), in order to illustrate how Boltzmann's ideas work, as well as to provide counterexamples to the rather widespread belief that, for irreversible behaviour to take place, the system must be ergodic or mixing. The first example, already discussed in [2], is a simple discrete model due to Mark Kac, the so-called ring model (Section 4). In that model, one can see exactly how the emergence of irreversible behaviour, on appropriate time scales, follows from the law of large numbers without appealing to notions like ergodicity. In the following Section, I will discuss approach to equilibrium in a mechanical model, but a very simple one, namely a set of uncoupled anharmonic oscillators. In fact, as will be seen in that Section, mixing does play a role in the explanation of irreversible behaviour, but not the one that is usually attributed to it.

In the last two sections, which will be more technical, I will discuss the conceptual significance of special probability distributions used in chaotic dynamical systems (the SRB measures) and in Bohmian mechanics (the $|\psi|^2$ distribution).

Some of the claims made below are conjectural or controversial. My goal is to raise some conceptual questions that have bothered me for some time, concerning the use of probabilities in physics, not necessarily to provide a satisfactory answer to them. Finally, the mathematical statements made here will be rather elementary and the proofs will not be given.

2 Two Notions of Probability

There are, traditionally, at least two different meanings given to the word 'probability' in the natural sciences. These two meanings are well-known but, since much confusion arises from the fact that the same word is used to denote two very different concepts, let me start by recalling them and explain how I see the connection between the two. First, I say 'in the natural sciences', because I do not want to discuss the purely mathematical notion of probability as measure (the notion introduced by Kolmogorov and others). This is of course an important branch of mathematics and the notion of probability that I will use, when formalized, will coincide with that mathematical notion, but I want to focus here on the role played by probabilities in our physical theories, which is not reducible to a purely mathematical concept (defined through a set of 'axioms').

So, the first notion that comes to mind is the so-called 'objective' or 'statistical' one, i.e. the view of probability as something like a 'theoretical frequency': if I say that the probability of the event E under condition X, Y, Z equals p, I mean that if I reproduce the conditions X, Y, Z sufficiently often, the event E will appear with frequency p. Of course, 'sufficiently often' is vague and this is the source of much criticism of that notion of probability. But, putting that objection aside for a moment and assuming that 'sufficiently often' can be given a precise meaning in concrete circumstances, probabilistic statements are, according to this view, factual statements that can be confirmed or refuted by observations or experiments.

By contrast, the 'subjective' or Bayesian use of the word probability refers to a form of reasoning and not (at least not directly) to a factual statement. Used in that sense, assigning a probability to an event expresses a (rational) judgment on the likelihood of that single event, based on the information available at that moment. Note that one is not interested here in what happens when one reproduces many times the 'same' event, as in the objective approach, but in the probability of a single event. This is of course very important in practice: when I wonder whether I need to take my umbrella because it will rain, or whether the stock market will crash next week, I am not mainly interested in the frequencies with which such events occur but with what will happen here and now; of course, these frequencies may be part of the information that is used in arriving at a rational judgment, but they are not typically the only information available.

How does one assign subjective probabilities to an event? In elementary textbooks, a probability is defined as the ratio between the number of favorable outcomes and the number of 'possible' ones. While the notion of favorable outcome is easy to define, the one of possible outcome is much harder. Indeed, for a Laplacean demon, nothing is uncertain and the only possible outcome is the actual one; hence, all probabilities are zeroes or ones. But *we* are not Laplacean demons[1] and it is here that ignorance enters. We try to reduce ourselves to a

[1] This was of course Laplace's main point in [12], although this is is often misunderstood. Laplace emphasized that human intelligence will forever remain 'infinitely distant' from the one of his demon.

series of cases about which we are 'equally ignorant', i.e. the information that we do have does not allow us to favour one case over the other, and that defines the number of 'possible' outcomes. The standard examples include the throwing of a dice or of a coin, where the counting is easy, but that situation is not typical.

At the time of Laplace, this method was called the 'principle of indifference'; its modern version is the maximum entropy principle. Here one assigns to each probability distribution $\boldsymbol{p} = (p_i)_{i=1}^N$ its Shannon (or Gibbs) entropy, given by:

$$S(\boldsymbol{p}) = -\sum_{i=1}^{N} p_i \ln p_i . \tag{1}$$

One then chooses the probability distribution that has the maximum entropy, among those that satisfy certain constraints that incorporate the information that we have about the system[2].

The rationale, like for the indifference principle, is not to introduce bias in our judgments, namely information that we do not have (like people who believe in lucky numbers). And one can reasonably argue that maximizing the Shannon entropy is indeed the best way to formalize that notion[3].

In practice, one starts by identifying a space of states in which the system under consideration can find itself and one assigns a prior distribution to it (maximizing entropy), which is then updated when new information becomes available[4].

Note that probabilistic statements, understood subjectively, are forms of reasoning, although not deductive ones[5]. Therefore, one cannot check them empirically. If someones says: Socrates is an angel; all angels are immortal; therefore Socrates is immortal, it is a valid (deductive) reasoning. Likewise, if I say that all I know about a coin is that it has two faces and that it looks symmetric, therefore the probability of 'head' is one half, it is a valid probabilistic reasoning; throwing the coin a thousand times with a result that is always tails does not disprove the reasoning; it only indicates that the initial assumption (of symmetry) was probably wrong (just as watching Socrates dead leads one to reconsider the notion that he is an angel or that the latter are immortal); the main point of Bayesianism is to give rules that allow to update one's probabilistic estimates, given previous observations.

Let me now discuss several objections or questions that are often raised about this 'subjective' view of probability which is more misunderstood and misrepresented than it ought to be.

1. *Commitment to subjectivism.* Some people think that a Bayesian view of probabilities presupposes of some form of subjectivism, meant as a doctrine

[2] In the simplest examples, the constraint is that the average of certain quantities (like the energy or the density) is fixed.

[3] See e.g. [3].

[4] For an exposition of how Bayesian updating is done, see e.g. [8–10].

[5] Indeed, they are a paradigmatic example of inductive reasoning. The calculus of probabilities is deductive, of course, but the Bayesian justification of the axioms of probability is not; nor is the choice of the prior distributions.

in philosophy or philosophy of science. To make matter worse, Bayesians sometimes talk as if all of science was about 'information' and never about facts or laws. Moreover, Bayesians often stress the idea that probabilities reflect our ignorance or quantify our ignorance and that makes some physicists uneasy: putting aside parapsychology, our knowledge or our ignorance do not play a causal role in the physical world; so, why should they enter in a fundamental way in our physical theories?

But there is no logical connection here: a subjectivist about probabilities may very well claim that there are objective facts in the world and laws governing them, and consider probabilities as being a tool used in situations where our knowledge of those facts and those laws is incomplete[6]. In fact, one could argue that, if there is any connection between Bayesianism and philosophical subjectivism, it goes in the opposite direction; a Bayesian should naturally think that one and only one among the 'possible' states is actually realized, and that there is a difference between what really happens in the world and what we know about it. But the philosophical subjectivist position often starts by *confusing* the world and our knowledge of it (for example, much of loose talk about everything being information often ignores the fact that 'information' is ultimately information about something which itself is not information). Thus, Bayesians should not be thought of as natural fellow-travellers of philosophical subjectivists.

Besides, ignorance does enter in the computations of probabilities but, as we will see in the next section, when we discuss the connection between probabilities and physics, this does not mean that either knowledge or ignorance are assumed to play a fundamental role in physics.

2. *Commitment to determinism.* At the other extreme, subjectivists are sometimes accused of holding a deterministic view of the world, which denies the possibility of free will or of intrinsic randomness (since all probabilities are 'subjective' or 'epistemic'). But, again, this is not necessary. A Bayesian simply tends to be agnostic concerning the issue of intrinsic randomness and will point out that it is difficult to find an argument showing the presence of intrinsic randomness in nature; indeed, it is well-known that some deterministic dynamical systems (the 'chaotic' ones) pass all the statistical tests that might indicate the presence of 'randomness'. So, how can we know, when we observe some irregular and unpredictable phenomenon, that this phenomenon is 'intrinsically random' rather than simply governed by unknown deterministic laws?[7].

---

[6] Which was the attitude adopted by Laplace in [12].

[7] The sad story of the so-called 'no hidden variable' theorems in quantum mechanics, that were widely believed to show the presence of intrinsic randomness in Nature, but that prove nothing of the sort, should strenghten the Bayesian skepticism (see [1] for a discussion of these theorems). Of course, in actual fact, the opposite happened: the widespread belief that quantum mechanics has shown that intrinsic randomness exists in nature is partly responsible for the loss of appeal of Bayesianism in the 20th century.

3. *Practical problems.* Another problem is that it is often hard to assign unambiguously a subjective probability to an event. It is easy, of course, for coin tossing or similar experiments where there are finitely many possible outcomes, which, moreover, are related by symmetry. In general, one may use maximum entropy principles, but then, one may encounter various problems: how to choose the right set of variables, how to assign an a priori distribution on those, corresponding to maximal ignorance, how to incorporate the "knowledge that we have". Many people consider such problems as objections to the subjectivist approach; but that misses the point of what this approach tries to achieve: finding the best thing to do in a bad situation, namely one where we are ignorant. Of course, the greater the ignorance, the worse the situation. To take an extreme case, we may not even realize that we are ignorant; a fortiori, there are situations where we do not know the right variables or how to put a natural prior probability distribution on them. But, so what? Bayesianism does not promise miracles; the Bayesian answer to that sort of criticism is to say: what else do you suggest to do (apart from giving up) ?
4. *(Ir)relevance to physics.* Yet another objection is: what is the physical meaning of probability one half for a single event? Bayesian thinking may be useful in bets or in various practical situations where decisions have to be made, but what role does that have in physical theories, which are supposed to describe the world as it is and not to deal with problems of practical rationality? I'll try to answer that last question in the next section.

3 Typicality, Scientific Explanations, and Statistical Mechanics

A way to make a connection between the two views on probability goes through the law of large numbers: the calculus of probabilities – viewed now as part of deductive reasoning – leads one to ascribe subjective probabilities close to one for certain events that are precisely those that the objective approach deals with, namely the frequencies with which some events occur, when we repeat many times the 'same' experiment[8]. So, rather than opposing the two views, one should carefully distinguish them, but regard the objective one as, in a sense, a special case of the subjective one (i.e. when the law of large numbers leads to subjective probabilities sufficiently close to one).

Let me now state the law of large numbers, using a terminology that will be useful when I turn to statistical mechanics below. Consider the simple example of coin flipping. Let 0 denote 'head' and 1, 'tail'. The 'space' of results of any single flip, $\{0,1\}$, will be called the 'physical space' while the space of all possible results of N flips, $\{0,1\}^N$, will be called the 'phase space'. The variables N_0, N_1

[8] Of course, when we say that we repeat the 'same' experiment, or that the results of different experiments are 'independent' of each other, we also try to quantify the knowledge that we have, i.e. that we do not see any differences or any causal connections between those experiments.

that count the number of heads (0) or tails (1) will be called macroscopic. Here we introduce an essential distinction between the macroscopic variables, or the macrostate, and the microstate. The microstate, for N flips, is the sequence of results for all the flips, while the macrostate simply specifies the values of N_0 and N_1. Although this example is trivial, let us draw the following analogy with statistical mechanics: N_0 and N_1 count, for a given point in phase space, corresponding to many 'particles' (i.e. to many flips), the number of 'particles' that belong to a given subset (0 or 1) of physical space.

Now, fix $\varepsilon > 0$ and call a configuration such that $|\frac{N_0}{N} - \frac{1}{2}| \leq \varepsilon$ *typical*, for that given ε, and *atypical* otherwise. Then, (a weak form of) the law of large numbers states that $\forall \varepsilon > 0$, $P(T(\varepsilon)) \rightarrow 1$ as $N \rightarrow \infty$ where $T(\varepsilon)$ is the set of typical configurations and P the product measure that assigns independent probabilities $\frac{1}{2}$ to each outcome of each flip. A more intuituive way to say the same thing is that, if we simply count the number of microstates that are typical, we find that they form a fraction of the total number of microstates close to 1, for N large.

Here we may already discuss the connection between probabilities and scientific explanations, the latter notion being obviously hard to state precisely[9]. A first form of scientific explanation is given by *laws*. If state A produces state B, then the occurence of B can be explained by the occurrence of A. If A is prepared in the laboratory, this kind of explanation is rather satisfactory. Of course, if B is some natural phenomena, then A itself has to be explained, and that leads to a potentially infinite regress[10]. But, in many situations, we do not have strict laws, e.g. in coin tossing, and thus we have to see what role probabilities play in our notion of explanation. Observe first that, if we toss a coin many times and we find approximately half heads and half tails, we do not feel that there is anything special to be explained. If, however, the result deviates strongly from that average, we'll look for an explanation (e.g. by saying that the coin is biased). This leads me to make the following suggestion; first, as discussed above, probabilities enter situations where our knowledge is incomplete and Bayesian methods allow us to make the most rational predictions in those situations. Now, suppose we want to explain some phenomenon when our knowledge *of the past* is such that this phenomenon could not have been predicted with certainty. I will say that our knowledge, although partial, is *sufficient* to explain that phenomenon if we would have predicted it using Bayesian computations and the information we had about the past. That notion of 'explanation' incorporates, of course, as a special case, the notion of explanation based on laws. Also, it fits with our intuition concerning the coin-tossing situation discussed above: being ignorant of any properties of the coin leads us to predict a fraction of heads or tails around

[9] One sometimes even hears that the goal of science is to describe phenomena or to predict them, not to explain them. But that view is difficult to maintain: human being demand explanations of various events, and if science does not give them, what will?

[10] Going back to the beginning of the universe, as discussed in Goldstein's contribution to this volume.

one-half. Hence, such a result is not surprising or, in other words, does not "need to be explained", while a deviation from it requires an explanation.

Turning to physics, consider for example the Maxwellian distribution of velocities, for a gas of N particles of mass m. Let $\Delta(\boldsymbol{u})$, be the cubic cell of size δ^3 centered around $\boldsymbol{u} \in (\delta\mathbb{Z})^3$. Let $\boldsymbol{v} = (\boldsymbol{v}_1, \dots, \boldsymbol{v}_N) \in \mathbb{R}^{3N}$ be an element of the 'phase space' of the system (where the coordinate part of the space are ignored), i.e. a configuration of velocities for all the particles.

Define the macrostate by the variables
$N_{\boldsymbol{u}}(\boldsymbol{v}) = |\{\alpha | \boldsymbol{v}_\alpha \in \Delta(\boldsymbol{u}), \alpha = 1, \dots, N\}|$.
Let $T(\varepsilon, \delta)$ be the set of typical points for ε, δ, namely those $\boldsymbol{v}$'s for which

$$\left| \frac{N_{\boldsymbol{u}}(\boldsymbol{v})}{N} - \frac{\exp(-\frac{\beta m |\boldsymbol{u}|^2}{2})}{(2\pi m k T)^{3/2}} \right| \leq \varepsilon. \tag{2}$$

for all cells $\Delta(\boldsymbol{u})$. $N_{\boldsymbol{u}}(\boldsymbol{v})$ is called the *empirical distribution* corresponding to the phase space point $\boldsymbol{v}$.

Let us consider the uniform measure on a constant energy surface of energy E (i.e. the restriction of the Lebesgue measure to that surface). Then a variant of the law of large numbers shows that the set of typical points on that surface has, for N large, a measure close to one (when T in (2) is related to E by $kT = \frac{2E}{3N}$).

Now, let us ask: how does one explain the occurrence of a Maxwellian distribution? The Bayesian answer is basically that there is nothing to explain, because this is analogous to the situation of coin flippings where the fractions of heads and tails are close to one half. Given that we know that the energy is conserved, symmetry considerations show that the uniform measure is the most natural one and, since the Maxwellian distribution is the empirical distribution corresponding to most phase points (relative to that measure), it is exactly what we would expect if we know nothing more about the system[11]. In fact, the only thing that would lead us *not* to predict the Maxwellian distribution would be some additional knowledge about the system (e.g. that there are some constraints or some external forces acting on it).

This answers the often heard question: "how does one justify the choice of the equilibrium measure?"; answering that question can vary from being trivial[12] (it is the natural choice on Bayesian grounds), to being impossible, if some other justification – for example, a dynamical justification that would not itself use probabilistic considerations – is demanded by those who ask the question.

[11] I am leaving aside here the problem of the choice of variables on which to put a uniform measure. I would argue that the natural variables are those (the positions and momenta) that characterize the state of the system or, in other words, the simplest ones that determine uniquely a solution to the equations of motion. Note that this is the only place in the reasoning where dynamical considerations enter.

[12] From the point of view defended here, what is not trivial are the mathematical consequences of that assumption, like (2) and all other such relations derived in equilibrium statistical mechanics. But the basic probabilistic assumption does not need a "deeper" justification.

However, one can ask a related question, which is less trivial: how does one explain the approach to equilibrium for a closed system which starts in an atypical (i.e. a nonequilibrium) configuration? This is the question that Boltzmann's analysis answers and is the topic of the next section.

4 A Simple Example: The Kac Ring Model

I will now analyze a simple model, due to Mark Kac ([11] p.99, see also Thompson ([13] p.23)), which nicely illustrates Boltzmann's solution to the problem of irreversibility, and shows how to avoid various misunderstandings and paradoxes.

I shall describe a slightly modified version of the model and state the relevant results, referring to [11] for the proofs (all the quotes in this section come from [11]).

"On a circle we consider n equidistant points"; m of the intervals between the points are marked and form a set called S. The complementary set (of $n-m$ intervals) will be called $\bar{S}$.

"Each of the n points is a site of a ball which can be either white (w) or black (b). During an elementary time interval each ball moves counterclockwise to the nearest site with the following proviso".

If the ball crosses an interval in S, it changes color upon completing the move but if it crosses an interval in $\bar{S}$, it performs the move without changing color.

"Suppose that we start with all white balls; the question is what happens after a large number of moves" (below, after (6), we shall also consider other initial conditions).

Here is the analogy with mechanical laws: the balls are described by their positions and their (discrete) "velocity", namely their color. One of the simplifying features of the model is that the "velocity" does not affect the motion. The only reason I call it a "velocity" is that it changes when the ball collides with a fixed "scatterer", i.e. an interval in S. Scattering with fixed objects tends to be easier to analyze than collisions between moving particles. The "equations of motion" are given by the counterclockwise motion, plus the changing of colors (see eqs (7, 8) below). These equations are obviously deterministic and reversible: if after a time t, we change the orientation of the motion from counterclockwise to clockwise, we return after t steps to the original state[13]. Moreover, the motion is strictly periodic: after $2n$ steps each interval has been crossed twice by each ball, hence they all come back to their original color. This is analogous to the Poincaré cycles that occur in mechanical systems[14], except that, here, the length of the cycle is the same for all configurations (there is no reason for this to hold in general mechanical systems). Finally, it is easy to find special configurations

[13] There is a small abuse here, because I seem to change the laws of motion by changing the orientation. But I can attach another discrete "velocity" parameter to the particles, having the same value for all of them, and indicating the orientation, clockwise or counterclockwise, of their motion. Then, the motion is truly reversible.

[14] For an isolated mechanical system, most points in any neighbourhood of a given point in phase space have a trajectory that will visit infinitely often that neighbourhood.

which obviously do not tend to equilibrium, because they do not tend to anything: start with all white balls and let every other interval belong to S (with $m = \frac{n}{2}$). Then, after two steps, all balls are black, after four steps they are all white again, etc... The motion is periodic with period 4.

Before dealing with those apparent paradoxes, note that, from a Bayesian viewpoint, one will assign a probability distribution that is uniform over configurations (of balls and of scatterers) compatible with whatever information we have. Thus, if we do not know anything, we assign a uniform distribution over all configurations. For N large, this distribution becomes concentrated on those configurations for which the number of black and of white balls are more or less equal, and this defines equilibrium.

What one would like to show is that, if we know initially that the number of white and black balls are different, and if we therefore assign a uniform distribution over the configurations compatible with that information, then, if one starts with a "typical" configurations relative to that distribution, the dynamics will evolve it towards an equilibrium one.

This is what the Boltzmann approach does; I will start with the analogue of Boltzmann's equation, and then turn to his more general ideas.

Analog of the Classical Solution of Boltzmann.

Let us first introduce the macrostate in this system which is defined by $N_w(t)$ i.e. by the total number of white balls at time t (i.e., after t moves; t being an integer). Of course, we could also take the total number of black balls $N_b(t) = N - N_w(t)$. On the other hand, let $N_w(S;t)$ ($N_b(S;t)$) be the number of white (black) balls which are going to cross an interval in S at time t.

"We have the immediate conservation relations:

$$\begin{aligned} N_w(t+1) &= N_w(t) - N_w(S;t) + N_b(S;t) \\ N_b(t+1) &= N_b(t) - N_b(S;t) + N_w(S;t) \end{aligned} \tag{3}$$

Now to follow Boltzmann, we introduce the assumption ("Stosszahlansatz" or "hypothesis of molecular chaos"[15])

$$\begin{aligned} N_w(S;t) &= mn^{-1}N_w(t) \\ N_b(S;t) &= mn^{-1}N_b(t)\text{''} \end{aligned} \tag{4}$$

Of course, if we want to solve (3) in a simple way, we have to make some assumption about $N_w(S;t)$, $N_b(S;t)$. Otherwise, one has to write equations for $N_w(S;t)$, $N_b(S;t)$ that will involve new variables and lead to an infinite regress.

The intuitive justification for this assumption is that each ball is "uncorrelated" with the event "the interval ahead of the ball belongs to S", so we write $N_w(S;t)$ as equal to $N_w(t)$, the total number of white balls, times the density $\frac{n}{m}$ of intervals in S. This assumption looks completely reasonable. However, upon

[15] The word "chaos" here has nothing to do with "chaos theory", and, of course, Boltzmann's hypothesis is much older than that theory.

reflection, it may lead to some puzzlement (just as the hypothesis of "molecular chaos" does): what does "uncorrelated" exactly mean? Why do we introduce a statistical assumption in a mechanical model? Fortunately here, these questions can be answered precisely and we shall answer them later. But let us first return to the Boltzmannian story.

"One obtains

$$N_w(t+1) - N_b(t+1) = (1 - 2mn^{-1})(N_w(t) - N_b(t))$$

Thus

$$\begin{aligned} n^{-1}[N_w t) - N_b(t)] &= (1 - 2mn^{-1})^t n^{-1}[N_w(0) - N_b(0)] \\ &= (1 - 2mn^{-1})^t \end{aligned} \tag{5}$$

and hence if

$$2m < n \tag{6}$$

(as we shall assume in the sequel) we obtain a *monotonic* approach to equipartition of white and black balls." Note that we get a monotonic approach for *all* initial conditions $(N_w(0) - N_b(0))$ of the balls.

The general approach of Boltzmann to irreversibility.

We can see here that Boltzmann's equation can at best be an approximation. The assumption (4) cannot hold for all times and for all configurations, because it would contradict the reversibility and the periodicity of the motion. In order to better understand what goes on, we need to reexamine the model at the microscopic level, first mechanically and then statistically. For each $i = 1, \cdots, n$, we introduce the variable

$$\epsilon_i = \begin{cases} +1 \text{ if the interval in front of } i \in \bar{S} \\ -1 \text{ if the interval in front of } i \in S \end{cases}$$

and we let

$$\eta_i(t) = \begin{cases} +1 \text{ if the ball at site } i \text{ at time } t \text{ is white} \\ -1 \text{ if the ball at site } i \text{ at time } t \text{ is black} \end{cases}$$

Then, we get the "equations of motion"

$$\eta_i(t) = \eta_{i-1}(t-1)\epsilon_{i-1} \tag{7}$$

whose solution is

$$\eta_i(t) = \eta_{i-t}(0)\epsilon_{i-1}\epsilon_{i-2}\cdots\epsilon_{i-t} \tag{8}$$

(where the subtractions in the indices are done modulo n). So we have an explicit solution of the equations of motion at the microscopic level.

We can express the macroscopic variables in terms of that solution:

$$N_w(t) - N_b(t) = \sum_{i=1}^{n} \eta_i(t) = \sum_{i=1}^{n} \eta_{i-t}(0)\epsilon_{i-1}\epsilon_{i-2}\cdots\epsilon_{i-t} \tag{9}$$

and we want to compute $n^{-1}(N_w(t) - N_b(t))$ for large n, for various choices of initial conditions $(\{\eta_i(0)\})$ and various sets S (determining the ϵ_i's). It is here that "statistical" assumptions enter. Namely, we fix an arbitrary initial condition $(\{\eta_i(0)\})$ and consider all possible sets S with $m = \mu n$ fixed (one can of course think of the choice of S as being part of the choice of initial conditions). Then, for each set S, one computes the "curve" $n^{-1}(N_w(t) - N_b(t))$ as a function of time. The result of the computation, done in [11], is that, for any given t and for n large compared to t, the overwhelming majority of these curves will approach $(1 - 2\frac{m}{n})^t = (1 - 2\mu)^t$, i.e. what is predicted by (3). (to fix ideas, Kac suggests to think of n as being of the order 10^{23} and t of order 10^6). The fraction of all curves that will deviate significantly from $(1 - 2\mu)^t$, for fixed t, goes to zero as $n^{-\frac{1}{2}}$, when $n \to \infty$.

Of course when I say "compute" I should rather say that one makes an estimate of the fraction of "exceptional" curves deviating from $(1 - 2\mu)^t$ at a fixed t. This estimate is similar to the law of large number and (7) is indeed of the form of a sum of (almost independent) variables (remembering that n is large compared to t).

I do not want to overemphasize the interest of this model. It has many simplifying features (for example, there is no conservation of momentum because the scatterers here are "fixed"). However, it has *all* the properties that have been invoked to show that mechanical systems cannot behave irreversibly, and therefore it is a perfect counterexample that allows us to refute all those arguments (and to understand exactly what is wrong with them): it is isolated (the balls plus the scatterers), deterministic, reversible, has Poincaré cycles and is not ergodic (see Remark 2 below).

This result, obtained in the Kac model, is exactly what one would like to show for general mechanical systems, in order to establish irreversibility. It is obvious why this is very hard. In general, one does not have an explicit solution (for an n-body system !) such as (7, 8), in terms of which the macroscopic variables can be expressed, see (9). It is also clear in this example what is exactly the status of our "ignorance". If we prepare the system many times and if the only variables that we can control are n and m, then we indeed expect to see the irreversible behaviour obtained above, simply because this is what happens *deterministically* for the *vast majority* of microscopic initial conditions corresponding to the macroscopic variables that we are able to control (i.e. for what we called a typical initial condition above). Note that we may, if we wish, say that we "ignore" the initial conditions, but there is nothing "subjective" here. Simply our ignorance does not prevent us from making, on Bayesian grounds, the right prediction and thus we say that the phenomenon of approach to equilibrium is explained.

I will conclude this section with some more technical remarks about this model.

Remarks

1. The Poincaré recurrence and the reversibility "paradoxes" are easy to understand: each curve studied is periodic of period $2n$. So that, if we did not

fix t and let $n \to \infty$, we would not observe "irreversible" behaviour. But this limit is physically correct. The recurrence time (n) is enormous compared to any physically accessible time. As for the reversibility objection, let us consider as initial condition a reversed configuration after time t. Then we know that, for that configuration and *that set* S, $n^{-1}(N_w(t) - N_b(t))$ will not be close to $(1-2\mu)^t$ at time t (since it will be back to its initial value 1). But all we are saying is that, for the vast majority of S's this limiting behaviour will be seen. For the reversed configuration, the original set S happens to be exceptional. But, for that configuration, it is still the case that the vast majority of sets S will lead to a typical behaviour. The same remark holds for the configuration with period 4 mentioned in the beginning.
Note also that, if we consider the set of configurations for which $n^{-1}(N_w(t) - N_b(t))$ is close to $(1-2\mu)^t$ for *all times*, then this set is empty, because of periodicity.

2. This model, although perfectly "irreversible", is not ergodic! Indeed, since it is periodic, no trajectory can "visit" more than $2n$ microscopic configurations. But the "phase space" contains 2^n configurations (two possibilities -black or white- at each site). So, only a very small fraction of the phase space is visited by a trajectory. This nicely illustrates the fact that ergodicity is not necessary for irreversibility. What is used here is only the fact that the vast majority of configurations give to the macroscopic variables a value close to their equilibrium one (as emphasized in [7]).

5 A Mechanical Example

In order to illustrate Boltzmann's ideas in a truly mechanical example, as well as to indicate the (limited) role that dynamical properties such as mixing plays in those ideas, consider the following simple system[16]: N uncoupled anharmonic oscillators of identical mass, whose Hamiltonian is

$$H = \sum_{i=1}^{N} \frac{p_i^2}{2m} + V(q_i) \tag{10}$$

where

$$V(q) = \frac{q^2}{2} + \lambda q^4 \tag{11}$$

with $\lambda > 0$.

Each such oscillator moves in the $\mathbb{R}^2$ plane along a closed constant energy curve determined by the initial conditions $(q(0), p(0))$. The frequency of the motion on each such curve depends on the energy, hence on the initial conditions (except for $\lambda = 0$ in (11), i.e. except for the harmonic oscillator whose oscillations are isochronous).

[16] This example was suggested to me by Pierre Collet.

Let us use the coordinates[17] given by the energy E and the angle φ. The equations of motion give $E =$ constant, $\varphi(t) = \varphi(0) + \omega(E)t$.

On Bayesian grounds, the equilibrium distribution is the microcanonical one, i.e. the uniform distribution on a constant energy surface, $\delta(H-E)\prod_{i=1}^{N} \mathrm{d}p_i \mathrm{d}q_i$, assuming that we know the energy of the system, which is the only quantity which is easy to know, in general (if we knew only the average energy, we would use the canonical distribution, or more general distribution if we didn't even know that).

To understand how convergence to equilibrium occurs, let ϱ be the density of an absolutely continuous measure on $\mathbb{R}^2$, and take initial data at "random" with respect to that measure and independently of each other. In other words, we consider a point $\underline{x}(0) \in \mathbb{R}^{2N}, \underline{x}(0) = (E_1(0), \varphi_1(0), \ldots, E_N(0), \varphi_N(0))$ which is 'typical' with respect to the measure

$$\mu = \prod_{i=1}^{N} \varrho(E_i, \varphi_i)\, \mathrm{d}E_i\, \mathrm{d}\varphi_i,$$

or, in other words, we are interested in properties that hold for all points in a set whose μ-measure is close to one. This measure might be the maximum entropy measure corresponding to some information about the initial state of the sytem, e.g. the average distribution of the angles (but we will not use this assumption). Finally consider a macroscopic quantity of the form

$$F(\underline{\varphi}) = \frac{1}{N} \sum_{i=1}^{N} f(\varphi_i) \tag{12}$$

with, say, f smooth.

Letting $\underline{\varphi}(t) = (\varphi_i(0) + \omega(E_i)t)_{i=1}^{N}$, it is easy to show that, in some sense which will be made precise soon,

$$F(\underline{\varphi}(t)) \text{ ``}\rightarrow\text{'' } \int_0^{2\pi} f(\varphi), d\varphi \tag{13}$$

and the RHS is just the equilibrium value of F, meaning the value obtained by integrating $F(\underline{\varphi})$ with the microcanonical measure on $\mathbb{R}^{2N}$ $\delta(H-E)\prod_{i=1}^{N} \mathrm{d}p_i \mathrm{d}q_i$ (as can be seen by changing variables to E_i, φ_i). To see where (13) comes from, and its precise meaning, apply the law of large numbers to show that, for N large,

$$\begin{aligned} F(\underline{\varphi}(t)) &\simeq \int F(\underline{\varphi}(t)) \prod_{i=1}^{N} \rho(\varphi_i, E_i) \mathrm{d}\varphi_i \mathrm{d}E_i \\ &= \int f(\varphi + \omega(E)t) \varrho(\varphi, E) \mathrm{d}\varphi \mathrm{d}E. \end{aligned} \tag{14}$$

[17] The action-angle variables (I, φ) are more conventional in mechanics but they are related to the coordinates used here by $\frac{\mathrm{d}E}{\mathrm{d}I} = \omega$ where $\omega = \omega(E)$ is the angular frequency of the motion of energy E.

for typical's $\varphi(t)$. Now, expanding f and ϱ into Fourier series and using the fact that $\frac{\mathrm{d}\omega(E)}{\mathrm{d}E} \neq 0$ for $\lambda \neq 0$ in (11), one can show that the RHS of (14) tends to $\int f(\varphi)\mathrm{d}\varphi$ as $t \to \infty$ (by assumption, $\int \varrho(\varphi, E)\mathrm{d}\varphi\mathrm{d}E = 1$). So, we have again, at least for quantities of the form (12), "convergence to equilibrium".

Note however that, for N fixed, the motion of $F(\underline{\varphi}(t))$ is quasiperiodic and so (13) does not really hold for $t \to \infty$ but only for t large but not too large compared to N. This is because, although the RHS of (14) converges to the RHS of (13), $F(\underline{\varphi}(t))$ is close to the RHS of (14) only for N large. To be mathematically precise, one should, as in the Kac model, first take $N \to \infty$ and then $t \to \infty$. The lack of convergence for fixed N is similar to the Poincaré recurrences.

The dynamical system on a constant energy surface in $\mathbb{R}^{2N}$ is not ergodic and, a fortiori, not mixing: the energy of each oscillator is conserved. Accordingly, a "convergence to equilibrium" like (13) would not hold, in general, if f in (12) depended also on E. This is of course a drawback of this model, which is due to its extreme simplicity.

There is nevertheless a sense in which "mixing" holds here and is relevant: in order for (14) to converge to $\int f(\varphi)\mathrm{d}\varphi$, we need to use the fact that ω depends on E (in the case of the harmonic oscillator, where this property does not hold, $F(\underline{\varphi}(t))$ will oscillate periodically and will not satisfy (13)). To see the role that this dependence plays, let ρ be a uniform density on a disk D in $\mathbb{R}^2$, of radius a, centered around $(q, 0)$ with $q > a$. Consider the annulus-like region A bounded by the constant energy curves $E_- = V(q-a)$ and $E_+ = V(q+a)$. Since the disk D is foliated by constant energy curves $E \in [E_-, E_+]$ and since, on each curve, the angular speed $\omega(E)$ is different, one can see that, under the time evolution, D will evolve into a elongated disk $D(t)$ which, while remaining of constant size (by Liouville's theorem), will 'fill'[18] the region A. This is precisely analogous to the kind of mixing *in physical space* (here, $\mathbb{R}^2$) like the ink drop in a glass of water considered by Gibbs (see [7]). But this is quite different from mixing *in phase space* ($\mathbb{R}^{2N}$ here) that does not take place in this system and that is not relevant for the "approach to equilibrium". The confusion between these two sorts of mixing properties is rather widespread.

Let me conclude with a general suggestion, based on this example, which however I do not know how to formulate precisely: namely, that a form of mixing is important for approcah to equilibrium to take place (after all, for the harmonic oscillator we have neither approach to equilibrium nor any form of mixing), but only in some kind of reduced phase space ($\mathbb{R}^2$ here), determined by the macroscopic variables.

6 Probabilities and Chaotic Dynamical Systems

In this section and in the next one, I will consider the 'naturalness', from a Bayesian viewpoint, of certain measures that enter either in classical dynamical systems or in quantum (Bohmian) mechanics.

[18] In the sense that $Area(A \cap D(t))$ will converge to $Area(A)Area(D)$, as $t \to \infty$, for measurable sets A.

Let me start with the Sinai-Ruelle-Bowen (SRB) measures in chaotic dynamical systems (see e.g. [6]). They are rather special objects, which are often singular (continuous) relative to the Lebesgue measure and that would probably never have been considered if we had not studied dynamical systems. How should one justify their physical relevance? One might invoke their stationarity. But they are far from unique in that respect. Not only do we have, in many cases, infinitely many stationary (discrete) measures concentrated on periodic orbits, we also have infinitely many stationary singular continuous measures[19]. There is, however, one property that singles out the SRB measures : they are the forward images of the Lebesgue measure (or, at least, of the restriction of that measure to the basin of attraction of the support of the SRB measure). This means that, for a typical initial condition x_0 in the basin of attraction and for a measurable set B,

$$\lim_{N\to\infty} \frac{1}{N} \sum_{n=0}^{N-1} \chi(f^n(x_0) \in B) = \int_B \mathrm{d}\mu(x), \tag{15}$$

where f is the map defining the dynamical system, μ is the SRB measure and $\chi(E)$ is the indicator function of the event E. Here 'typical' means "except for a set of points of zero Lebesgue measure".

Note that these chaotic dynamical systems may have few degrees of freedom and thus the usual distinction between microscopic and macroscopic quantities cannot be formulated. However, one may argue that the role of the macroscopic quantities is played here by frequencies of visits of the trajectory of a typical point to measurable sets and those frequencies are simply given by the SRB measure of that set. Indeed, these frequencies are reproducible quantities in these systems (unlike individual trajectories that are not reproducible because of chaoticity) and they can be observed in computer or real experiments.

Anyway, the point that I want to emphasize here is that the *initial* measure, the one with respect to which typicality of the initial conditions is defined, is the Lebesgue measure which, in general, is not the SRB measure and is not even stationary. And, what recommand this measure as a natural measure of typicality is a form of the principle of indifference or of maximum entropy[20].

---

[19] When there exists a Markov partition, the map defining the dynamical system can be conjugated to the shift on a space of symbol sequences. Then, any translation (or shift) invariant measure on that space (e.g. any translation invariant Gibbs state) will give rise to a stationary measure for the original dynamical system when it is transported back to the phase space of that system. And such stationary measures can be (singular) continuous.

[20] If we perform computer simulations, the behaviour of trajectories with "random" initial points will be the one described by the SRB measure, which can be understood because the 'randomness' of the initial point, i.e. our ignorance about it, leads us again to consider the Lebesgue measure, because it is uniform over initial configurations.

7 Bohmian Mechanics and Quantum Equilibrium

In this section, I will make some remarks on the naturalness of the "quantum equilibrium" hypothesis, i.e. of the choice of an initial distribution of particle positions coinciding with the square of the absolute value of the wave function (see [5]). I will assume some familiarity with Bohmian mechanics as well as with the idea of quantum equilibrium. The authors of [4] want to justify this choice by invoking its equivariance, which they view as a natural generalization of stationarity, the latter property being offered as a justification of the Lebesgue measure, i.e. of the microcanonical ensemble, in statistical mechanics. As explained in section 3, it seems to me that the use of the Lebesgue measure in statistical mechanics should and could be justified on Bayesian grounds, rather than invoking stationarity. But, before discussing further this argument, let me emphasize that, whatever disagreement I may have with that particular justification, I do not regard it in any way as a criticism of Bohm's theory itself. Indeed, I consider that, at present, Bohm's theory is, by far, the most satisfactory theory that we have of quantum phenomena and that most objections that are raised against it are either totally misguided or simply reflect the fact that this theory is basically the only one which is sufficiently well formulated so that objections to it can be raised[21]. This preliminary remark should not be necessary and is only made because Bohm's has been unfairly criticized and misrepresented to an extent that is truly extraordinary, and totally unusual in the history of science (even more than Bayesian ideas).

Coming back to the justification of the probability distribution on initial conditions, here are some problems, of various degree of importance, that I see in using stationarity as a naturalness condition for a measure on initial conditions.

1. First, a familiar way to think of stationary measures[22] is as being those that describe the large time asymptotics of trajectories; but why should that imply that those measures are natural when applied to *initial conditions*? Moreover, there is vague sense in which this "begs the question" because it assumes that stationarity (or equilibrium) has been reached from the start.
2. The notion of stationarity is too sophisticated mathematically; that, usually is not a problem, but remember that here we try to nail down what we intuitively mean by scientific explanation. And a mathematical notion like stationarity is far away from any intuition that we may have about explanation.

[21] Here, I realize that I am slightly unfair to spontaneous collapse theories, of the GRW type; but what I have in mind is e.g. the widespread belief that orthodox, Copenhagen-like, quantum theories have no problem incorporating special relativity, while Bohmian mechanics does. Indeed, whatever the Copenhagen interpretation really means, it must somewhere introduce a collapse or reduction of the state vector or an intervention of the observer or some – usually verbal – substitute for that. And, to my knowledge, the Lorentz invariance of that part of the "theory" is never discussed. And, if it was, problems caused by EPR-type experiments, that are the source of the difficulty in obtaining a Lorentz invariant Bohmian theory, would soon become evident.

[22] Which, however, is not used in [4].

Of course, maximum entropy principles and the like are also sophisticated, in a sense. But they are much closer to our everyday intuition; in fact, it is likely that even animals make Bayesian 'computations' before taking 'decisions'. There is a difference between maximum entropy principles, which try to formalize our pre-scientific intuitions and stationarity, which, it seems to me, can only appeal to the mathematically trained scientist.

3. If we look at the second picture in Figure 1 of [7], and consider *that configuration* as our initial conditions, we could regard it as being 'typical' for a certain distribution, and use that distribution to infer the future behaviour of the system (i.e. the last two figures). But that distribution will not be stationary; in fact, even the first picture corresponds to typical configuration of a distribution that is not stationary, once the hole between the two half-boxes is open[23]. And, in both cases, the corresponding distribution can probably be justified on Bayesian grounds, as being the maximum entropy distribution contrained by the density profile that we observe.
4. As discussed before, there are often many stationary measures and, in the case of SRB measures at least, they cannot be singled out using stationarity[24].
5. The most basic objection is that probabilities are not physical quantities, but tools used to predict or explain. In particular, there is no good reason why one should always let probabilities change in time according to the evolution induced by the physical laws (like the Liouville evolution in classical physics)[25]. They certainly do not obey such laws when one does Bayesian updating. But if that is granted, stationarity, or any other dynamical property, does not seem to be a particularly natural requirement to impose on probabilities.

The upshot of this discussion is that the proper way to justify the use of the Lebesgue measure in statistical mechanics is not to invoke stationarity, but rather to appeal to Bayesian arguments.

Finally, to see what difference arises, in Bohmian mechanics, between these two notions of naturalness (stationarity and maximum entropy), consider the following example (which is artificially simple, but by no means unique)[26].

Take a 'universe' consisting of non interacting particles in a one-dimensional box, all of which are in their ground state. The latter can be taken to be proportional to a sine function that vanishes on the boundary of the box. Let us call ϕ that function (following the notation of [5]). The wave function of the

[23] Of course, if the hole is closed, the uniform distribution is stationary.

[24] Or stationarity plus absolute continuity with respect to Lebesgue measure, since the SRB measures is not, in general, absolutely continuous.

[25] Once one realizes that, the "paradox" posed by the constancy of the Gibbs entropy under the Liouville evolution simply disappears.

[26] The point of this example is to discuss the notion of quantum equilibrium; however, I will not explain or use the nice analysis of the concept of effective wave function which is needed to understand this concept more deeply; see [4,5].

"universe", Ψ, is a product of such functions, one for each particle[27]. To define quantum equilibrium, let us proceed as with the Maxwellian distribution of velocities, see (2). Let $\Delta(u)$, be the interval of size δ centered around $u \in \delta\mathbb{Z}$. Let $Q = (\mathbf{Q}_1, \ldots, \mathbf{Q}_N) \in \mathbb{R}^{3N}$ be an element of the configuration space of the system.

Define the macrostate by the variables $N_u(Q) = |\{\alpha | \mathbf{Q}_\alpha \in \Delta(u), \alpha=1, \ldots, N\}|$. Let $T(\varepsilon, \delta)$ be the set of typical points *relative to* $|\Psi|^2$ for ε, δ, namely those Q's for which

$$\left| \frac{N_u(Q)}{N} - |\phi(u)|^2 \right| \leq \varepsilon. \tag{16}$$

for all cells $\Delta(u)$ ($N_u(Q)$ is a coarse grained version of the empirical distribution denoted $\rho_{\text{emp}}^{N,\phi}$ in [5] and (16) corresponds to equation (9) there).

The idea of quantum equilibrium in this example is that, again because of the law of large numbers, the set of typical points, as defined above, will have a $|\Psi|^2$-measure close to 1, for N large. So, if one picks a microscopic configuration of the universe Q that is typical relative to $|\Psi|^2$, it will give rise, because of (16), to an empirical distribution satisfying Born's statistical law[28].

So, argue the authors of [4], we have shown that at least one initial configuration of the universe gives rise to Born's law, which is already non-trivial. In fact, many configurations do the job: almost all those in the support of the $|\Psi|^2$-measure. And the authors of [4] rightly invoke the notion of *typicality* in order to claim that quantum equilibrium and therefore Born's law is actually very natural. But, for that last claim to be right, one needs to argue that the measure with respect to which the configurations under discussion are typical is itself "natural" (every configuration is typical with respect to at least one measure – the delta measure concentrated on itself).

However, note that the empirical distribution in (16) must be peaked away from the boundaries of the box, because of the vanishing of $|\phi|^2$ (i.e. of the sine function) there.

But, when we open the box, what would we naturally expect to find? On Bayesian grounds, I would expect a uniform distribution. Of course, if we knew that there is a force, for example, repelling the particles from the boundaries of the box, we might revise our expectations. But the properties of the ground state here are not of that genre; in fact, they imply nothing for the dynamics of the system since here nothing moves[29]. In fact the only 'explanation' of the fact that we obtain a $|\phi|^2$ distribution rather than a uniform distribution is probably

---

[27] Putting aside here questions of statistics as well as the role of measuring devices outside the box.

[28] In more general situations, this holds also, but, to show that, one needs to invoke the equivariance of the dynamics, see [4,5].

[29] In this particular example, stationarity with respect to the dynamics does not single out *any* measure; Indeed, the laws of Bohmian mechanics imply that, in a ground state, nothing moves, so that all distributions are stationary. Whatever empirical distribution God (or whatever replaced him) picked at the begining of times, would be found by us, if the box had remained isolated until we open it.

that God likes quantum equilibrium and Born's law and so put it there at the beginning of times.

The upshot of this discussion is that quantum equilibrium, in Bohmian mechanics, should, in my view, be presented as a postulate, independent of the other ones, rather than as somehow being the only natural or reasonable choice. It is not a particularly unnatural choice and it is true that quantum equilibrium is still far less mysterious than classical non equilibrium at the origin of the universe (i.e. what Goldstein calls the hard problem of irreversibility in [7]). But one should not be present it as more natural than it is.

Acknowledgments. I would like to thank P. Collet, D. Dürr, S. Goldstein, R. Tumulka and N. Zanghì for interesting discussions.

References

1. J. Bell, *Speakable and Unspeakable in Quantum Mechanics* (Cambridge University Press, Cambridge 1987)
2. J. Bricmont, Science of chaos, or chaos in science? Physicalia Magazine **17**, pp. 159-208 (1995), and in: *The flight from science and reason*, ed. by P.R. Gross, N. Levitt, and M.W. Lewis; Annals of the New York Academy of Sciences, **775** pp. 131-175 (1996). Available from: http://dogma.free.fr/txt/JB-Chaos.htm.
3. R. T. Cox, Amer. Journal of Phys. **17**, pp 1-133 (1946)
4. D. Dürr, S. Goldstein, N. Zanghì, Journal of Stat. Phys. **67**, pp. 843-907 (1992)
5. D. Dürr, this volume; available from http://www.mathematik.uni-muenchen.de/~duerr/
6. J-P. Eckmann, D. Ruelle, Rev. Mod. Phys. **57**, pp. 617-656 (1985).
7. S. Goldstein, this volume; available from http://www.math.rutgers.edu/~oldstein
8. E.T. Jaynes, *Papers on Probability, Statistics and Statistical Physics*, ed. by R. D. Rosencrantz (Reidel, Dordrecht 1983)
9. E.T. Jaynes, *Probability Theory as Extended Logic*, available from http://bayes.wustl.edu/
10. E.T. Jaynes, E. T., Bayesian Methods: General Background, in: *Maximum-Entropy and Bayesian Methods in Applied Statistics*, ed. by J. H. Justice, (Cambridge University Press, Cambridge 1986) available from http://bayes.wustl.edu/
11. M. Kac, *Probability and Related Topics in the Physical Sciences* (Interscience Pub., New York 1959)
12. P.S. Laplace, *A Philosophical Essay on Probabilities*, Transl. by F. W. Truscott and F. L. Emory, (Dover Pub., New York, 1951). Original: *Essai philosophique sur les probabilités*, (C. Bourgeois, Paris 1986, text of the fifth edition, 1825).
13. C. J. Thompson, *Mathematical Statistical Mechanics*, (Princeton University Press, Princeton, 1972)

Part II

Classical Statistical Mechanics

The Rise of Statistical Mechanics

Carlo Cercignani

Dipartimento di Matematica, Politecnico di Milano, I–20133 Milano, Italy

Abstract. The atomic theory of matter is reviewed from the times of the early greeks until the times of Boltzmann and Gibbs. The modern kinetic theory of gases including the Boltzmann equation is discussed from a historical point of view.

The question whether matter is atomistically constituted or continuous is already present in the writings of Greek philosophers. The first atomic theory is actually credited to Democritus of Abdera who lived in the 5th Century B.C.. We can echo here what Maxwell said in a popular lecture, if we modify the number of centuries:

> "We do not know much about the science organisation of Thrace twenty-*three* centuries ago, or of the machinery then employed for diffusing an interest in physical research. There were men, however, in those days, who devoted their lives to the pursuit of knowledge with an ardour worthy of the most distinguished members of the British Association; and the lectures in which Democritus explained the atomic theory to his fellow-citizens of Abdera realised, not in golden opinions only, but in golden talents, a sum hardly equalled even in America".

The atomic theory was also maintained by other philosophers such as Leucippus (5th Century B.C.) and through Epicurus (341–270 B.C.) it was transmitted to Romans. The most complete exposition of the view of the ancients is the famous poem of Lucretius (99–55 B.C.), *De Rerum Natura* ("On the Nature of Things"). Apart from being just qualitative, this theory was inaccurate on a basic point: the law of inertia was of course not available and thus the molecular motion was ascribed to gravity (the vertical direction being thought of as absolute). In addition, collisions were replaced by a stochastic change of direction, named *clinamen* ("change of slope").

In medieval times some Arabian thinkers accepted the atomic theory, which was, on the contrary, fiercely fought by the scholastic theologians in the West, who maintained that it conflicted with the dogma of transubstantiation. During the Renaissance period, ideas related to atomism occur in the writings of Giordano Bruno (1548–1600), Galileo Galilei (1564–1642), and Francis Bacon (1561–1626). Later the French philosopher Petrus Gassendi (1592–1655) considered the idea of the atomic constitution of matter as a basic point of his philosophy.

In more recent times Descartes held that it is the basic property of space to be occupied by matter and denied the existence of atoms. Leibniz, on the other hand, regarded his monad as the ultimate element of everything.

Thus, from the start, modern science found itself confronted by two opposite views of reality, which can be regarded as a continuum or as made up of discrete particles. Sometimes both descriptions can be applied with the same results. The continuum theory makes use of a small number of continuous quantities, such as the density, the bulk velocity, and the pressure of a gas, whose evolution in time is ruled by essentially irreversible laws, which imply dissipation and growth of entropy. These laws are in general asymmetric with respect to the time variable, the "time-arrow" pointing from the past into future, *i.e.* with an orientation coinciding with that of increasing entropy. The atomic theory, on the other hand, conceives matter as discontinuous and made up of an extremely large number of particles, moving according to the time-symmetric laws of mechanics.

Since, as we have just said, these two descriptions frequently yield the same results, we must explain the fact that regular physical processes, described by continuous variables, emerge at the macroscopic level of everyday life from the extremely complex motions of a huge number of particles and the fact that the passage to a continuum description is accompanied by the break in the time-symmetry and leads from reversible microscopic motions to irreversible macroscopic phenomena.

Many scientists used the concept of atom to deduce results about the macroscopic behavior of matter, but the first systematic attempt must be credited to Roger Joseph Boscovich (1711–1787), whose theory may be taken as an example of the purest monadism, but was somewhat qualitative.

Daniel Bernoulli (1700–1782) introduced the idea of the atomic structure of matter as we understand it today. He explained the origin of pressure by means of an early example of a statistical calculation in mechanics and gave birth to the kinetic theory of gases. The new idea was that the mechanical effect of the impact of these moving molecules when they strike against a solid constitutes what is commonly called the pressure of the gas. In fact, if we were guided solely by the atomic hypothesis, we might assume the pressure to be produced by the repulsions of the molecules (this seems to be Boscovich's explanation). Although Bernoulli's scheme [1] was able to account for the elementary properties of gases (compressibility, tendency to expand, rise of temperature during compression and fall during expansion, trend toward uniformity or space homogeneity), no definite opinion could be passed on it until it was investigated quantitatively. The actual development of the kinetic theory of gases accordingly took place much later, in the nineteenth century.

The fact that gases were the first subject of statistical theories of mechanics can easily be understood; in fact, the behavior of a gas can be easily understood on a statistical basis. The word "gas" (or "gaz") was indeed coined by the chemist J.B. van Helmont in 1620 after the word "chaos" (which had been used by Paracelsius to denote a gas).

Although the rules generating the dynamics of an ensemble of molecules (originally thought of as tiny rigid bodies, a surprisingly good model) are easy to prescribe, the associated phenomena are not so simple. They are actually rather difficult to understand, especially if one is interested in the trend of the system for long periods of time (the so-called ergodic properties) or in the case when the number of spheres is very large (kinetic and hydrodynamical limits).

The theory proposed by Bernoulli was soon afterwards brought forward independently by George-Louis Lesage of Geneva (1724–1803), who, however, devoted most of his work to the explanation of gravitation as due to the impact of atoms. Then John Herapath (1790–1869), in his *Mathematical Physics*, published in 1847, made a much more extensive application of the theory. Apparently without knowing Bernoulli's work, he went further and was the first to state the idea of an absolute temperature depending on molecular velocities, although he used a rather unusual temperature (essentially the square of our absolute temperature). He gave the first explicit value for the average speed of a molecule in a gas. James Prescott Joule (1818–1889), to whom this priority is usually attributed, seems to have based his calculations on Herapath's.

Another British scientist who rediscovered elementary kinetic theory and was not acknowledged by his contemporaries is John James Waterston (1811–1883). Very few people paid attention to his work even in Britain. Refence to his papers is most remarkably absent from Maxwell's writings. Rankine did quote him twice in 1864 and 1869. His work was rediscovered, rather too late, by Lord Rayleigh in 1891.

Thus the glory of discovering the first ideas about modern kinetic theory shifted to Germany. A not particularly exciting paper by August Karl Krönig (1822–1879) is however frequently quoted because it might have had the important role of drawing the attention of Rudolf Clausius (1822–1888) to the subject. Clausius had just moved to the Eidgenössische Polytechnikum in Zürich when Krönig's paper appeared, but it seems plausible that he had done some unpublished work on the subject earlier.

Clausius took kinetic theory into a mature stage, with the explicit recognition that thermal energy is but the kinetic energy of the random motions of the molecules and the explanation of the first law of thermodynamics in kinetic terms.

Clausius had been the first to formulate the Second Law of thermodynamics and was, later, to discover the hidden concept of entropy. His first paper on the kinetic theory of gases, entitled *Über die Art der Bewegung, welche wir Wärme nennen* (The kind of motion we call heat), which appeared in 1857 in Poggendorff's *Annalen*, defined the scope of most nineteenth century work in kinetic theory. His earlier experience in atmospheric physics and in thermodynamics prompted him to find an answer to the naïve objections to the idea that molecules travel at very large speeds. Thus he laid down the program and some concepts of the new theory. One of the first scientists to react to Clausius's first paper was the Italian chemist Stanislao Cannizzaro (1826–1910), who had revived the chemical atomic theory in its modern form based on the hypothesis of Amedeo Avogadro (1776–1856). His widely known *Sunto di un corso di Filosofia Chimica* (published in 1858 and reprinted as *Sketch of a Course of Chemical Philosophy*, The Alembic Club, Edinburgh, 1961) was distributed at a meeting in Karlsruhe in 1860 and quoted the new researches ("from Gay-Lussac to Clausius") to support Avogadro's views.

Clausius introduced the concept of the mean free path (*normale mittlere Weglänge*), in 1858 [2]. In 1860, two years after Clausius's paper, on the basis of this concept, James Clerk Maxwell (1831–1879) developed a preliminary theory of transport processes, such as heat transfer, viscous drag, diffusion. He also remarked that the approach of assuming all the molecules at rest and one moving was too simplistic and one needed to compute how the molecules distribute with respect to their speeds. He thus introduced the concept of a distribution function. In the same paper he gave a (very) heuristic derivation of the velocity distribution function that bears his name [3].

Maxwell almost immediately realized that the concept of mean free path was inadequate as a foundation for kinetic theory and in 1867 developed a much more accurate method [4], based on the so-called transfer equations, and realized the particularly simple properties of a molecular model, according to which the molecules are mass points (therefore not hard spheres) interacting at a distance with a repelling force inversely proportional to the fifth power of the distance (these ficticious molecules are nowadays commonly called Maxwell molecules). In the same paper he gave a better justification of his formula for the velocity distribution function for a gas in equilibrium.

With his transfer equations, Maxwell had come very close to an *evolution equation for the distribution function*, but this last step [5] must be credited to Ludwig Boltzmann (1844–1906). The equation under consideration is usually called the Boltzmann equation and sometimes the Maxwell–Boltzmann equation (to recognize the important role played by Maxwell in its discovery).

It remained, in fact, the unsolved important problem of deducing the Second Law of thermodynamics. As is well-known by elementary physics, this principle is often subdivided into two parts, according to whether we consider just reversible processes or irreversible processes as well.

The modern idea of irreversibility in physical processes is based upon the second law of thermodynamics in its most general form. This law was put forward for the first time by Sadi Carnot in 1824 in a paper on the efficiency of the steam engines [6].

In the scientific literature before 1850 one finds scattered statements about something that is lost or dissipated when heat is used to produce mechanical work, but only in 1852 William Thomson (later to become Lord Kelvin) asserted the existence of "A universal tendency in nature to the dissipation of mechanical energy" [7]. The consequences of Thomson's Principle of Dissipation were elaborated by Hermann von Helmholtz, who, two years later, described the "heat death" of the universe, the consequence of the transformation of all energy into heat [8]. It is to be stressed that Clausius had already remarked in 1850 [9] that, although Carnot's argument can be reconciled with the equivalence of work and heat, through a slight modification, something more than the impossibility of perpetual motion had to be invoked as a postulate. In fact both the First Law (equivalence of heat and work) and Carnot's argument do not show any feature of irreversibility, whereas heat "always shows a tendency to equalize temperature differences and therefore to pass from hotter to colder bodies" [9]. A more accurate and succinct statement of the Second Law can be found in another paper by Clausius, appeared in 1856 [10]:

> "Heat can never pass from a colder to a warmer body without some other change, connected therewith, occurring at the same time".

The modern statement of the Principle of Dissipation is based upon the notion of entropy, introduced by Clausius in 1865 [11]. Although Clausius's formulation did not add any new physical content, merely the fact of choosing a new name for something which had previously been represented only by mathematical formulae and rather heavy circumlocutions, had an undoubted influence upon the subsequent developments of the matter.

In the description of matter as a collection of molecules, instead of a continuum, questions related to reversibility are presented for the first time with the invention, almost as a joke, of what is now known as "Maxwell's demon".

The invention of the demon shows how clearly Maxwell saw the heart of the matter and we can but be disappointed by the fact that his writings on the statistical character of the second law are penetrating but fragmentary.

The first attempts at explaining the Second Law on the grounds of kinetic theory are due to Rankine [13]. He assumed atomic trajectories which do not change during a thermodynamical transformation. For this reason, he was thus criticized by Boltzmann [14], since this assumption is incompatible with the deformation that the system undergoes during the said transformation.

Boltzmann himself makes his first appearance in the field in 1866 (the concept of entropy had been introduced the year before!) with a paper [15] where he tries to prove the Second Law starting from purely mechanical theorems, under the rather restrictive assumption that the molecular motions are periodic, with period τ, and the awkward remark, which might perhaps be justified, that "if the orbits do not close after a finite time, one may think that they do in an infinite one". Essentially, Boltzmann remarks that temperature may be thought of as the time average of kinetic energy, while heat can be equated to the average increase of kinetic energy; if we compute the unspecified period from one of the relations and replace the result into the other, it turns out that the heat divided by the temperature is an exact differential. This part of the paper appears to be a rather primitive justification of the first part of the Second Law; as for the second part Boltzmann's argument belongs more to pure thermodynamics than to statistical mechanics and leads to concluding that entropy must increase in an irreversible process.

In 1866 Boltzmann was practically nobody and the paper received little attention. It is thus not surprising to find that Clausius did not read it and published a completely analogous paper four years later [16]. Boltzmann was then ready to publish a comment [17], that quoted word by word about ten pages of his previous paper and concluded by saying:

> "I think I have established my priority. At the end I wish to express my pleasure because an authority like Dr. Clausius contributes to the dissemination of the ideas contained in my papers on the mechanical theory of heat".

Clausius recognized Boltzmann's priority [18], apologizing because of the scarce time he had had in the last few years to keep abreast with the subject. He also added that, in his opinion, Boltzmann's argument were not as general as his own. Boltzmann did not reply to this remark; by then his scientific interests had taken another path.

Maxwell observed this and later disputes with detachment – and amusement. In his words:

> "It is rare sport to see those learned Germans contending for the priority of the discovery that the 2nd law... is the Hamiltonsche Princip."

Why was Maxwell so witty about "those learned Germans"? Because the prize for which they were contending was an illusion. He knew already, as his discussion of the demon named after him shows, that if heat is motion, then the Second Law "is equivalent to a denial of our power to perform the operation just described [to transform heat into ordinary motion], either by a train of mechanism, or by any method yet discovered. Hence, if the heat consists in the motion of its parts, the separate parts which move must be so small that we cannot in any way lay hold of them to stop them" [19]. In other words, the Second Law expresses a limitation on the possibility of acting on those tiny objects, the atoms, with our usual macroscopic tools.

As for Clausius, he always stuck to his more mechanical viewpoint; his use of probabilities seems to be essentially restricted to the use of mean values. It seems fitting to quote here Gibbs's words, in his obituary of Clausius:

> "In reading Clausius we seem to be reading mechanics; in reading Maxwell, and in much of Boltzmann's most valuable work, we seem rather to be reading in the theory of probabilities." [20]

Boltzmann had not discovered the simple but deep truth about the statistical nature of the second principle as yet. His path toward his own statistical approach and especially toward a clear understanding of the profound meaning of his own successful discoveries was long and painful. On the other hand, Maxwell seems to have also missed the importance of Boltzmann's main result, the H-theorem. *If* a scientist of the stature of Maxwell missed the importance of a result offering a way of measuring our inability to transform heat into ordinary motion, then we should be sympathetic with our contemporaries, when they are unable to understand the meaning of Boltzmann's discovery.

The problem of irreversibility came to the forefront in kinetic theory, with Boltzmann, who in 1872 [5] not only derived the equation that bears his name, but also introduced a definition of entropy in terms of the distribution function of the molecular velocities and proved that, as a consequence of his equation, the entropy that he had defined must always increase, or, at most, in a situation of statistical equilibrium, remain constant.

Before writing this paper, Boltzmann had learned to master Maxwell's techniques [4], which we have already alluded to. In fact, by 1868 already he had extended Maxwell's distribution to the case when the molecules are in equilibrium in a force field with potential [21], including the case of polyatomic molecules as well [22]. The energy equipartition theorem was also extended by him to the case of polyatomic molecules [23]. In the mean time he had established contacts with Kirchhoff and Helmholtz.

Boltzmann interprets Maxwell's distribution function in two different ways, which he seems to consider as *a priori* equivalent: the first way is based on the fraction of a time interval sufficiently long, during which the velocity of a specific molecule has values within a certain volume element in velocity space, whereas the second way (quoted in a footnote to paper [21]) is based on the fraction of molecules which, at a given instant, have a velocity in the said volume element. It seems clear that Boltzmann did not, at that time, feel any need to analyse the equivalence, implicitly assumed, between these two meanings, which are so different. He soon realized, however (footnote to paper [23]), that it was necessary to make a "not improbable" assumption for real bodies made of molecules that are moving because they possess "the motion that we call heat". This assumption, according to which the coordinates and the velocities of the molecules take on, in an equilibrium state, all values compatible with the assigned total energy of the gas, became later familiar as ergodic hypothesis, the name given to it by Paul and Tatiana Ehrenfest [24].

Before writing his basic paper of 1872, in 1871 Boltzmann felt ready for a new attempt to understand the Second Law [25], starting from the equilibrium law that he had obtained in his previous papers and illustrating the difference between heat and work. He also proceeded to computing it explicitly for a perfect gas and for a simple model of a solid body, thus finding, in the first case, a result well known in thermodynamics, in the second an expression, from which he easily succeeded in obtaining the Dulong–Petit formula for specific heats (the first application of statistical mechanics to solids).

The derivation by Boltzmann is impeccable, if one grants that the equilibrium distribution is what is called nowadays Maxwell–Boltzmann, and is now more or less standard. It was, however, a treatment that excluded irreversible phenomena and it could not have been otherwise since the said distribution holds only for equilibrium states.

But Boltzmann was by then ready for the last step, *i.e.* the extension of the statistical treatment to irreversible phenomena, on the basis of his integrodifferential equation, with a memoir of almost 100 pages [5] with the strange title, "Further researches on the thermal equilibrium of gas molecules", chosen to present a wealth of new results.

By means of his equation, Boltzmann showed not only that the Maxwell distribution is a steady solution of the equation, but that no other such solution can be found. This goal is achieved by introducing a quantity, that turns out to be, apart from a constant factor, the opposite of the entropy; the possibility of expressing the entropy in terms of the distribution function, though in a certain sense not unexpected, does not cease to stand as a remarkable fact, that must have produced a deep impression on Boltzmann's contemporaries. In fact, as remarked by Boltzmann himself, it implied an entirely different approach to the proof of the Second Law, that showed not only the existence of an entropy function for the equilibrium states, but also permitted to study its increase in irreversible processes.

The paper goes on with an alternative derivation based on a model with discrete energies, in such a way that the integrodifferential equation for the distribution function becomes a system of ordinary nonlinear differential equations. The use of discrete energies has always appeared "much clearer and intuitive" [5] to Boltzmann. This statement of his may sound like a naïvety, but might also indicate a surprising intuition about the difficulties of a rigorous proof of the trend to equilibrium. As a matter of fact, these difficulties in the proof of the trend just mentioned disappear if one has to deal with a discrete, finite system of equations, since the unknown f is, at any time instant, a finite set of numbers, instead of a function (in mathematical jargon, we would say that we are dealing with a finite-dimensional space, rather than with a function space); this simplification permits to make use of a property, already known in Boltzmann's days (the so-called Bolzano–Weierstrass theorem) in order to deduce the trend under consideration without particularly refined mathematical arguments.

Many historians of science have underlined the circumstance that these discrete models used by Boltzmann led Planck to the discovery of his energy *quanta*, as Planck himself acknowledged [26].

Just a few pages of the voluminous memoir by Boltzmann concern the calculation of the transport properties in a gas. It is in these pages, however, that Boltzmann laid down his equation in the most familiar form for us, where the distribution function depends upon time, velocity and position.

Extending Boltzmann's treatment to liquids and solids proves extremely difficult. In fact, except for sparse results, the only case dealt with for systems more complicated than the case of gases considered so far is equilibrium statistical mechanics.

Once more it must be ascribed to Boltzmann the merit of having started this branch of statistical mechanics with a basic paper [27], written in 1884 and much less frequently quoted than his other contributions. In this paper he formulated the hypothesis that some among the possible steady distributions can be interpreted as macroscopic equilibrium states. This fundamental work by Boltzmann was taken up again, widened and expounded in a classical treatise by Gibbs [28] and it is the terminology introduced by Gibbs that is currently used. As a matter of fact a statistical ensemble (in Gibbs's terminology) is called *monode* by Boltzmann. The question posed in the abovementioned paper [27] is the following:

what statistical families of steady distributions have the property that, when producing an infinitesimal change in their parameters, the infinitesimal changes of the average total energy of the system, E, of the pressure p and the volume V are such that $(dE+pdV)/T$ (where T is the average kinetic energy per particle) is an exact differential (at least in the thermodynamic limit, when $V \to \infty$, $N \to \infty$, whereas N/V remains bounded)? These families are called *orthodes* by Boltzmann. The answer given by Boltzmann to his own question is that there are at least two ensembles of this kind, the *ergode* (Gibbs's *microcanonical ensemble*) and the *holode* (Gibbs's *canonical* ensemble).

Although Boltzmann originated [27] the study of equilibrium states for more general situations than that, already considered by Maxwell, of a dilute gas in the absence of external forces, it is not to his name but to Gibbs's that one usually associates the methods of this area (the most completely developed one) of statistical mechanics. Even the terminology (microcanonical, canonical, grand canonical ensembles) is the one due to Gibbs, while the first two ensembles were clearly defined (with different names) and used by Boltzmann. It is then beyond doubt, in the words of Klein [29], that

> "it was Boltzmann, and not Maxwell or Gibbs, who worked out precisely *how* the second law is related to probability, creating the subject of statistical mechanics."

We may add, for the sake of clarity, that Gibbs invented the name of the subject.

Gibbs was, of course, aware of Boltzmann's priority. Why, then, beyond some generic acknowledgements to Boltzmann, it is Gibbs's name that emerges? It is an interesting question, to which one may be tempted to give an easy answer, the fact that not many people read the lengthy papers written by Boltzmann.

At the time when he developed his esposition [28] of statistical mechanics, Gibbs was at the end of a scientific career devoted to the study and application of thermodynamics, in which abstract thoughts were illuminated by several geometric representations, but not by images based on mechanical models typical of atomism. The energeticists, who opposed and even scorned the use of molecular ideas, particularly valued Gibbs for his having avoided these ideas in his work. In particular, G. Helm praised his thermodynamic writings, because they "established the strict consequences of the two laws with no hankering or yearning after mechanics" [30].

The most widely known paper by Gibbs, concerned with the application of the principles of thermodynamics to the study of thermodynamic equilibrium, is the long memoir, written in 1874, "On the Equilibrium of Hetereogeneous Substances" [31]. But in the midst of his thermodynamics, Gibbs turned up the paradox on diffusion that still bears his name. He concluded his discussion of this apparent paradox by remarking: "In other words, the impossibility of an uncompensated decrease of entropy seems to be reduced to improbability". This succinct statement was used by Boltzmann, years later, as the motto for his *Lectures on the Theory of Gases.*

Boltzmann's paper [27] was practically forgotten and Gibbs's treatise became the standard reference for equilibrium statistical mechanics.

In his paper [32], where Boltzmann's paper is considered with detail and competence, Gallavotti claims that the reason for its oblivion is rather due to the fact, hinted at in previous chapters, that Boltzmann's work is known only through the popularization of the Ehrenfests encyclopedia article [24], which is as good a treatise on the foundations of statistical mechanics as little it has to do with many key ideas of Boltzmann. In fact, people who have had a chance of talking with illustrious physicists, who had been students of Paul Ehrenfest, may realize, by reading Boltzmann and the encyclopedia article [24] that, when they said that they were reporting Boltzmann's views, they were really talking about that article [24]. Gallavotti [32] should be mentioned here, not only because he appears to be the second author, after Klein, to underline the importance and the basic role of Boltzmann's paper [27], but also because he proposes a new etymology of the words *monode*, *ergode*, *holode*. This has some relevance, even if these terms have disappeared from common usage, because the second one has originated the much used adjective *ergodic*.

I have tentatively added [33] a third explanation for the oblivion of Boltzmann's work, in addition to Gallavotti's and the fact that Boltzmann's rather lengthy treatment was superseded by Gibbs's concise presentation. This explanation is related to the role played by Niels Bohr. He was probably the most influential scientist of this century (in the words of Max Born, when proposing both Einstein and him as foreign members of the Göttingen Academy of Sciences: "His influence on theoretical and experimental research of our time is greater than that of any other physicist"). Bohr had a great opinion of Gibbs, expressed over and over, because he had introduced the statistical ensembles, but did not think highly of Boltzmann, and he may have induced a mistrust in Boltzmann's work in a large number of physicists of this century. It is clear that he had not read Boltzmann, who had been the originator of the first two kinds of ensemble. This is confirmed by his coworker L. Rosenfeld in an interview on Bohr, released to T. Kuhn and J. L. Heilbron, and kept in the Niels Bohr Archive in Copenhagen. These are the relevant sentences of Rosenfeld:

> "I don't think that he had ever read Boltzmann, or at any rate not Boltzmann's papers. He had read, I think, and studied very carefully Boltzmann's Lectures on gas theory"

and

> "He said that Boltzmann had spoiled things by insisting on the properties of mechanical systems."

Apparently Bohr was not alone. Einstein once said to one of his students: "Boltzmann's work is not easy to read. There are great physicists who have not understood it." [34]

Bohr's opinions go back to his youth, because he had expressed reservations about Boltzmann's views already in 1912, when lecturing on the statistical foundations of thermodynamics in Copenhagen [35].

We remark that the lengthy and extremely well-documented treatment of the ergodic hypothesis by Brush [36] mentions in passing that Boltzmann had discovered the method of ensembles before Gibbs, but he devotes no discussion to this extremely important point.

The influence of Boltzmann's ideas on the physics of twentieth century is obvious, even if it often appears as interpreted in the writings of different authors, mainly through those of Gibbs, Planck, Einstein and the Ehrenfests. To the arguments developed above, we can add the fact that in the last years of his life Boltzmann devoted himself more to defending his theories and his viewpoint on theoretical physics than to expounding systematically his discoveries and methods or applying these methods to new areas that were opening up (theory of electrons, black-body radiation, Brownian motion). Even his published lectures [37], and in particular the second volume, are devoted more to showing the usefulness of the basic concepts of the kinetic theory of gases than to underlining the role played by their author in the development of the same theory. As we have hinted at above, some applications, of which Boltzmann was well aware, were never developed by him. Perhaps the most evident case is that of Brownian motion; Boltzmann mentions this motion in a couple of places, but he does not develop the consequences of some of his remarks such as the following one [38]:

> "...likewise, it is observed that very small particles in a gas execute motions which result from the fact that the pressure on the surface of the particles may fluctuate"

It was left to Einstein to arrive at a theory, which was to constitute the starting point to ascertain the atomic structure of matter beyond any doubt. Einstein himself mentioned his own astonishment at the fact that this result had not been obtained by Boltzmann, saying in a conversation that

> "it is puzzling that Boltzmann did not himself draw this most perspicuous consequence, since Boltzmann had laid the foundations for the whole subject" [39].

Let us remark here that Einstein had certainly read Boltzmann's lectures before he started his own work on statistical mechanics. It seems doubtful whether he was aware of Boltzmann's papers; he certainly covered this gap later and as a result gained a high opinion of the Austrian scientist. Concerning the first point we can quote from Einstein's autobiographical sketch:

> "Not acquainted with the earlier investigations of Boltzmann and Gibbs which had appeared earlier and which actually exhausted the subject, I developed the statistical mechanics and the molecular-kinetic theory of thermodynamics which was based on the former. My major aim in this was to find facts which would guarantee as much as possible the existence of atoms of definite finite size. In the midst of this I discovered that, according to atomistic theory, there would have to be a movement of suspended microscopic particles open to observation, without knowing that observations concerning Brownian motion were already long familiar." [40]

A very early judgement by Einstein on Boltzmann, based presumably on Boltzmann's lectures [37], can be found in a letter of his to his girlfriend, Mileva Mariċ, which says:

> "The Boltzmann is magnificent. I have almost finished it. He is a masterly expounder. I am firmly convinced that the principles of the theory are right, which means that I am convinced that in the case of gases we are really dealing with discrete mass points of definite finite size, which are moving according to certain conditions. Boltzmann very correctly emphasizes that the hypothetical forces between the molecules are not an essential component of the theory, as the whole energy is of the kinetic kind. This is a step forward in the dynamical explanation of physical phenomena." [41]

In a letter to Marcel Grossmann one can read:

> "Lately I have been engrossed in Boltzmann's work on the kinetic theory of gases and these last few days I wrote a short paper myself that provides the keystone in the chain of proofs that he started. [...] I'll probably publish it in the Annalen" [41]

Concerning the second point (i.e. the fact that Einstein got a more detailed knowledge of Boltzmann's work and a great admiration for him), it seems sufficient to recall that, in an exposition to a wide public of the theory of relativity, a subject not at all related to Boltzmann's work, Einstein writes in the introduction:

> "I adhered scrupulously to the precept of that brilliant theoretical physicist L. Boltzmann, according to whom matters of elegance ought to be left to the tailor and the cobbler." [42]

Planck's contributions to the black-body theory – which took place between the end of the nineteenth and the beginning of the twentieth century – open the new era in physics that followed the period of Boltzmann's most important activity took place, i.e. the thirty years which close the nineteenth century. And Planck's work is indeed the connecting link between Boltzmann and quantum mechanics.

This is a good point to stop, not without commenting that last century's contributions were not only in terms of techniques but also of conceptual framework. This undoubtedly played a role also in the development and interpretation of quantum mechanics.

References

1. D. Bernoulli: *Hydrodynamica*, Argentorati, Strassburg (1738).
2. R. Clausius: "Über die mittlere Länge der Wege, welche bei der Molekularbewegung gasförmiger Körper von den einzelnen Molekülen zurückgelegt werden; nebst einigen anderen Bemerkungen über die mechanische Wärmetheorie", Poggendorff Ann. **105**, 239-258 (1858).

3. J.C. Maxwell: "Illustration of the dynamical theory of gases. Part I: On the motions and collisions of perfectly elastic spheres. Part II: On the process of diffusion of two or more kinds of moving particles among one another. Part III: On the collisions of perfectly elastic bodies of any form", Phil. Mag. (4), 19-32, (20) 21-37 (1860).
4. J.C. Maxwell: "On the dynamical theory of gases", Phil. Trans. Roy. Soc. **157**, 49-88, (1867).
5. L. Boltzmann: "Weitere Studien über das Wärmegleichgewicht unter Gasmolekülen", Sitzungsberichte Akad. Wiss., II, **66**, 275-370 (1872).
6. S. Carnot: *Reflexions sur la puissance motrice du feu et sur les machines propre à développer cette puissance*, Bachelier, Paris (1824).
7. W.Thomson (Lord Kelvin): "On a universal tendency in nature to the dissipation of mechanical energy", Phil. Mag. (4) **4**, 304 (1852)
8. H. von Helmholtz: *Über die Erhaltung der Kraft*, Berlin (1847).
9. R. Clausius: "Über die bewegende Kraft der Wärme, und die Gesestze, welche sich daraus für die Wärmelehre selbst ableiten lassen", Pogg. Ann. **79**, 368-500 (1850) (English translation by W. F. Magie in: S. Carnot, *Reflections on the Motive Power of Fire and Other Papers on the Second Law of Thermodynamics*, E. Mendoza, ed., Dover, New York (1960)).
10. R. Clausius: "On a modified form of the second fundamental theorem in the mechanical theory of heat", Philosohical Magazine **12**, 86 (1856).
11. R. Clausius: "Über verschiedene für die Anwendung bequeme Formen der Hauptgleichungen der mechanischen Wärmetheorie", Pogg. Ann. **125**, 353 (1865).
12. W. J. M. Rankine: "On the second law of the thermodynamics", Phil. Mag. (4) **30**, 241-245 (1865)
13. W. J. M. Rankine: "Sur la nécessité de vulgariser la seconde loi de la thermodynamique", Ann. Chim. (4) **12**, 258-266 (1867)
14. L. Boltzmann: "Theorie der Wärme", Fortschritte der Physik **26**, 441-504 (1870)
15. L. Boltzmann: "Über die mechanische Bedeutung des zweiten Hauptsatzes der Wärmetheorie", Wien. Ber. **53**, 195-220 (1866)
16. R. Clausius: "Über Zurückführung des zweiten Hauptsatzes der mechanischen Wärmetheorie auf allgemeine mechanische Prinzipien", Pogg. Ann. **142**, 433-461 (1871).
17. L. Boltzmann: "Zur Priorität der Auffindung der Beziehung zwischen dem zweiten Hauptsatze der mechanischen Wärmetheorie und dem Prinzip der kleinsten Wirkung", Pogg. Ann. **143**, 211-230 (1871).
18. R. Clausius: "Bemerkungen zu der Prioritätsreclamation des Hrn. Boltzmann", Pogg. Ann. **144**, 265-280 (1871).
19. J. C. Maxwell: *Theory of Heat*, Longmans, London, 1871, Chapter XXII.
20. J. W. Gibbs: "Rudolf Julius Emmanuel Clausius", Proc. Am. Acad. **16**, 458-465 (1889).
21. L. Boltzmann: "Studien über das Gleichgewicht der lebendingen Kraft zwischen bewegten materiellen Punkten ", Sitzungsberichte Akad. Wiss. **58**, 517 (1868).
22. L. Boltzmann: "Über das Wärmegleichgewicht zwischen mehratomigen Gasmolekülen", Sitzungsberichte Akad. Wiss. **63**, 397 (1871).
23. L. Boltzmann: "Einige allgemeine Sätze über Wärmegleichgewicht", Sitzungsberichte Akad. Wiss., **63**, 679 (1871).
24. P. and T. Ehrenfest: "Begriffliche Grundlagen der statistischen Auffassung in der Mechanik", in *Enzyklopädie der mathematischen wissenschaften*, Vol. IV, Part 32 (Lipsia, 1911).

25. L. Boltzmann: "Analytischer Beweis des zweiten Hauptsatzes der mechanischen Wärmetheorie aus den Sätzen über das Gleichgewicht der lebendingen Kraft", Sitzungsberichte Akad. Wiss. **63**, 712 (1871).
26. M. Planck: *Wissenschaftliche Selbstbiographie*, Leipzig (1948).
27. L. Boltzmann: "Über die Möglichkeit der Begründung einer kinetischen Gastheorie auf anziehende Kräfte allein", Wien. Ber. **89**, 714 (1884).
28. J. W. Gibbs: *Elementary principles in statistical mechanics, developed with special reference to the rational foundations of thermodynamics* , Yale University Press (1902).
29. M. J. Klein: "The Development of Boltzmann's Statistical Ideas", in *The Boltzmann Equation. Theory and Application*, p. 53-106, E. G. D. Cohen and W. Thirring, eds., Springer-Verlag, Vienna (1973).
30. G. Helm: *Die Energetik nach ihrer geschichtlichen Entwicklung*, Veit, Leipzig (1898).
31. J. W. Gibbs: "On the equilibrium of heterogeneous substances", in *The Scientific Papers of J. Willard Gibbs*, H. A. Bumstead and R. G. Van Name, eds., Vol. I, pp. 55-353, Longmans, Green, and Co., New York (1906).
32. G. Gallavotti: "Ergodicity, ensembles, irreversibility in Boltzmann and beyond", Jour. Stat. Phys. **78**, 1571-1590 (1995).
33. C. Cercignani: *Ludwig Boltzmann. The Man Who Trusted Atoms*, Oxford University Press, Oxford (1998).
34. C. Seelig: *Albert Einstein*, p. 176, Europa Verlag, Zürich (1960).
35. N. Bohr: *Collected works*, Vol. 6, p. 320, North-Holland, Amsterdam (1972).
36. S. Brush: *The Kind of Motion We Call Heat*, Elsevier, Amsterdam (1976).
37. L. Boltzmann: *Vorlesungen über Gastheorie*, 2 voll., J. A. Barth, Leipzig,(1895-1898).
38. L. Boltzmann: "Entgegnung auf die wärmetheoretischen Betrachtungen des Hrn. E. Zermelo". Ann. Phys. **57**, 773-784 (1896). (English translation in S. G. Brush, *Kinetic theory. Vol. 2 Irreversible Processes*, 218-228, Pergamon Press, Oxford (1966)).
39. A. Sommerfeld: "Zum Andenken an Marian von Smoluchowski", Phys. Zeitschr. **18**, 533-539 (1917).
40. A. Einstein: in *Albert Einstein, Philosopher-Scientist*, P.A. Schilpp ed., Tudor, New York (1949).
41. A. Beck and P. Havas, eds.: *The collected papers of Albert Einstein. Vol. I. The early years*, English translation by A. Beck, p. 149, Princeton University Press, Princeton (1987).
42. A. Einstein: *On the Special and the General Relativity Theory, a Popular Exposition*, Methuen, London (1920). [translation of *Über die spezielle und allgemeine Relativitätstheorie (gemeinveständlich)*, Vieweg, Braunschweig, (1917)].

Boltzmann's Approach to Statistical Mechanics

Sheldon Goldstein

Departments of Mathematics and Physics
Rutgers University
Piscataway NJ 08854, USA

Abstract. In the last quarter of the nineteenth century, Ludwig Boltzmann explained how irreversible macroscopic laws, in particular the second law of thermodynamics, originate in the time-reversible laws of microscopic physics. Boltzmann's analysis, the essence of which I shall review here, is basically correct. The most famous criticisms of Boltzmann's later work on the subject have little merit. Most twentieth century innovations – such as the identification of the state of a physical system with a probability distribution ϱ on its phase space, of its thermodynamic entropy with the Gibbs entropy of ϱ, and the invocation of the notions of ergodicity and mixing for the justification of the foundations of statistical mechanics – are thoroughly misguided.

1 Introduction

I shall focus here on Boltzmann's approach to the problem of the arrow of time: the origin of irreversible macroscopic laws, e.g., and most importantly, the second law of thermodynamics – the law of increasing entropy – in the reversible laws of microscopic physics. I shall assume, as of course did Boltzmann, a classical framework, for the most part ignoring quantum mechanics.

As a matter of fact, it is widely believed that the transition from classical mechanics to quantum mechanics requires no essential modification of Boltzmann's ideas. This may well be so, but I believe that the traditional formulation of quantum theory is much too vague to permit any definitive conclusions on this matter. (For a non-traditional formulation of quantum mechanics that avoids its conceptual incoherence, a formulation which I quite naturally believe is well worth very serious consideration, see the contribution of Detlef Dürr to this volume.)

For a more detailed presentation of much of what is discussed here, the reader should consult the excellent papers of Joel Lebowitz [1] on this subject, as well as Jean Bricmont's sharp critique [2] of some recent proposals. See also the contributions of Bricmont and Herbert Spohn to this volume.

Most macroscopic phenomena are irreversible: They would look quite different, and in fact usually quite astonishing, if run backwards in time. Consider for example the breaking of a glass, the boiling of an egg, birth and death. At a certain time a glass sits on a table; at another time, a *later* time, what had been that glass now lies on the floor in a mess of tiny fragments. And corresponding to such irreversible phenomena there are irreversible equations and laws, for example the diffusion or heat equation, the Navier-Stokes equation, Boltzmann's equation, and, perhaps most important for the issue at hand, the second law of thermodynamics, which can be regarded as lying behind the irreversibility of the irreversible macroscopic equations.

This irreversible behavior and these irreversible laws must somehow be a consequence of the (more) fundamental microscopic laws governing the behavior of the constituents of the systems obeying the irreversible laws. But these microscopic laws are symmetric under time reversal, and their solutions, run backwards in time, are also solutions. The correct detailed resolution of this apparent paradox was provided by Boltzmann more than a century ago. And the essential idea was understood, by Maxwell and Lord Kelvin, even earlier. Here is Lord Kelvin [3], writing in 1874:

> If, then, the motion of every particle of matter in the universe were precisely reversed at any instant, the course of nature would be simply reversed for ever after. The bursting bubble of foam at the foot of a waterfall would reunite and descend into the water ... Boulders would recover from the mud the materials required to rebuild them into their previous jagged forms, and would become reunited to the mountain peak from which they had formerly broken away. And if the materialistic hypothesis of life were true, living creatures would grow backwards, with conscious knowledge of the future, but no memory of the past, and would become again unborn. But the real phenomena of life infinitely transcend human science ... Far otherwise, however, is it in respect to the reversal of the motions of matter uninfluenced by life, a very elementary consideration of which leads to a full explanation of the theory of dissipation of energy.

The adequacy of Boltzmann's resolution was controversial at the time and remains so even today. Nonetheless, the detailed solution found by Boltzmann, based in part on his understanding of the microscopic meaning of entropy, was well appreciated and admired by many of his contemporaries. For example, towards the end of his life Schrödinger [5] reported that "no perception in physics has ever seemed more important to me than that of Boltzmann – despite Planck and Einstein." In a more detailed assessment, Schrödinger [6] declared that

> The spontaneous transition from order to disorder is the quintessence of Boltzmann's theory ... This theory really grants an understanding and does not ... reason away the dissymmetry of things by means of an a priori sense of direction of time ... No one who has once understood Boltzmann's theory will ever again have recourse to such expedients. It would be a scientific regression beside which a repudiation of Copernicus in favor of Ptolemy would seem trifling.

However, as I've already indicated, the issue has never been entirely settled to everyone's complete satisfaction. In fact Schrödinger concluded the statement that I've just quoted with the observation that

> Nevertheless, objections to the theory have been raised again and again in the course of past decades and not (only) by fools but (also) by fine

> thinkers. If we ... eliminate the subtle misunderstandings ... we ... find ... a significant residue ... which needs exploring ...

Notice Schrödinger's parenthetical comments, suggesting that he did not have a very high opinion of most of the objections that continued to be raised against Boltzmann's ideas. Note also his separation of a residual core of confusion from what was regarded – or should have been – as settled.

Roughly corresponding to this separation, the problem of irreversibility can be regarded as having two parts: an easy part and a hard part. For many physicists who find Boltzmann's solution inadequate, it is only the "hard part" that causes them trouble. However, for a great many others even the "easy part" seems hard, and generates an enormous amount of confusion.

Moreover, even with the solution to both parts of the problem, there are still questions, of a philosophical character, that remain. These questions, touched upon in Section 4, could be regarded as defining the really hard part of the problem. But it would be better to see them, not so much as part of the problem of irreversibility, but rather as general questions about the meaning and nature of scientific explanation.

2 The Easy Part

Boltzmann's great achievement was to arrive at an understanding of the meaning of entropy and why it should tend to increase, almost never decreasing. This understanding is such as also to explain why systems evolve to equilibrium states, of maximal entropy.

Consider, for example the sequence of snapshots of a gas in a box illustrated in Fig. 1. We see here the transition from a low entropy nonequilibrium state on

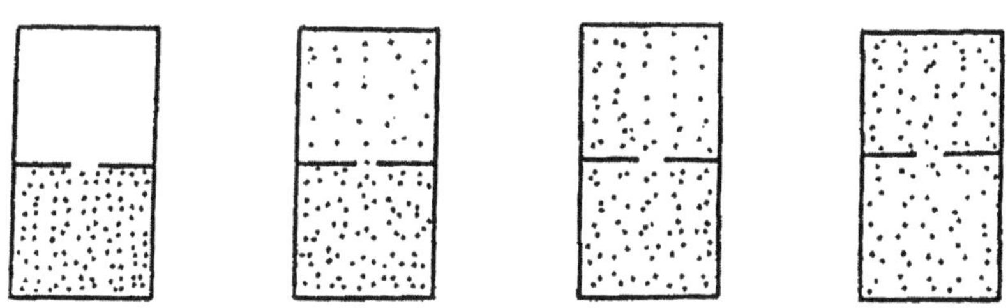

Fig. 1. Snapshots of a gas in a box

the left, with the gas entirely on the bottom of the box, through states of higher entropy, with the gas expanding into the full box, until at the right we have a high entropy equilibrium state with the gas uniformly distributed throughout the box.

The question that Boltzmann addressed is quite simple and straightforward: Why does the gas prefer to be in equilibrium – that is, to look like the snapshot on the right? The answer that he found is, perhaps, even simpler but also, somehow, rather subtle.

A complete description of the state of the gas is provided by its *phase point* X, a point in the phase space of possible microscopic states of the gas. Many different phase points correspond to each of the snapshots in Fig. 1. There is nothing particularly special about any specific *equilibrium phase point* – a phase point corresponding to the snapshot on the right, a system in equilibrium. The dynamics prefers a given equilibrium phase point neither more nor less than it prefers any other given phase point, even a specific far-from-equilibrium phase point, corresponding say to the leftmost snapshot.

There are, however, for a system at a given energy E, far more equilibrium phase points than nonequilibrium phase points, overwhelming more, in fact, than the totality of nonequilibrium phase points at that energy – corresponding to all possible ways the system can fail to be in equilibrium, and described, for example, by the various density distributions perceptibly different from the uniform one. The only relevant sense in which the equilibrium phase points could be regarded as special is that there are vastly more of them.

I shall now sketch Boltzmann's account of why this should be so, indicating the central role of entropy and its meaning in providing a quantitative foundation for what I've just described. A crucial ingredient in this account is the vast separation of scales between the microscopic and the macroscopic levels, without which irreversible phenomena could not possibly emerge from reversible microscopic laws.

Boltzmann's Entropy

Consider the microscopic state $X = (\boldsymbol{q}_1, \ldots, \boldsymbol{q}_N, \boldsymbol{p}_1, \ldots, \boldsymbol{p}_N)$ of a classical system consisting of a large number N of identical particles forming a gas in a box Λ, with positions $\boldsymbol{q}_i \in \Lambda$ and momenta $\boldsymbol{p}_i$. The evolution of the system is determined, via Hamilton's equations of motion $\mathrm{d}\boldsymbol{q}_i/\mathrm{d}t = \partial H/\partial \boldsymbol{p}_i$, $\mathrm{d}\boldsymbol{p}_i/\mathrm{d}t = -\partial H/\partial \boldsymbol{q}_i$, by a Hamiltonian function $H(\boldsymbol{q}_1, \ldots, \boldsymbol{q}_N, \boldsymbol{p}_1, \ldots, \boldsymbol{p}_N)$, the energy function for the system, normally assumed to be the sum of a kinetic energy term and a potential energy term but whose form shall not much concern us here. Since the energy is a constant of this motion, we may take as the relevant set of possible states, not the full phase space $\Omega(\Lambda)$ of the system under consideration but the energy surface $\Omega_E = \{X \in \Omega(\Lambda) \mid H(X) = E\}$ corresponding to the value E of the energy of the system.

Each snapshot in Fig. 1 corresponds to a subset Γ of Ω_E, namely the set of all phase points that "look like" the snapshot. Every phase point X belongs to a macrostate $\Gamma(X)$ consisting of phase points that are macroscopically similar to X. More precisely, partition the 1-particle phase space (the $\boldsymbol{q}, \boldsymbol{p}$ – space) into macroscopically small but microscopically large cells Δ_α and specify (approximately, by providing intervals) the number n_α of particles in each cell. Each such specification determines a *macrostate*, the set of all phase points whose particles are distributed in the manner specified. Different specifications yield different macrostates and the set of all such macrostates defines a partition of our phase space Ω_E, into macrostates. $\Gamma(X)$ is then the macrostate – the element of this partition – to which X belongs. All points in $\Gamma(X)$ have, in particular, similar spatial particle densities (and similar bulk-velocity profiles).

Different macrostates typically have vastly different sizes, and this difference is conveniently quantified by *Boltzmann's entropy*:

$$S(X) = k \log |\Gamma(X)| \ , \tag{1}$$

where k is Boltzmann's constant, expressing the relationship between macroscopic and microscopic units, and $|\ \ |$ denotes volume, given by the time-invariant projection of the Liouville (Lebesgue) measure onto Ω_E. (The definition (1) is correct up to an additive constant, depending on N, that is important but shall not concern us here.) Following Boltzmann, let us identify Boltzmann's entropy with the thermodynamic entropy, to which the Second Law refers. Then that we should expect entropy to increase – but for exceedingly rare exceptions never decreasing – more or less directly follows from the very concept (1) of entropy once the vast separation of scales between the microscopic and the macroscopic levels is taken into account.

This is because the entropy S is an extensive variable, proportional to the number N of particles in the system. In fact, the "log" in (1) suggests that the various possible values of $|\Gamma(X)|$ need to be measured on a logarithmic scale, these values differing so greatly that without taking the logarithm it would be difficult to place them on the same scale or graph. With the logarithm, an extensive quantity is obtained, and after dividing by the spatial volume of the system we obtain an entropy per unit volume whose differences, for different macrostates, are of order unity.

What all of this means is that typical values of $|\Gamma(X)|$, and, more to the point, typical ratios of such values, for example between the value of $|\Gamma(X)|$ for the equilibrium state depicted in the snapshot on the right of Fig. 1 and the value for any nonequilibrium state, say the one depicted on the left, are of "order" 10^N, or, more explicitly, of order $10^{10^{20}}$ for a system with 10^{20} particles. (That this is the correct order is easily seen by noting that when the condition that all particles be on the left side of the box is dropped, the volume of the set of all possible phase points consistent with the constraints increases by a factor of 2^N – a factor of 2 for each of the particles. Note also that in terms of the sense of "order" appropriate for comparing quantities as vast as those under consideration here, "of order $10^{10^{20}}$" should not be regarded as different from "of order $2^{10^{20}}$.")

Thus Ω_E consists almost entirely of phase points in the equilibrium macrostate Γ_{eq}, with ridiculously few exceptions whose totality has volume of order $10^{-10^{20}}$ relative to that of Ω_E. For a nonequilibrium phase point X of energy E, the Hamiltonian dynamics governing the motion X_t arising from X would have to be ridiculously special to avoid reasonably quickly carrying X_t into Γ_{eq} and keeping it there for an extremely long time – unless, of course, X itself were ridiculously special.

The Objections of Zermelo and Loschmidt

Zermelo and Loschmidt claimed that Boltzmann's analysis, showing how irreversible macroscopic behavior emerges from reversible microscopic laws, was – and had to be – inadequate. Loschmidt noted that since the classical equations of motion are time-reversible, we may obtain solutions X_t to these equations that violate the macroscopic laws by time-reversing solutions that obey them, thereby obtaining, for example, a solution which runs through the snapshots in Fig. 1 in time-reversed order, right to left. And Zermelo pointed out that there is another reason why such anti-thermodynamic solutions must exist: Poincaré recurrence, which guarantees that most solutions (the exceptions have, in fact, at most measure 0) that initially belong to the macrostate depicted in the left snapshot will eventually return to that macrostate. It thus follows that anti-thermodynamic behavior is just as consistent with the microscopic laws as thermodynamic behavior (which is of course obvious in view of the reversibility of those laws), so that neither, in fact, could be a consequence of those laws.

However, no genuine conflict with the analysis of Boltzmann follows from these objections. Boltzmann did not (finally) claim that all phase points should behave properly, but only that by far most of them – in any given macrostate and in the sense of volume relative to that of the macrostate – should. Some may behave improperly, and the arguments of Zermelo and Loschmidt merely show that such bad phase points in fact exist. For example, from Loschmidt we know that we may obtain such points by reversing the velocities of the equilibrium phase points to which nonequilibrium phase points (reasonably quickly) evolve.

Here is part of Boltzmann's response to Zermelo [7]:

> I have ... emphasized that the second law of thermodynamics is from the molecular viewpoint merely a statistical law. Zermelo's paper shows that my writings have been misunderstood; ... Poincaré's theorem, which Zermelo explains at the beginning of his paper, is clearly correct, but his application of it to the theory of heat is not. ... Thus, when Zermelo concludes, from the theoretical fact that the initial states in a gas must recur – without having calculated how long a time this will take – that the hypotheses of gas theory must be rejected or else fundamentally changed, he is just like a dice player who has calculated that the probability of a sequence of 1000 one's is not zero, and then concludes that his dice must be loaded since he has not yet observed such a sequence!

The Relevance of Ergodicity and Mixing

The basic notions of ergodic theory, in particular ergodicity and mixing, are widely believed to play a crucial role in the foundations of statistical mechanics. Ergodicity, roughly speaking the absence of constants of the motion other than functions of the energy H, implies the equality of time-averages and phase-space averages, i.e., of the long-time average of a quantity as it changes under the dynamics and the uniform average of that quantity, over the relevant energy surface Ω_E, as the phase point at which it is evaluated varies over that surface. This is supposed to justify the use, to define the equilibrium values of thermodynamic quantities, of such phase-space averages with respect to the *microcanonical ensemble*, the uniform distribution over the energy surface, the idea being that the observed values of these quantities are time-averages, since measurement takes time.

This use of ergodicity is thoroughly misguided. Boltzmann's key insight was that, given the energy of a system, the overwhelming majority of its phase points on the corresponding energy surface are equilibrium phase points, all of which look macroscopically more or less the same. This means that the value of any thermodynamic quantity is, to all intents and purposes, constant on the energy surface, and averaging over the energy surface will thus reproduce that constant value, regardless of whether or not the system is ergodic.

For example, one characteristic shared by the equilibrium phase points, and thus by the great majority of phase points on the energy surface, is a common Maxwellian empirical distribution for the velocities. Here again is Boltzmann [7], expressing his frustration with Zermelo for failing, it seems, to appreciate that fact:

> [The Maxwell distribution] is characterized by the fact that by far the largest number of possible velocity distributions have the characteristic properties of the Maxwell distribution, and compared to these there are only a relatively small number of possible distributions that deviate significantly from Maxwell's. Whereas Zermelo says that the number of states that finally lead to the Maxwellian state is small compared to all possible states, I assert on the contrary that by far the largest number of possible states are "Maxwellian" and that the number that deviate from the Maxwellian state is vanishingly small.

There is another problem with this use of ergodicity, a mismatch of time scales. The time scale appropriate for the ergodicity of a gas in a box is, roughly speaking, the time necessary for a trajectory for the motion in phase space to wind all over the phase space, and this is at least as long as a Poincaré recurrence time, the time necessary for the gas, say after leaving the state depicted in the left snapshot in Fig. 1, to return to that state, a time as large as the corresponding microstate is small and hence of order $10^{10^{20}}$ in your favorite unit of time, a time far far larger than that believed to be the age of the universe, since the big bang. Thus ergodicity couldn't possibly be very relevant to an account of phenomena, such as those with which thermodynamics is concerned, taking place on reasonable time scales.

Mixing is supposed to explain why systems evolve to a state of equilibrium. The idea here is that since such a state is in a sense characterized by a special probability distribution, namely the microcanonical ensemble, evolution to equilibrium amounts to the convergence of a generic (nonequilibrium) distribution to the special one, under the dynamics on probabilities arising from the phase-space motion. But this, made suitably precise, amounts to the assertion that the system is (a) *mixing* (system), this terminology referring to the fact that the notion is usually defined by the following (equivalent) property of the dynamics: The points of any region R – of nonvanishing volume, no matter how small – of the energy surface will evolve under the dynamics, after a suitably long time t, to points that fill out a region R_t that is distorted and convoluted in such a way as to be spread more or less uniformly throughout the energy surface, in the sense that for any reasonable function f, the uniform averages of f over Ω_E and over R_t are more or less the same.

Since the energy surface Ω_E consists almost entirely of a single macrostate, the equilibrium macrostate Γ_{eq}, the mixing property of the evolution on the energy surface pretty much amounts to the condition that (even small) subregions of Γ_{eq} become uniformly spread over Γ_{eq} after a sufficiently long time. But this could not possibly be relevant to the problem of approach to equilibrium, since it concerns only phase points that describe a system that is already in equilibrium. Approach to equilibrium concerns the passage of a nonequilibrium phase point, lying outside of Γ_{eq}, into Γ_{eq}, and this tends to happen, in fact typically rather quickly, merely because Γ_{eq} is so extremely large. (Note, however, that although mixing is a much stronger property of the dynamics than ergodicity, the mixing time scale could be much smaller than that for ergodicity, and could in fact be of the same order as the time required for a system to reach equilibrium.)

This abuse of mixing is so obviously wrong-headed that one can't help wondering about the sources of the confusion. Here are two possibilities:

One of the most widely mentioned metaphors for mixing, invoked by Gibbs [8] when he introduced the notion, is the spreading of a tiny ink drop when it is placed in a glass of water and the water is stirred. The spreading of the ink all over the water is also an example of approach to equilibrium. However, it is important to bear in mind that insofar as it illustrates mixing, this example should be regarded as having as its phase space the points of the liquid, a three-dimensional set, whereas the relevant phase space for the approach to equilibrium in the example is the set of configurations of the liquid, describing how the ink is distributed in the water, a space of enormous dimension. The sort of mixing that can be relevant to approach to equilibrium takes place in physical space, not in phase space.

A related point: there is a quantity of a probabilistic character whose approach to equilibrium for a low density gas does reflect, famously and correctly, that of the system. I have in mind here, of course, the one-particle Boltzmann function $f(\boldsymbol{q},\boldsymbol{p})$, whose approach to equilibrium is governed by Boltzmann's equation and which describes the empirical distribution, or coarse-grained density, of the particles of the gas in the one-particle phase space. It is worth stressing that $f(\boldsymbol{q},\boldsymbol{p})$, since it is an empirical distribution, is determined by the state of the system, given by a point X in its full phase space. $f(\boldsymbol{q},\boldsymbol{p})$ can in fact be regarded as describing a macrostate of the system by using it to specify, up to a constant of proportionality, the numbers n_α defining the macrostate; see the paragraph preceding the one containing equation 1.

This legitimate association of approach to equilibrium for the system with the approach of f to an equilibrium distribution, Maxwellian in velocity and spatially homogeneous, has unfortunately suggested to many that approach to equilibrium for a more general gas should be identified with the convergence of the full n-particle distribution function to one that is constant on the relevant energy surface, describing the microcanonical ensemble – in a word, mixing. But the n-particle distribution function is not an empirical distribution and, unlike $f(\boldsymbol{q}, \boldsymbol{p})$, is not merely a characteristic of the actual state X of the gas.

Another source of confusion lies in the widespread tendency to identify states of a physical system with probability measures on its phase space, a tendency partly owing to the very success achieved by statistical mechanics through the use of statistical methods in general, and the standard ensembles in particular; and partly owing to the baleful influence of quantum mechanics, one of the main lessons of which is all too widely believed to be that a detailed description of a quantum mechanical system is fundamentally impossible, so that the state of such a system must be identified with an object of a statistical character, be it the wave function of the system or a positive linear functional on its algebra of observables.

Boltzmann's Entropy Versus the Gibbs Entropy

The identification of the state of a system with a probability measure, given, say, by a density ϱ on its phase space, has led to the widespread identification of the thermodynamic entropy of a system with its *Gibbs entropy*:

$$S_{\mathrm{G}}(\varrho) = -k \int \varrho(X) \log \varrho(X) \,\mathrm{d}X \,. \tag{2}$$

One of the most important features of the Gibbs entropy is that it is a constant of the motion: Writing ϱ_t for the evolution on densities induced by the motion on phase space, we have that $S_{\mathrm{G}}(\varrho_t)$ is independent of t; in particular it does not increase to its equilibrium value.

It is frequently asked how this can be compatible with the Second Law. The answer is very simple. The Second Law is concerned with the thermodynamic entropy, and this is given by Boltzmann's entropy (1), not by the Gibbs entropy (2). In fact, the Gibbs entropy is not even an entity of the right sort: It is a function of a probability distribution, i.e., of an ensemble of systems, and not a function on phase space, a function of the actual state X of an individual system, the behavior of which the Second Law – and macro-physics in general – is supposed to describe.

The widespread tendency to identify the thermodynamic entropy with the Gibbs entropy rather than with Boltzmann's entropy, while clearly misguided, is easy to understand. Boltzmann's entropy is a bit vague, since it depends upon somewhat arbitrary choices that go into the specification of the macrostates of a system, and, other things being equal, vagueness is bad. This vagueness, however, is of little practical consequence, and indeed upon reflection is quite appropriate for the problem of defining entropy, a concept that somehow relates the microscopic level of description and the (not sharply definable) macroscopic level. But the vagueness is there nonetheless.

Unlike Boltzmann's entropy, the Gibbs entropy is sharply defined, without arbitrariness. It is also a very natural functional of a probability distribution, having a great many useful and appealing features into which I shall not enter here. It is the basis of extremely important developments in information theory and in the ergodic theory of dynamical systems, and has proven to be of great value in the derivation of hydrodynamical laws from suitable microscopic first principles. It is a very fine concept and can be used to define a great many things, but thermodynamic entropy is not one of them.

Certainly contributing to the tendency to identify the thermodynamic entropy with the Gibbs entropy is the fact that for systems in equilibrium the Gibbs entropy agrees with Boltzmann's entropy. More precisely, and more generally, $S(X) = S_{\mathrm{G}}(\varrho)$ whenever ϱ is the uniform distribution over the macrostate to which X belongs. Moreover – and this is probably the origin of the confusion – Boltzmann showed that for a low density gas

$$S(X) \approx -kN \int f(\boldsymbol{q},\boldsymbol{p}) \ln f(\boldsymbol{q},\boldsymbol{p}) \, \mathrm{d}\boldsymbol{q}\mathrm{d}\boldsymbol{p} \tag{3}$$

whenever X belongs to the macrostate defined by the Boltzmann function $f(\boldsymbol{q},\boldsymbol{p})$.

This formula involves the Gibbs entropy for the one-particle distribution f and is appropriate, as mentioned, for a low density gas, in which correlations between particles are insignificant. This has suggested to many that to obtain an expression for entropy valid beyond the low density regime one should apply a similar formula to the full probability distribution, capturing all the correlations then present. The result, of course, is the Gibbs entropy (2). The mistake in doing so lies in failing to appreciate that it is the left hand side of (3), defined by (1), that is fundamental, not the right hand side, which is merely the result of a computation valid in a special situation.

It is widely believed that thermodynamic entropy is a reflection of our ignorance of the precise microscopic state of a macroscopic system, and that if we somehow knew the exact phase point for the system, its entropy would be zero or meaningless. But entropy is a quantity playing a precise role in a formalism governing an aspect of the behavior of macroscopic systems. This behavior is completely determined by the evolution of the detailed microscopic state of these systems, regardless of what any person or any other being happens to know about that state. The widespread acceptance of the preposterous notion that how macroscopic systems behave could be affected merely by what we know about them is simply another instance of the distressing effect that quantum mechanics has had upon the ability of physicists to think clearly about fundamental issues in physics.

3 The Hard Part

In the previous section we addressed the easy part of the problem of irreversibility: Suppose a system, e.g., a gas in a box, is in a state of low entropy at some time. Why should its entropy tend to be larger at a later time? The reason is basically that states of large entropy correspond to regions in phase space of enormously greater volume than those of lower entropy. We now turn to the hard part of the problem: Why should there be an arrow of time in our universe, governed as it is, at the fundamental level, by reversible microscopic laws?

The problem here can be appreciated by focusing on the question: What is the origin of the low entropy initial states? If they are so "unlikely," why should systems find themselves in such states? In many cases, the answer is that we or an experimenter created them, from states of lower entropy still. If we continue to ask such questions, we come to the conclusion that the cause of low entropy states on earth, the source in effect of negative entropy, is our (very low entropy) sun, whose high energy photons are absorbed by the earth, which converts them to a great many low energy photons (having together much larger entropy), permitting entropy-decreasing processes to occur on our planet without violation of overall entropy non-decrease. And if we push further we eventually arrive at a cosmological low entropy state, in the distant past, for the universe as a whole.

And what about the origin of this state? Figure 2, taken from Roger Penrose's *The Emperor's New Mind* [9, page 343], illustrates the difficulty.

Penrose estimates the volume of the region of phase space corresponding to the possible initial states of the universe to be one part in $10^{10^{123}}$ of the entire relevant phase space. He bases this upon the Bekenstein-Hawking entropy of a 10^{80} baryon mass black hole, which, in 'natural units,' is 10^{123}, corresponding to a "Big Crunch state' of volume of order $10^{10^{123}}$, a reasonable estimate, Penrose argues, for the volume of the entire relevant phase space for a closed universe. (It does not much matter what we take as the entropy of the initial state, be it 10^{20} or 10^{80} or 10^{100} – the relevant ratio of volumes will always be of order $10^{10^{123}}$. As to whether $10^{10^{123}}$ is indeed a good estimate for the volume of the relevant phase space, it has been suggested that ∞ might be more on target; see the contribution of Michael Kiessling to this volume.) As to why the universe should have begun in such an exceedingly improbable macrostate, an answer that has often been suggested is that such a state arose from a fluctuation out of equilibrium. In fact, if the universal dynamics were ergodic, such a fluctuation must eventually occur, repeatedly, for all phase points with the possible exception of a set of measure 0.

Nonetheless, this answer is quite unsatisfactory; indeed, according to Feynman [10] it is "ridiculous." The problem is that if the explanation of entropy increase and the arrow of time in our universe is that they have emerged from a low entropy state that arose from a fluctuation, then that fluctuation should have been no larger than necessary – that is, to a state like the present state of the universe, and not to a state of much lower entropy as seems to have existed in

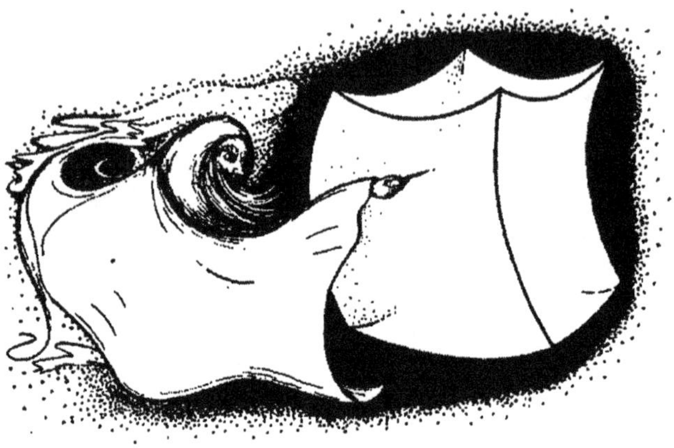

Fig. 2. In order to produce a universe resembling the one in which we live, the Creator would have to aim for an absurdly tiny volume of the phase space of possible universes – about $1/10^{10^{123}}$ of the entire volume, for the situation under consideration. (The pin, and the spot aimed for, are not drawn to scale!)

the past. Here is Feynman, referring to astronomy, to history books and history, and to paleontology:

> Since we always make the prediction that in a place where we have not looked we shall see stars in a similar condition, or find the same statement about Napoleon, or that we shall see bones like the bones that we have seen before, the success of all those sciences indicates that the world did not come from a fluctuation ... Therefore I think it is necessary to add to the physical laws the hypothesis that in the past the universe was more ordered ... than it is today – I think this is the additional statement that is needed to make sense, and to make an understanding of the irreversibility.

The view expressed here by Feynman may seem to be the view at which Boltzmann himself ultimately arrived:

> The second law of thermodynamics can be proved from the mechanical theory if one assumes that the present state of the universe, or at least that part which surrounds us, started to evolve from an improbable state and is still in a relatively improbable state. This is a reasonable assumption to make, since it enables us to explain the facts of experience, and one should not expect to be able to deduce it from anything more fundamental.

However, this statement, with which Boltzmann began his second response to Zermelo [11], probably should not be read as a repudiation of the fluctuation hypothesis ridiculed by Feynman, since towards the end of the very same article, in its §4, Boltzmann advocated this hypothesis.

Be that as it may, what we need to complete Boltzmann's account of irreversibility is a reasonable hypothesis on the initial state of the universe, in effect an additional physical law. This hypothesis must imply that this state had very low entropy, but, unlike Feynman's suggestion above, it need not explicitly stipulate that this be so. What is required is that the initial state not be too contrived, that it be somehow reasonable, indeed that it be natural.

Moreover, it seems that gravity, even classical gravity, affords just such a possibility. This is because the attractive nature of the gravitational interaction is such that gravitating matter tends to clump, clumped states having larger entropy. (For more on this see the contribution of Michael Kiessling to this volume.) This important difference from the behavior of ordinary matter, for which gravity can be more or less ignored, is well illustrated by Penrose in [9, page 338], see Fig. 3. What is important for our purposes here is that what is arguably

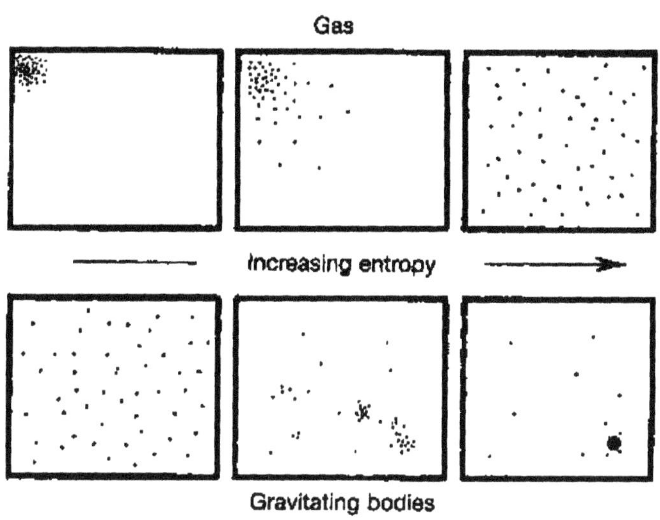

Fig. 3. For an ordinary gas, increasing entropy tends to make the distribution more uniform. For a system of gravitating bodies the reverse is true. High entropy is achieved by gravitational clumping – and the highest of all, by collapse to a black hole.

the most random, typical, natural, and least contrived initial state for a system of gravitating particles, one in which they are uniformly distributed over space (with, say, all of them at rest) also happens to be a state of very low entropy, exactly what is needed to complete Boltzmann's account of irreversibility.

4 Typicality and Explanation

According to the Boltzmannian scenario propounded here, the overwhelming majority, as measured by relative phase-space volume, of phase points in a (very small) initial macrostate of the universe evolve in such a way as to behave – for reasonable times, that are not too large on the time scale defined by the present age of the universe, since the big bang – thermodynamically, with suitably closed subsystems having increasing entropy, exhibiting irreversible behavior, and so on. In other words, *typical* phase space points yield the behavior that it was our (or Boltzmann's) purpose to explain. Thus we should expect such behavior to be prevail in our universe.

This raises the question, What is the force of such an explanation, based, as it is, merely on what would "typically" happen, not on what must inevitably happen? Now, as a scientist I see no problem here. What more could reasonably be expected by way of explanation? Boltzmann [11] makes a similar point when he complains that

> The applicability of probability theory to a particular case cannot of course be proved rigorously. ... Despite this, every insurance company relies on probability theory. ... The assumption that these rare cases are not observed in nature is not strictly provable (nor is the entire mechanical picture itself) but in view of what has been said it is so natural and obvious, and so much in agreement with all experience with probabilities, ...that any doubt on this point certainly cannot put in question the validity of the theory when it is otherwise so useful.
>
> It is completely incomprehensible to me how anyone can see a refutation of the applicability of probability theory in the fact that some other argument shows that exceptions must occur now and then over a period of eons of time; for probability theory itself teaches just the same thing.

However, as philosophers we might in fact be inclined to demand more. We might ask whether we've really explained a phenomenon (such as irreversibility, an arrow of time in our world, or whatever) if we have (merely!) shown that it is typical, but that exceptions, while in a sense extraordinarily rare, still exist in abundance – at least as logical and mathematical possibilities. But in doing so it seems to me that we will have to face very hard questions about just what is meant by scientific explanation, or explanation of any sort. We might conclude that the questions that we are asking are relevant, not only to the issue of the origin of irreversibility, but to most, if not all, problems – such as Hume's problem of induction [12], of why the future should at all be expected to resemble the past – about the nature of science and the status and justification of scientific theories.

The point is that in science, as in life, we must learn to cope with uncertainty. As Boltzmann noted in the above quotation, the mechanical picture itself is not strictly provable. Nor is any other scientific theory. No matter how strongly a particular theory seems to be supported by the evidence at hand, there are always logically possible alternative accounts of the very same evidence, however far fetched these may be. The account that we tend to believe – what we consider to be the inference to the best explanation – seems to us simpler, more elegant, more natural, and certainly less contrived, than the alternatives. Nonetheless, our expectations for future behavior, based on the best available evidence, can never be regarded as inevitable consequences of that evidence, but only – at best – as overwhelming likely.

With regard to 'probablistic' or 'statistical' explanations, involving uncertainty about initial conditions, we say that a phenomenon has been explained if it holds for *typical* initial conditions, that is with rare exceptions as defined by a suitable "measure" μ of typicality. The phenomenon has been explained if the set E of exceptional initial conditions satisfies $\mu(E) \ll 1$.

Of course it is essential that the measure of typicality be natural and not contrived. It should be an object that could, somehow, have been agreed upon before the phenomenon to be explained was even considered. For dynamical systems such as we are discussing here, the measure of typicality should be naturally related to the dynamics, and the most common such condition is that of stationarity. After all, our notion of typicality should not change with time. And for classical mechanics, for which symplectic or canonical structure plays a crucial role in the dynamics, the most natural measure is the volume measure defined by the symplectic coordinates, the measure we have been invoking throughout this article.

Here is a small point, but one worth making if we intend to worry a bit about the justification of explanation via typicality: While typicality is usually defined – as it was here – in terms of a probability measure, the basic concept is not genuinely probablistic, but rather a less detailed concept. A measure μ of typicality need not be countably additive, nor even finitely additive. Moreover, for any event E, if μ is merely a measure of typicality, there is no point worrying about, nor any sense to, the question as to the real meaning of say '$\mu(E) = 1/2$'. Distinctions such as between '$\mu(E) = 1/2$' and '$\mu(E) = 3/4$' are distinctions without a difference.

The only thing that matters for a measure μ of typicality is '$\mu(E) \ll 1$': a measure of typicality plays solely the role of informing us when a set E of exceptions is sufficiently small that we may in effect ignore it and regard the phenomenon in question, occurring off the set E, as having been explained. And our future expectation is for behavior that is typical in this sense. After all, our firm belief in the impossibility of perpetual motion machines is not based primarily on the fact that none has ever been achieved so much as on Boltzmann's account of why such a machine would be a practical impossibility.

What I've just described is the calculus of explanation via appeal to typicality. The rigorous justification of such a calculus is, as I've already indicated, another matter entirely, a problem, like so many similar problems involving crucial uncertainty, of extraordinary difficulty. To begin to address this problem we would have to investigate how explanation and typicality are related, which would of course require that we analyze exactly what is meant by 'explanation' and by 'typicality.' If typicality is not probability – the explication of which is itself a controversial problem of immense difficulty – it would help to be aware of this from the outset.

Much has been written about such matters of justification and nondeductive rationality (for a recent discussion see [13]). Much more should be written and, no doubt, will be written. But not here!

References

1. J.L. Lebowitz: Physics Today **46**, 32 (1993); 'Microscopic Reversibility and Macroscopic Behavior: Physical Explanations and Mathematical Derivations'. In: *25 Years of Non-Equilibrium Statistical Mechanics, Proceedings of the Sitges Conference in Barcelona, Spain, 1994*, Lecture Notes in Physics, ed. by J.J. Brey, J. Marro, J.M. Rubí, M. San Miguel (Springer, Heidelberg 1995). For a very recent comprehensive overview see 'Microscopic Origin of Irreversible Macroscopic Behavior', preprint, available on the server at xxx.lanl.gov and in the Texas Mathematical Physics Archives.
2. J. Bricmont: 'Science of Chaos or Chaos in Science?'. In: *The Flight from Science and Reason*, Annals of the New York Academy of Sciences **775**, ed. by P.R. Gross, N. Levitt, M.W. Lewis (The New York Academy of Sciences, New York 1996) pp. 131–175
3. W. Thomson: Proceedings of the Royal Society of Edinburgh **3**, 325 (1874); reprinted in [4]
4. S.G. Brush: *Kinetic Theory* (Pergamon, Oxford 1966)
5. E. Schrödinger: *What is Life? : The Physical Aspect of the Living Cell with Mind and Matter & Autobiographical Sketches* (Cambridge University Press, Cambridge 1992)
6. E. Schrödinger: *What is Life? and Other Scientific Essays* (Doubleday Anchor Books, New York 1965) section 6
7. L. Boltzmann: Annalen der Physik **57**, 773 (1896); reprinted and translated as Chapter 8 in [4]
8. J.W. Gibbs: *Elementary Principles in Statistical Mechanics* (Yale University Press, New Haven 1902) Chapter XII
9. R. Penrose: *The Emperor's New Mind* (Oxford University Press, New York, Oxford 1989)
10. R. Feynman: *The Character of Physical Law* (MIT Press, Cambridge 1967) Section 5
11. L. Boltzmann: Annalen der Physik **60**, 392 (1897); reprinted and translated as Chapter 10 in [4]
12. D. Hume: *An Enquiry Concerning Human Understanding* (Prometheus Books, Amherst, New York 1988)
13. L. Bonjour: *In Defense of Pure Reason: A Rationalist Account of A Priori Justification*, Cambridge Studies in Philosophy (Cambridge University Press, Cambridge 1997)

Microscopic Time Reversibility and the Boltzmann Equation

Herbert Spohn

Zentrum Mathematik and Physik Department
TU München, D-80290 München

Abstract. O.E. Lanford proves that in the Boltzmann–Grad limit the dynamics of hard spheres is governed by the Boltzmann equation (with a technical restriction on the time span). This result allows us to explain, with more precision according to usual standards, the emergence of an irreversible macroscopic equation out of the reversible microscopic dynamics.

The tension between time-reversibility on the microscopic level and the experienced irreversibility at the human scale has intrigued natural scientists and philosophers alike. I have no intention to even enter the stage. As a physicist I only want to advertise the pair classical Newtonian mechanics (which is invariant under time reversal) and the Boltzmann equation (which is not) for two reasons. Firstly, the pair is a realistic, experimentally well supported description of dilute gases. Secondly, the Boltzmann equation is derived with mathematical rigor [1–3], which is a welcome safeguard against incorrect claims.

I have nothing original to say. Despite, occasionally even dark, clouds of confusion, there have always been physicists who clearly understood the issue. My only excuse is that a concrete example will serve well and provides a secure basis for further understanding and enquiry.

Let me remind you: As the fundamental theory we assume that the dynamics of atoms (or molecules) are governed by *Newton's equation of motion*,

$$m\frac{\mathrm{d}^2}{\mathrm{d}t^2}q_j(t) = F_j\left(q_1(t), ..., q_N(t)\right), \qquad j = 1, ..., N. \tag{N}$$

For simplicity, the particles are all equal with common mass m. $q_j(t)$ is the position of molecule number j at time t. The force, F_j, acting on molecule j depends, in principle, on the positions of all other molecules. As Loschmidt noticed, the time reversal $t \to -t$ leaves (N) unaltered because $(-1)^2 = 1$. (N) *is invariant* under time reversal. To study the physical properties of low density gases, Boltzmann introduced the distribution function $f(\boldsymbol{r}, \boldsymbol{v}, t)$ which counts the number of molecules in the position volume element d^3r at $\boldsymbol{r}$ and the velocity volume element d^3v at $\boldsymbol{v}$ at time t. $f(t)$ is governed by the *Boltzmann equation*

$$\frac{\mathrm{d}}{\mathrm{d}t}f(t) = -\boldsymbol{v}\cdot\nabla_r f(t) + Q\left(f(t), f(t)\right). \tag{B}$$

The first term is the change in f through rectilinear, free motion of particles, whereas the second term, the collision term, accounts for binary collisions between molecules. When changing $t \to -t$ one also has to change the sign of the velocity $\boldsymbol{v}$. Thus (B) becomes

$$\frac{\mathrm{d}}{\mathrm{d}t} f(t) = -\boldsymbol{v} \cdot \nabla_r f(t) - Q\left(f(t), f(t)\right), \tag{aB}$$

sometimes called *anti-Boltzmann equation* because of the minus sign in front of the collision term. (B) *is not invariant* under time-reversal.

The irreversible character of the Boltzmann equation can be read off most easily by introducing the H-functional

$$H(t) = k_B \int \mathrm{d}^3\boldsymbol{r}\mathrm{d}^3\boldsymbol{v}\, f(\boldsymbol{r}, \boldsymbol{v}, t) \log f(\boldsymbol{r}, \boldsymbol{v}, t).$$

$H(t)$ is the negative of the entropy. Provided $f(t)$ is governed by (B) it decreases in time and eventually settles at an equilibrium value determined through the initial average density and average kinetic energy (given the gas is enclosed in a container). However under (aB), $H(t)$ increases without bound until f becomes negative and thus looses its physical interpretation. The arguments to be presented will be more transparent when discussed for the full Boltzmann equation (B) rather than for $H(t)$ only.

In his original 1872 paper Boltzmann seems to assert

Claim 1: (B) is a strict consequence of (N).

This led to Loschmidt's

Objection 1: (B) cannot follow from (N) because both equations transform differently under time reversal.

Note that, on face of it, Objection 1 is a mathematical, not an empirical issue. But lingering behind is the question of whether the irreversible evolution equations so well suited to describe the macroscopic world around us are *consistent* with reversible microscopic dynamics.

At this stage it may help to follow the distinction introduced by R. Penrose in his book "Shadows of the Mind" [4], page 237. He classifies

- *Z-mysteries or puzzle mysteries.* These are genuinely puzzling discoveries and insights which however become understood and commonly accepted once further evidence is available.

- *X-mysteries or paradox mysteries.* Such findings cannot be accounted for by current theories and will eventually lead to a radical change in the theoretical framework.

Of course, there is some arbitrariness in these definitions. But one accepted example of an X-mystery is the spectrum of the black body radiation before Planck. There is a certain tendency to up-grade one's own problem to the highest mystery category. Thus Objection 1 is usually called Loschmidt's reversibility paradox. In my opinion the Loschmidt argument was a genuine puzzle at the time. Up-grading it to a paradox has done more harm than good and did not enhance a clear understanding. In particular, the uninitiated is raised to a level of expectancy which can never be satisfied.

Equations (N) and (B) are incomplete. As mathematics tells us they have to be supplemented with initial conditions, conventionally taken at time $t = 0$. Since (N) is a second order differential equation, we need positions and velocities $(q_1(0), v_1(0), ..., q_N(0), v_N(0))$ as initial conditions, whereas for the Boltzmann equation one has to specify the initial distribution function $f(\boldsymbol{r}, \boldsymbol{v}, 0)$. While the Equation (N) is time-reversal invariant, particular solutions corresponding to given initial data certainly are not, leaving aside some obvious exceptions. Therefore Claim 1 is now refined to

Claim 2: If at time $t = 0$ $(q_1(0), v_1(0), ..., q_N(0), v_N(0))$ are close to $f(\boldsymbol{r}, \boldsymbol{v}, 0)$, then $(q_1(t), v_1(t), ..., q_N(t), v_N(t))$ (as solution of (N)) are close to $f(\boldsymbol{r}, \boldsymbol{v}, t)$ (as solution to (B)), at least over some reasonable time span.

"Close", here, means that the number of particles in the volume element $\mathrm{d}^3\boldsymbol{r}\mathrm{d}^3\boldsymbol{v}$ at time t is given by $Nf(\boldsymbol{r}, \boldsymbol{v}, t)$ up to some small error. This is like many pixels approximating a seemingly continuous picture, as we all know from TV. Such a notion of "close" can be defined more precisely, but it will not be needed here.

Objection 2: Claim 2 is not valid.

The argument is now a little bit more subtle, but again based on time-reversal. We assume that the initial data $(q_1(0), v_1(0), ..., q_N(0), v_N(0))$ well approximate $f(\boldsymbol{r}, \boldsymbol{v}, 0)$ and let the dynamics run for a time span $t_0 > 0$. According to Claim 2, at time t_0 positions and velocities of the particles are still close to the Boltzmann distribution function $f(t_0)$, in particular $(q_1(t_0), -v_1(t_0), ..., q_N(t_0), -v_N(t_0))$ is close to $f(\boldsymbol{r}, -\boldsymbol{v}, t_0)$. On this basis Claim 2 asserts the same situation to be found at time $2t_0$. However, after velocity reversal (N) traces back its own history, whereas (B) further proceeds towards equilibrium. Thus Claim 2 cannot be valid, QED.

Note that we did not make a physical velocity reversal. It is just used to construct a counterexample against the strong Claim 2.

The famous resolution of Boltzmann lies in admitting exceptional initial conditions which however have a small probability. In modern parlance

Assertion 3: For typical initial data $(q_1(0), v_1(0), ..., q_N(0), v_N(0))$ the assertion in Claim 2 holds.

Mathematical theorems [2,3] confirm Boltzmann's point of view and make the somewhat vexing notion of "typicality" precise. We might say that typical initial conditions break the time reversal invariance.

Interlude: What is a typical number in the interval [0,1]?

Well, it is just a number between 0 and 1 drawn at random, the one my computer would print upon typing RAN. Is 1/3 typical? What about $1/\sqrt{2}$? Clearly we need a criterion to decide, which I propose as follows. Let $x \in [0,1]$ with binary expansion $x = 0.\sigma_1\sigma_2\sigma_3...$, where either $\sigma_n = 0$ or $\sigma_n = 1, n = 1,2....$ x is *typical* if the fraction of singletons 0 (and of singletons 1) is $\frac{1}{2}$, the fraction of pairs 00 (and of pairs 01, 10, 11) is $\frac{1}{4}$, the fraction of triples 000 (and of triples 001, 010, 100, 011, 101, 110, 111) is $\frac{1}{8}$, etc.. A well known theorem states that the set of all typical numbers has length (Lebesgue measure) 1. In this sense almost all numbers in [0,1] are typical. A good random number generator will make the fractions as required at least for short sequences. $\frac{1}{3} = 0.0101...$ is not typical because the fraction of pairs is already wrong. $1/\sqrt{2} = 0.10110101000001...$? I do not know, but statistical tests up to $2 \cdot 10^5$ digits seem to indicate yes. Thus, despite their abundance to explicitly exhibit a typical number is not so easy. □

"Typical" in Assertion 3 means roughly that we ignore all initial data which are not close to $f(\boldsymbol{r},\boldsymbol{v},0)$. Of those which are, we make a random choice. More precisely, we sample q_1, v_1 according to $f(\boldsymbol{r},\boldsymbol{v},0)$. Then independently q_2, v_2 again according to $f(\boldsymbol{r},\boldsymbol{v},0)$, etc., 10^{23} times. If some q_n is too close to the previously sampled positions $q_j, j = 1,...,n-1$, i.e. $|q_j - q_n| \leq a$ with a the molecular diameter, then q_n is not accepted. In this fashion we construct the random initial data $(q_1, v_1, ..., q_N, v_N)$ close to $f(\boldsymbol{r},\boldsymbol{v},0)$. I emphasize that "typical" refers to a specific $f(\boldsymbol{r},\boldsymbol{v})$. Sometimes typical is taken as relative to some global equilibrium measure, as for example the uniform distribution on the surface of constant energy. This can be very misleading because the nonequilibrium initial conditions of interest would have only a tiny probability with respect to this measure. Note that the random choice is made only at $t = 0$ relative to $f(\boldsymbol{r},\boldsymbol{v},0)$. After that the dynamics proceed deterministically according to either (N) or (B). In Assertion 3 it would certainly be desirable to have a non probabilistic characterization of good initial data, i.e. of initial conditions for (N) which closely follow (B). Such a criterion is not in sight, which only reflects the fact that, despite intense efforts, the dynamics of many interacting particles is not a well understood subject.

In Newton's equation of motion we may well ask what happens for $t < 0$.

Assertion 4: The solution of (N) for a typical initial datum $(q_1(0), v_1(0), ..., q_N(0), v_N(0))$ close to $f(\boldsymbol{r},\boldsymbol{v},0)$ is at time $t < 0$ close to the solution of the anti-Boltzmann equation (aB) when solved *backward* in time.

This is the *real* puzzle, because it contradicts everyday experience. Sure enough: Let us consider a gas in a container and assume that at time $t = 0$ we observe the gas somewhat compressed to the right half of the vessel. Then in the future ($t > 0$) the gas will expand and become uniformly distributed. This is precisely what Assertion 3 states. In the past ($t < 0$) we certainly expect that the gas was even further compressed to the right until the time when an outside experimenter had carefully prepared the gas in one corner (which is easy). However, Assertion 4 (a valid consequence of (N)) states just the opposite, namely that the gas further expands for $t < 0$ as well. *So why does the method which works so well for the future fail so badly for the past?*

It may be useful to repeat the argument in other words. Suppose at time $-\tau, \tau > 0$, we prepared an initial configuration of molecules typical for $f(\boldsymbol{r},\boldsymbol{v},-\tau)$

corresponding to compression of the gas in the right-hand corner. Then at time $t = 0$ the molecule configuration (as obtained by the solution of (N)) will be *untypical* for $f(\boldsymbol{r}, \boldsymbol{v}, 0)$ (as obtained from the solution of (B) with initial datum $f(\boldsymbol{r}, \boldsymbol{v}, -\tau)$). The configuration $(q_1(0), v_1(0), ..., q_N(0), v_N(0))$ at time $t = 0$ is not a random configuration relative to $f(\boldsymbol{r}, \boldsymbol{v}, 0)$, because it has a very special past. This causes no harm to the future evolution since we can always refer back to time $-\tau$. However, if the knowledge at time $t = 0$ is all that is available, we would make the wrong retrodiction.

In the context of a lab situation the puzzle does not really apply. We use the empirical fact that constrained equilibrium states can be manufactured without difficulty. If the constraint is released the future time evolution proceeds according to Assertion 3, whereas the past time evolution is only hypothetical. However, irreversible processes occur everywhere around us – entropy is constantly generated – without careful arrangement of some initial constrained equilibrium state. Thus, in the spirit of Assertion 3, at some sufficiently early time initial conditions typical for some nonequilibrium state must have been present, the relaxation of which we witness today. How early? No doubt the sun has been a fantastically stable source of low entropy radiation. This empirical fact forces us to go back in time over cosmological scales.

I take Assertions 3 and 4 as a valid indication that the thermodynamic arrow of time is linked to the cosmological arrow of time. Of course, this does not mean that if the universe were one day to start to contract, heat would flow from cold to hot. Rather if we want to understand why we live now in a nonequilibrium world well governed by irreversible evolution equations (as the Boltzmann equation), we are forced to consider the universe as a whole.

I recommend the books by R.P. Feynman [5] and by R. Penrose [6], presenting very similar points of view, the book by C. Cercignani [7] for a historical background, and the masterly lecture by J.L.Lebowitz [8], from whom I learned.

References

1. O.E. Lanford, *Time Evolution of Large Classical Systems.* Lecture Notes in Physics Vol. 38, pp. 1-111, ed. J. Moser, (Springer, Berlin, 1975)
2. H. Spohn, *Large Scale Dynamics of Interacting Particles*, (Springer, Heidelberg 1991)
3. C. Cercignani, R. Illner, and M. Pulvirenti, *The Mathematical Theory of Dilute Gases*, (Springer, New York 1994)
4. R. Penrose, *Shadows of the Mind*, (Oxford University Press 1994)
5. R. P. Feynman *The Character of Physical Law*, (British Broadcasting Corporation, London 1965)
6. R. Penrose, *The Emperor's New Mind* (Oxford University Press, 1989)
7. C. Cercignani, *Ludwig Boltzmann, The Man Who Trusted Atoms*, (Oxford University Press 1998)
8. J. L. Lebowitz, *Macroscopic laws, microscopic dynamics, time's arrow and Boltzmann's entropy.* In: *Physica* **A 194**, *1 (1993)*

The Direction of Time *

Oliver Penrose

Heriot-Watt University, Riccarton, Edinburgh EH14 4AS, UK

Abstract. It is argued, using a relativistic space-time view of the Universe, that Reichenbach's "principle of the common cause" provides a good basis for understanding the time direction of a variety of time-asymmetric physical processes. Most of the mathematical formulation is based on a probabilistic model using classical mechanics, but the extension to quantum mechanics is also considered.

1 Does Time Have a Direction?

When we speak of "the direction of time" we are not really saying that time itself has a direction, any more than we would say that space itself has a direction. What we are saying is that the events and processes that take place in time have a direction (i.e. they are not symmetrical under time reversal) and, moreover, that this time direction is the same for all the events.

To see the distinction, imagine that you look into a body of water and see some fish all facing in the same (spatial) direction. You would not attribute the directedness of this fish population to any "direction of space"; it would be much more natural, assuming the body of water to be a river, to attribute the directedness of the fish population to the direction of a physical phenomenon which is external to them: that is, to the direction of the flow of the river.

An external influence of this kind is not the only mechanism by which a set of objects (or creatures) can line up along a particular direction: they may line up because of an interaction between them rather than some external influence. It might be that the water is stagnant, imposing no direction on the fish from outside, but that the fish happen to like facing in the same direction as their neighbours. In this case there is an interaction between neighbouring fish, but the interaction itself is invariant under space reversal: if two fish are facing the same way and you turn both of them around, they should be just as happy about each other as before. In such cases, when an asymmetrical arrangement arises out of a symmetrical interaction, we speak of spontaneous symmetry breaking.

In the case of the time asymmetry, we are surrounded by examples of processes that are directed in time, in the sense that the time-reversed version of the same process would be virtually impossible (imagine raindrops rising through the air and attaching themselves to clouds, for example); moreover, as we shall

* This paper is dedicated to the memory of Dennis Sciama (1926–1999), the sadly missed friend who taught me so much about cosmology, the foundations of physics, and scientific writing.

see in more detail later, the time direction of each process is the same every time it happens (imagine a world where raindrops fell on some planets but rose on others[1]). Just as in the case of the asymmetrical spatial arrangement of fishes, there are two possible explanations of such a large-scale asymmetry: either it is due to some overarching asymmetry which separately causes the time direction of each of the individual processes, or else it is due to some linkage between different events which, though itself symmetric under time reversal, leads most or all of them to adopt the same time direction as their neighbours so that an asymmetry can arise through spontaneous symmetry breaking.

The most obvious physical linkage between different processes comes from the fact that the same material particles may participate in different processes at different times; the states of the same piece of matter at different times are linked by dynamical laws, such as Newton's laws of motion which control the motion of the particles composing that matter. Processes involving different particles can also be linked, because of interactions between the particles, and the interactions are subject to these same dynamical laws. The dynamical laws are symmetrical under time reversal; nevertheless, in analogy with our example of the fishes, this symmetry is perfectly consistent with the asymmetry of the arrangement of events and processes that actually takes place.

In this article I will look at and characterize the time directions of various types of asymmetric processes that we see in the world around us and investigate whether all these asymmetries can be traced back to a single asymmmetric cause or principle, or whether, on the other hand, they should be attributed to some form of symmetry breaking. The conclusion will be that there is, indeed, a single asymmetric principle which accounts for the observed asymmetry of most physical processes. Many of the ideas used come from an earlier article written with I. C. Percival [17], but others are new, for example the extension to quantum mechanics.

2 Some Time-Directed Processes

Let's begin by listing some physical processes that are manifestly asymmetrical under time reversal.

The subjective direction of time. Our own subjective experience gives us a very clear distinction between the future and the past. We remember the past, not the future. Through our memory of events that have happened to us in the past, we know what those events were; whereas a person who "knew" the future in such a way would be credited with supernatural powers. Although some future events can be predicted with near certainty (e.g. that the Sun will rise tomorrow), we do not know them in the same sense as we know the past until they have actually happened. And many future events are the subject of great uncertainty: future scientific discoveries, for example. We just do not know in advance what the new discoveries will be. That is what makes science so fascinating – and so dangerous.

[1] This possibility may not be quite so fanciful as it sounds. See [20]

Part of our ignorance about the future is ignorance about what we ourselves will decide to do in the future. It is this ignorance that leads to the sense of "free will". Suppose you are lucky enough to be offered a new job. At first, you do not know whether or not you will decide to accept it. You feel free to choose between accepting and declining. Later on, after you have decided, the uncertainty is gone. Because your decision is now in the past, you know by remembering what your decision was and the reasons for it. You no longer have any sense of free will in relation to that particular decision. The (temporary) sense of free will which you had about the decision before you made it arose from your ignorance of the future at that time. As soon as the ignorance disappeared, so did the sense of free will.

It is not uncommon to talk about the "flow" of time, as if time were a moving river in which you have to swim, like a fish, in order to stay where you are. Equivalently one could look at this relative motion from the opposite point of view, the fish swimming past water that remains stationary; in the latter picture there is no flow as such, but instead a sequence of present instants which follow one another in sequence. The time direction of this sequence does not come from any "flow", but rather comes from the asymmetry of memory mentioned earlier. At each instant you can remember the instants in the past that have already happened, including the ones that happened only a moment ago, but you cannot remember or know the ones in the future. The time asymmetry of the "flow" is a direct consequence of the asymmetry of memory; calling it a "flow" does not tell us anything about the direction of time that we do not already know from the time asymmetry of memory.

Recording devices. Why is it, then, that we can only remember the past, not the future? It is helpful to think of devices which we understand better than the human memory but which do a similar job although they are nothing like as complex and wonderful. I have in mind recording devices such as a camera or the "memory" of a computer. These devices can only record the past, not the future, and our memories are time-asymmetric in just the same way.

Irreversible processes in materials. Various macroscopic processes in materials have a well-defined time direction attached. Friction is a familiar example: it always slows down a moving body, never accelerates it. The internal friction of a liquid, known as viscosity, similarly has a definite time direction. Other examples are heat conduction (the heat always goes from the hotter place to the colder, never the other way) and diffusion. The time directions of all the processes in this category are encapsulated in the Second Law of Thermodynamics, which tells us how to define, for macroscopic systems that are not too far from equilibrium, a quantity called entropy which has the property that the entropy of an isolated system is bound to increase rather than to decrease. The restriction to systems that are not too far from equilibrium can be lifted in some cases, notably gases, for which Boltzmann's H-theorem [3] provides a ready-made non-decreasing quantity which can be identifed with the entropy.

Radio transmission. When the electrons in a radio antenna move back and forth, they emit expanding spherical waves. Mathematically, these waves are described by the retarded solutions of Maxwell's equations for the electromagnetic field (i.e. the solutions obtained using retarded potentials). Maxwell's equations also have a different type of solution, the so-called advanced solutions, which describe contracting spherical waves; but such contracting waves would be observed only under very special conditions. Although Maxwell's equations are invariant under time reversal, the time inverse of a physically plausible expanding-wave solution is a physically implausible contracting-wave solution.

The expanding universe. Astronomical observation tells us that the Universe is expanding. Is the fact that our Universe is expanding rather than contracting connected with the other time asymmetries mentioned above, or is it just an accident of the particular stage in cosmological evolution we happen to be living in?

Black holes. General Relativity theory predicted the possibility of black holes, a certain type of singularity in the solution of Einstein's equations for space-time, which swallows up everything that comes near to it. The time inverse of a black hole is a white hole, an object that would be spewing forth matter and/or radiation. It is believed [18] that black holes do occur in our Universe, but that white holes do not.

3 The Dynamical Laws

The astonishing thing about the processes listed above is that, although they are all manifestly asymmetrical under time reversal, every one of them takes place in a system governed, at the microscopic level, by a dynamical law which is *symmetrical* under time reversal. For many of these processes, a perfectly adequate dynamical model is a system of interacting particles governed by Newton's laws of motion – or, if we want to use more up-do-date physics, by the Schrödinger equation for such a system. In the case of the radio antenna, the dynamical model is provided by an electromagnetic field governed by Maxwell's equations, and in the cosmological examples it is a space-time manifold governed by Einstein's equations. All these dynamical laws are invariant under time reversal.

To provide a convenient way of discussing the laws of dynamics and properties of these laws such as invariance under time reversal, I'll use a purely classical model of the Universe. The possibility of generalizing the model to quantum mechanics will be discussed in Section 7. The model assumes a finite speed of light, which no moving particle and indeed no causal influence of any kind can surpass, and is therefore compatible with relativity theory.

Imagine a space-time map in which the trajectories of all the particles (atoms, molecules, etc.) in the model universe are shown in microscopic detail. In principle the electromagnetic field should also be represented. The space-time in which the map is drawn will be denoted by U. Fig. 1 shows a one-dimensional map of this kind, containing just one trajectory. For generality, we shall use (special) relativistic mechanics, with a finite speed of light, so that the space in Fig. 1 represents Minkowski space-time. (Similar diagrams could be drawn for non-flat cosmological models such as the Einstein-de Sitter model [21].)

For each space-time point X, we define its *future zone* to comprise all the space-time points (including X istelf) that can be reached from X by travelling no faster than light. A signal or causal influence that starts from X cannot reach any space-time point outside the future zone of X.

The *past zone* of X comprises those space-time points (including X itself) from which X can be reached by travelling no faster than light. Signals starting from space-time points outside the past zone of X cannot influence what happens at X. Since real particles travel slower than light, any particle trajectory passing through X stays inside the past and future zones of X.

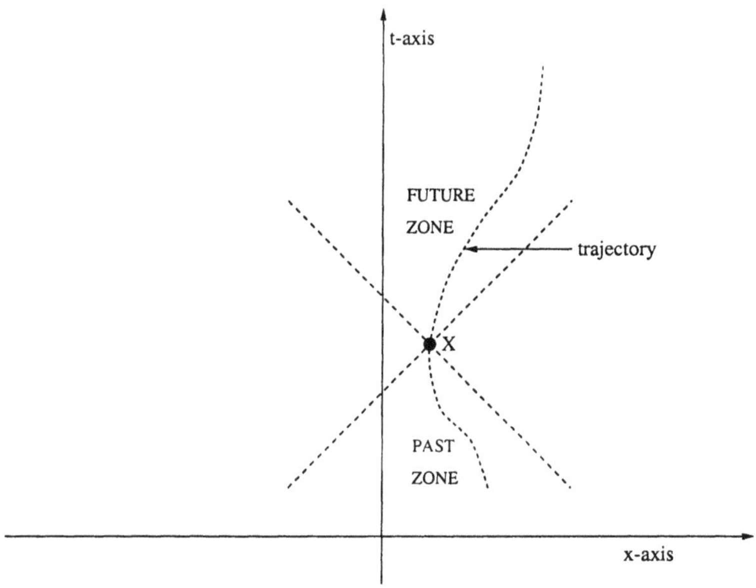

Fig. 1. Typical particle trajectory in a universe with one space dimension. The future and past zones of the space-time point X on the trajectory are shown as quadrants. The two dotted half-lines bounding the future (past) zone of X are the trajectories of the two possible light rays starting (ending) at X; their equations are $x = ct$ and $x = -ct$ where c is the speed of light – which, for the purpose of drawing the diagram, has been taken to be 1 unit.

By the *history* of the model universe being considered, I mean (disregarding the electromagnetic field for simplicity) the totality of all the trajectories in it; since each trajectory is a set of points in U, the history is also a certain set of points in U. This history will be denoted by the symbol ω. Given a space-time region A within U, we shall use the term the *history of* A to mean the intersection of A with the history of U; that is to say, the totality of all pieces of trajectory that lie inside A. The history of A will be denoted by $\omega \cap A$ or just a. See Fig. 2.

The laws of dynamics place certain restrictions on the histories; for example, if the model universe were to consist entirely of non-interacting particles, then all the trajectories would be straight lines. We shall not assume that all the trajectories are straight lines, but we shall make two specific assumptions, both of which are satisfied by Newton's equations for the mechanics of a system of particles and by Maxwell's equations for the electromagnetic field: these assumptions are that the dynamical laws are (i) invariant under time reversal and (ii) deterministic. (We could also require relativistic invariance but this is not essential). By time-reversal invariance we mean that the time inverse of any possible history (obtained by inverting the space-time map of that history) is also a possible history – or, more formally, that if R denotes the time-reversal transformation in some inertial frame of reference, implemented by reversing the time co-ordinate in that frame of reference, and if ω is any history that is compatible with the laws of dynamics, then $R\omega$ is also compatible with the laws of dynamics.

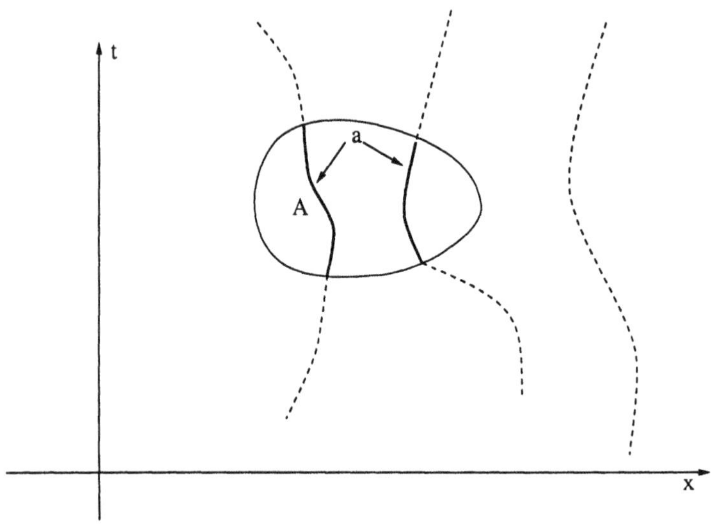

Fig. 2. The history a of the egg-shaped space-time region A consists of the the two part-trajectories which lie inside A.

The following definitions will help us to formulate the concept of determinism in this theory. By the *past zone* of a region A, denoted $PZ(A)$, I mean the union of the past zones of all the points in A. The region A istelf is part of $PZ(A)$. If A and B are two disjoint regions, it may happen that the part of $PZ(A)$ lying outside B consists of two (or more) disconnected parts, as in Fig. 3. In that case, we shall say that B *severs* $PZ(A)$. In a similar way, one region can sever the future zone of another.

Now we can formulate what we mean by determinism in this theory. If a region B severs the past zone of a region A, then any particle or causal influence affecting the history of A must pass through B. The history of B includes all these causal influences, and so it determines the history of A. In symbols, this means that there exists a function f_{AB} from the history space of B to that of A

such that the laws of dynamics require

$$a = f_{AB}(b). \tag{1}$$

For deterministic dynamics, such an equation will hold for every pair of regions A, B such that B severs the past zone of A.

Since we are assuming the laws of dynamics to be time-symmetric, the time inverse of (1) also holds. That is to say, if there is a region B' which severs the *future* zone of A then the dynamical state of A is completely determined by that of B'. In other words, there exists a function $g_{AB'}$ such that the laws of dynamics requre

$$a = g_{AB'}(b'), \tag{2}$$

and such an equation holds for every pair of regions such that B' severs the future zone of A.

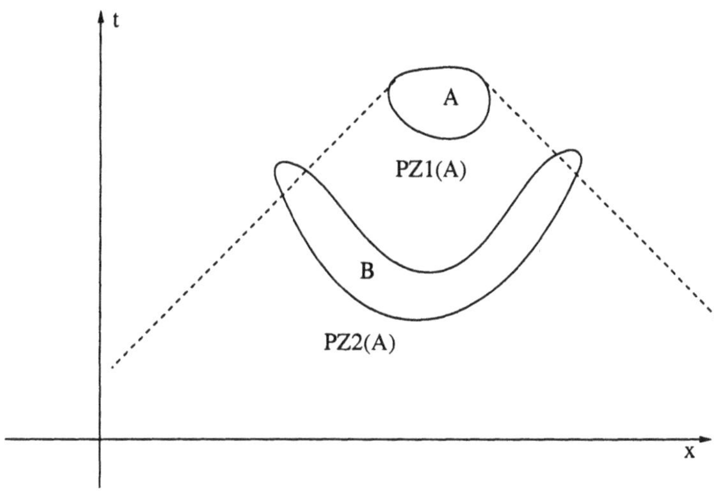

Fig. 3. Example where B severs the past zone of A. The past zone of A consists of three separate parts: one part inside B, and two disconnected parts outside B, marked $PZ1(A)$ and $PZ2(A)$. By definition, A itelf is a part of $PZ1(A)$

4 Probabilities: The Mathematical Model

As we have seen, time direction is not a property of the laws of motion themselves. The laws of motion are too general to solve the problem of time asymmetry, since they do not distinguish between a physically reasonable motion and its physically unreasonable time inverse. To characterize the time asymmetry we need to say something about which solutions of the dynamical equations are likely to occur in the real world. This can be done using probability concepts. The mathematical model I shall use will be based on the following postulates:

i. there is a probability measure on the space Ω consisting of of all conceivable microsopic histories of the model universe U (including histories that do not obey the dynamical laws);
ii. the measure is invariant and ergodic under spatial shifts (translations);
iii. the actual history of U was selected at random using this measure, so that any property which holds with probability 1 under the measure (and whose definition does not mention the actual history) is a property of the actual history.

To keep the mathematical formulation of postulate (i) as simple as possible, let us assume that Ω is discrete, and that the set Ω_A of conceivable histories of any given (finite) space-time region A is finite. (The extension to the more realistic case of history spaces that are not discrete involves only standard methods of probability theory.) For each A, there is a probability distribution over the histories in Ω_A, that is to say a set of non-negative numbers $p_A(a)$, the probabilities, such that $\sum_{a \in \Omega_A} p_A(a) = 1$. The probability distributions for different regions must satisfy certain consistency relations, arising out of the fact that if one region is a subset of another then the probability distribution for the smaller region is completely determined by that for the larger; it will not be necessary to give these consistency relations explicitly.

There is no contradiction here between determinism and the use of probabilities: we are dealing with a probability distribution over a space of deterministic histories. But determinism does impose some conditions on the probability distribution, the most obvious of which is that any trajectory in Ω_A violating the dynamical laws (e.g. Newton's equations of motion) has probability zero. Determinism also imposes conditions relating the probability distributions in different regions; thus from formula (1) it follows that if B severs the past zone of A then there is a function f_{AB} such that

$$p_A(a) = \sum_{b \in \Omega_B} \delta(a, f_{AB}(b)) p_B(b) \tag{3}$$

where $\delta(\cdot,\cdot)$ is defined to be 1 if its two arguments are the same and 0 if they are different.

To formulate the first part of postulate (ii) mathematically, we assume that there is a group of spatial shifts[2] T such that if A is any region in the universe U then TA is also a region in U, and that if a lies in Ω_A (i.e. if it is a possible history of A) then its translate Ta is a possible history of TA. If A is a union of disjoint parts, then TA is obtained by applying T to every part. Then the postulate of shift invariance can be written

$$p_{TA}(Ta) = p_A(a) \quad (a \in \Omega_A) \tag{4}$$

holding for all space-time regions A and all spatial shifts T.

[2] Unlike the dynamical model, our probability model cannot be relativistically invariant, since the group of spatial shifts is not the full symmetry group of (special) relativistic space-time. But the lower symmetry of the probability model should not be reckoned as a disadvantage, since the whole purpose of the probability model is to identify deviations from this full symmetry, in particular deviations from time-reversal symetry.

To a very good approximation, we would expect probabilities to be invariant not only under space shifts but also under time shifts, provided that the size of the time shift is not too big. Invariance under shifts of a few seconds, days, years or even millennia is fine, but to postulate invariance under arbitrarily large time shifts, comparable with the age of the universe or greater, would be to commit oneself to the steady-state cosmological theory, which is no longer in fashion.

The second part of postulate (ii), ergodicity, means that a law of large numbers holds with respect to space shifts: given any event E that can occur in a space-time region A, the fraction of the space translates of A in which the (spatially shifted) event E actually occurs is almost surely equal to the probability of E. In symbols, if E is a subset of Ω_A (for example E could consist of just one history) then the following statement is true with probability 1:

$$\lim_{n\to\infty} \frac{1}{n}\sum_{i=1}^{n} \chi_n(\omega) = p_A(E) \tag{5}$$

where χ_n denotes the indicator function for the occurrence of the event T^nE in the region T^nA; i.e. $\chi_n(\omega)$ is defined to be 1 if $\omega \cap T^nA \in T^nE$ and to be 0 if not.

By postulate (iii) eqn (5), being true with probability 1, is true in the actual Universe. It is this property that connects the mathematical probability model with observable properties of the real world, enabling us to equate the probabilities in it with frequencies of real events. In practice, it is easier to use time shifts than spatial ones, reproducing the same situation at different times in the same place rather than in different places at the same time. Since the time shifts used in estimating the average on the left side of (5) by this method are likely to be very small compared to the age of the Universe, the use of time shifts when formula (5) is used to estimate probabilities should be a very good approximation.

As an example, A could be any space-time region consisting of a cube with side 1 metre lasting for 1 second, starting now, and E could denote the event in that the cube is empty of particles throughout its lifetime Then $p_A(a)$ is the probability that a 1 metre cube, randomly chosen at this moment somewhere in the Universe, is empty of particles during the second after the moment when it is chosen. Knowing something about the density and temperature of the of gas in outer space, one could give a reasonable numerical estimate of this probability.

5 Physical Probabilities

The probabilities we usually deal with in science refer to events not in outer space but here on Earth. In the present formalism, these are conditional probabilities, conditioned on some particular experimental set-up. For example, suppose you throw a spinning coin into the air and note which side is up when it lands. The probability of the outcome 'heads' can be written as a conditional probability:

$$\mathrm{Prob}(\mathcal{H}) = p_{A\cup B}(\mathcal{H}|\mathcal{B}) = \frac{p_{A\cup B}(\mathcal{H}\times\mathcal{B})}{p_B(\mathcal{B})} \tag{6}$$

where $\mathcal{H}$ represents the 'heads' outcome and $\mathcal{B}$ represents the launching of the coin and the other requirements that make the tossing of a coin possible, for example the presence of a floor and a gravitational field. The left side of 6 is to be thought of as a physical property of the coin, measurable by replicating the experimental condition $\mathcal{B}$ many times and counting the fraction of occasisions when the outcome is $\mathcal{H}$.

In our probability model, the macroscopic event $\mathcal{H}$ is represented by a set of histories for a space-time region A which includes the place and time where the coin comes to rest on the floor, and $\mathcal{B}$ is represented by a set of histories for an earlier space-time region B which is disjoint from A and includes the place and time where the coin is launched. The notation $\mathcal{H} \times \mathcal{B}$ denotes the set of histories $\varpi \in \Omega_{AB}$ such that $(\varpi \cap A) \in \mathcal{H}$ and $(\varpi \cap B) \in \mathcal{B}$. Note that B comes earlier than A, not later: the coin is spun before the time when it is observed on the floor, not after. Moreover, to ensure that all the influences that might influence the motion of the coin are properly controlled, B should sever the past zone of A, as in Fig. 3.

By itself, equation (6) is just a definition; it contains no information about the behaviour of real coins. We can put some empirical information into it, however if we take account of something that every probability theorist (though not every gambler) believes to be true, namely that there are many features of the macroscopic state $\mathcal{A}$ of the region A upon which Prob($\mathcal{H}$) depends very slightly, if at all. The sex of the experimenter, for example, makes no difference to the probability of $\mathcal{H}$; nor does it matter (so the probability theorists believe) how many times the 'heads' outcome occurred on the previous occasions when the coin was spun. It is this independence of irrelevant features of the macroscopic state of A (a feature that is given the name "statistical stability" in [14]) that makes it possible to think of Prob($\mathcal{H}$) as a physical quantity, which can be measured in many different laboratories to give the same answer.

Our experience of the statistical stability of physical probabilities indicates that, if we define the macroscopic states in the right way, the probabilities of macroscopic events should be independent of what happened before the region B got into the macroscopic state $\mathcal{B}$. In symbols, we expect the equation

$$p_{A\cup B}(\mathcal{H}|\mathcal{B}) = p_{A\cup B\cup C}(\mathcal{H}|\mathcal{B} \times \mathcal{C}) \qquad (7)$$

to hold for all macroscopic events $\mathcal{C}$ in the history space of an extra space-time region C which is arbitrary except that it must sever the past zone of B, as shown in Fig. 4. Thus, if C is any space-time region that contains the experimenter just before the coin is spun, $\mathcal{C}$ could be the event "the experimenter is a woman"; or if C is a space-time region containing all the previous occasions when the coin was spun, $\mathcal{C}$ could be the event "all the previous spins gave the result 'heads'".

In the mathematical theory of probability, equation (7) is part of the definition of a Markov chain; so (7) can be regarded as saying that it is possible to choose the macroscopic states such as $\mathcal{B}$ in such a way that their probabilities have a Markovian structure. In [14], such a Markovian structure is taken as one of the main postulates in a deductive treatment of the foundations of statistical mechanics, and it is shown to lead to an equation expressing transition probabilities such as $p_{A\cup B}(\mathcal{H}|\mathcal{B})$ in terms of purely dynamical quantities. That discussion does not use space-time maps, however, being geared to the standard methods of statistical mechanics where the system we are interested in is considered to be isolated from the rest of the world during the period when its time evolution is studied. The approach described here arose in part from trying to avoid this restriction.

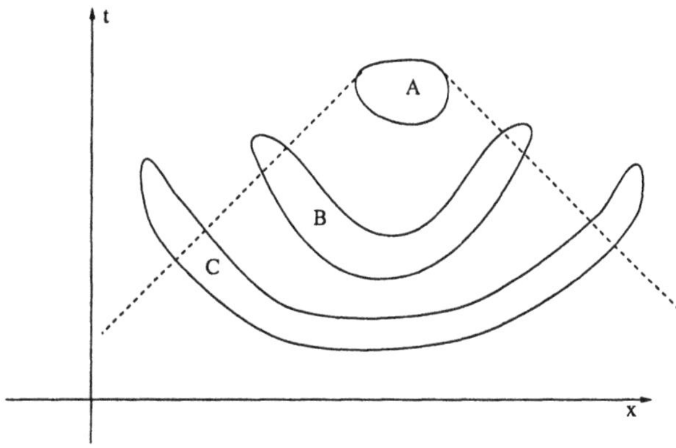

Fig. 4. The arrangement of regions A, B, C for equation (7). The region B severs the past zone of A and C severs that of B.

What about the time direction of physical probabilities? At first sight, equation (6) appears to contain a very clear time asymmetry, since in the experiments done to measure physical probability we always prepare the system before we observe it, not after: the region B in (6) always comes before A, not after it. But is this really a time asymmetry of the equation? Using the definition of conditional probability, and replacing $\mathcal{H}$ by a more general macroscopic event $\mathcal{A}$ in A, (7) can be rewritten as

$$\frac{p_{A\cup B}(\mathcal{A}\times\mathcal{B})}{p_B(\mathcal{B})} = \frac{p_{A\cup B\cup C}(\mathcal{A}\times\mathcal{B}\times\mathcal{C})}{p_{B\cup C}(\mathcal{B}\times\mathcal{C})}$$

which can be rearranged to give

$$p_{B\cup C}(\mathcal{C}|\mathcal{B}) = p_{A\cup B\cup C}(\mathcal{C}|\mathcal{B}\times\mathcal{A}). \tag{9}$$

Thus the conditional probability of the earlier event $\mathcal{C}$ given the later one $\mathcal{B}$ is independent of the even later event $\mathcal{A}$. The Markovian condition (7), despite its asymmetrical formal appearance, is in fact symmetrical under time reversal as far as the time order of the three events $\mathcal{A}, \mathcal{B}, \mathcal{C}$ is concerned. One might be tempted to conclude from the reversed Markovian condition (9) that the conditional probability of an earlier macroscopic event given a later one has a similar type of statistical regularity to that of the later event given the earlier one, in which case we could use the left side of (9) to define a "time-reversed physical probability" for the earlier event given the later one.

Nevertheless, such a conclusion would be wrong. The "time-reversed physical probability" corresponding to the left side of (6), namely $p_{B \cup A}(\mathcal{B}|\mathcal{H})$, would be the probability, given that a coin is found on the floor with its 'heads' side uppermost, that the coin arrived there as the result of a coin-tossing experiment (rather than, for example, as the result of somebody's dropping it on the floor by mistake). Such probabilities are well-defined in the model, but physicists cannot measure them without going outside their laboratories, nor philosophers without going outside their studies; they need outside information to tell them, for example, how often people actually do drop coins on the floor by mistake. In short, the time-inverse probability $p_{B \cup A}(\mathcal{B}|\mathcal{H})$ cannot be measured by laboratory experiments and therefore, unlike the left side of (7), is not a physically measurable property of the coin.

The reason for the wrong conclusion discussed in the preceding paragraph is that the Markovian condition resides not only in the formula (7) (with $\mathcal{H}$ now replaced by a general macroscopic event $\mathcal{A}$) but also in the geometrical relation between the space-time regions A, B, C appearing in it. Unlike the formula itself this geometrical relation, illustrated in Fig. 4, is not symmetrical under the symmetry operation of reversing time and interchanging the labels A and C: thus, C severs the past zone of A, but A does not sever the future zone of C.

So the Markovian condition (7), if true for suitably specified macroscopic events, does after all give us a direction of time, its asymmetry under time reversal deriving not from any algebraic property of the formula, but from the geometrical asymmetry of the relation between the space-time regions A and B illustrated in Fig. 3: if we reversed this relation, making B sever the future zone of A instead of the past, and C sever the future zone of B, then there would be no reason to expect either the formula (7) or the equivalent version (9) to hold.

6 The Common Cause Principle

The time asymmetry we found in the preceding section is not a completely satisfactory answer to the problem of characterizing the time asymmetry of probabilities. It is not clear that the macroscopic states such as $\mathcal{B}$ can be defined in such a way that the Markovian condition (7) is satisfied; moreover, there is the difficulty (pointed out to me by G. Sewell [22]) of proving consistency of the theory by showing that the dynamical consequences of the formula (7), studied in detail in [14] for isolated systems, are compatible with the dynamical laws. In the present section we look at a different approach, which uses microscopic histories rather than macroscopic events, and is much more easily reconciled with the dynamical laws.

In this discussion it will be assumed that there was an initial time, let us call it t_0, at which the Universe as we know it began. In order not to prejudge the time direction problem by making the intrinsic structure of the Universe temporally asymmetric quite apart from what is happening inside it, we suppose for the time being that there is will also be a final time t_1 at which the Universe as we know it will end. Such a time is indeed a feature of some cosmological models, see for example [21], although of course these models use curved space-time so that the straight light rays in our diagrams would have to be replaced by curved ones. We shall find that the value of t_1 plays no part in the discussion (except that it must be greater than t_0); so the results will apply equally well to a model universe in which $t_1 = +\infty$. The case $t_0 = -\infty$, which arises in the steady-state cosmological model, can also be treated (see [17]) but will be ignored here for simplicity.

The idea of referring back the present condition of the Universe, via deterministic mechanical laws, to its condition at time t_0 goes back at least to to Boltzmann who writes, when discussing the time asymmetry or "uni-directedness" of his H theorem,

> "The uni-directedness of this process is obviously not caused by the equations of motion of the molecules, for those do not change when the direction of time is changed. The uni-directedness lies uniquely and solely in the initial conditions. However, this is not to be understood in the sense that for each experiment one would have to make all over again the special assumption that the initial conditions are just particular ones and not the opposite, equally possible ones; rather, a unified fundamental assumption about the initial constitution of the world suffices, from which it follows automatically with logical necessity that whenever bodies engage in interaction then the correct initial conditions must prevail". [4]

Boltzmann's proposal for achieving this was to "conceive of the world as an enormously large mechanical system ... which starts from a completely ordered initial state, and even at present is still in a substantially ordered state" [5]. According to the usual interpretation of entropy as disorder, Boltzmann's remark means that the Universe started in a state of low entropy, and the entropy has been increasing ever since but is still quite low. Excellent elucidations of Boltzmann's proposal are given in [12,18]. Boltzmann's insight about the entropy of the initial state of the Universe is not the whole story, however. The law of increasing entropy is a property of certain processes in materials, and such proceses are only one item in our list of time-asymmetric processes in section 2. Moreover, it is hard to see how the value of just one number, the entropy, at the one time $t = t_0$ can control the subsequent evolution of the Universe with such exquisite precision as to determine the time direction of all the physical processes that happen everywhere for ever after. At the very least some information about the probabilities at time $t = t_0$ seems necessary.

As a step towards advancing Boltzmann's programme a bit further, I shall make use of Reichenbach's "principle of the common cause" [19] to formulate a reasonable hypothesis about the initial probabilities. Reichenbach's principle asserts that "if an improbable coincidence has occurred there must exist a common cause". Following Reichenbach himself, we may interpret "improbable coincidence" to mean simply a correlation, or more precisely a deviation from the product formula for the joint probability of two events or histories in two spatially separated space-time regions A and B. As for the "common cause", like all causes it takes place before its effect, and must therefore be some event in the past zones of both A and B, that is to say in their intersection $PZ(A) \cap PZ(B)$, which I shall call the *common past* of A and B. Thus, in the one-dimensional Universe illustrated in Fig. 5, if A and B are correlated, i.e. if $p_{A \cup B}(a \times b) \neq p_A(a)p_B(b)$, then Reichenbach's principle leads us to seek the cause of the correlation in some event that takes place in the region marked C.

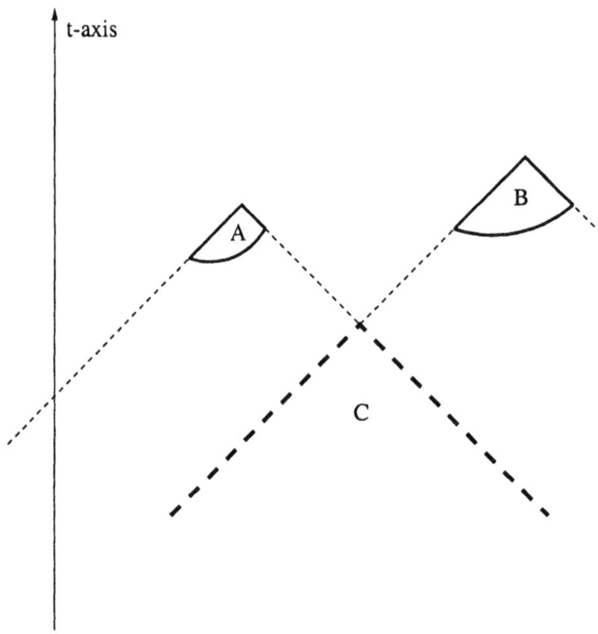

Fig. 5. Space-time regions for Reichenbach's common cause principle and the law of conditional independence. The region below the heavier dashed lines is $C = PZ(A) \cap PZ(B)$, the common past of A and B.

It is an elementary consequence of Reichenbach's principle that if the space-time regions A and B are so far apart that their past zones are disjoint, i.e. if $PZ(A) \cap PZ(B)$ is empty, as in Fig. 6, then the two regions are uncorrelated, that is to say the formula

$$p_{A \cup B}(a \times b) = p_A(a)p_B(b) \tag{10}$$

holds for all a in Ω_A and all b in Ω_B.

Equation (10) can be applied to the case where the two "regions" are subsets of the manifold $t = t_0$, such as the two segments marked M and N in Fig. 6; this leads us to the conclusion that disjoint parts of the $t = t_0$ manifold are uncorrelated[3].

[3] In [16] this independence of different parts of the $t = t_0$ manifold was taken as an axiom

From the independence of disjoint parts of the $t = t_0$ manifold we can derive a formula, which may be called [17] the *law of conditional independence*, relating the probabilities in two regions whose past zones are not disjoint. It states that if A and B are any two space-time regions, and C is the common past of A and B, then A and B are conditionally independent given the history of C. In symbols[4]

$$p_{A \cup B \cup C}(a \times b|c) = p_{A \cup C}(a|c) p_{B \cup C}(b|c) \tag{11}$$

for all $a \in \Omega_A, b \in \Omega_B, c \in \Omega_C$.

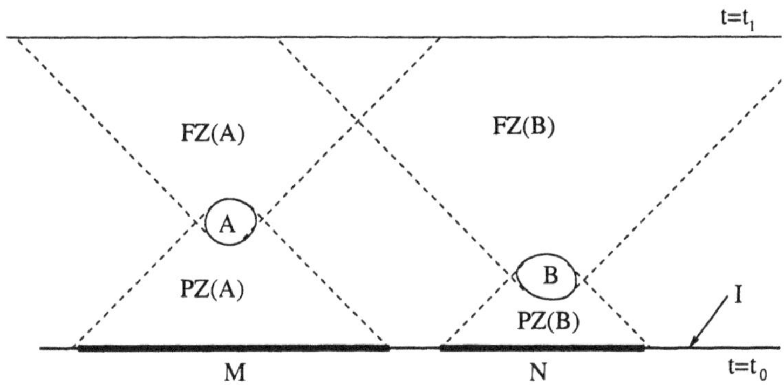

Fig. 6. The initial time is t_0, the final time t_1. The regions A and B are uncorrelated, because their past zones do not overlap (even though their future zones do overlap). For the same reason, the segments M and N are uncorrelated.

To prove (11), define I to be the initial manifold, on which $t = t_0$, and define two subsets M, N of I (see Fig. 7) by the formulas

$$\begin{aligned} M &= I \cap (PZ(A)) - C) \\ N &= I \cap (PZ(B)) - C), \end{aligned} \tag{12}$$

and let m, n denote their respective histories. By an argument similar to the one based on Fig. 6, the three regions M, N, C are uncorrelated with one another.

It can be seen from Fig. 7 that the definitions (12) etc. imply that $C \cup M$ severs the past zone of A, and that $C \cup N$ severs the past zone of B. Hence, by the determinism condition (1), there exist functions f, g such that

$$\begin{aligned} a &= f(m, c) \\ b &= g(n, c). \end{aligned} \tag{13}$$

[4] The same equation is given in [17] but the condition used there to characterize C appears to be too weak to ensure the truth of (11). Equation (11), together with a diagram equivalent to Fig. 5, also appears in Bell's paper [1] about the impossibility of explaining quantum non-locality in terms of local "beables".

Applying the formula (3) we find that

$$\begin{aligned} p_{A \cup C}(a \times c) &= \sum_{m \in \Omega_M} p_{M \cup C}(m \times c)\delta(a, f(m,c)) \\ &= p_C(c) \sum_{m \in \Omega_M} p_M(m)\delta(a, f(m,c)) \end{aligned} \tag{14}$$

where in the last line we have used the fact that the regions M and C are uncorrelated. From (14) and the definition of conditional probability we obtain

$$p_{A \cup C}(a|c) = \sum_{m \in \Omega_M} p_M(m)\delta(a, f(m,c)). \tag{15}$$

Similar formulas can be worked out for $p_{B \cup C}(b|c)$ and $p_{A \cup B \cup C}(a \times b|c)$, and using all three formulas in (11) we find that the two sides of (11) are equal. This completes the proof.

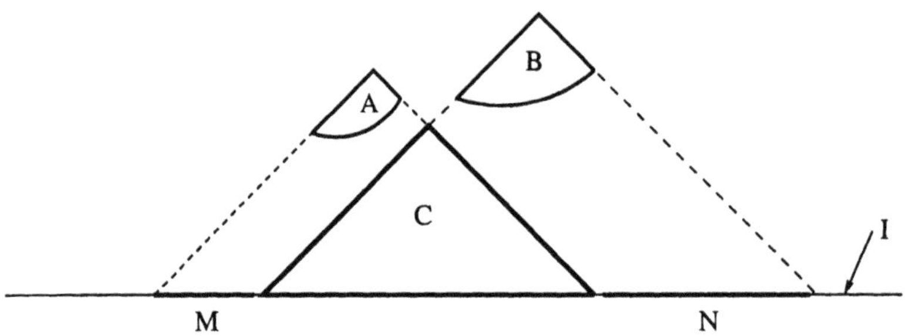

Fig. 7 Space-time regions used in proving the law of conditional independence.

Before going on to the extension of these ideas to quantum mechanics, a word about Bohmian mechanics [2] is in order. From our point of view, Bohmian mechanics is a deterministic classical theory, but since it contains simultaneous action at a distance the speed of light in our treatment would have to be taken infinite. This would make the dotted lines representing the light rays in the diagrams horizontal. Our formulation of Reichenbach's common cause principle would no longer work, but its consequence that disjoint pieces of the $t = t_0$ manifold are uncorrelated might still be adopted as an axiom in its own right. Initial conditions for Bohmian mechanics are treated in detail in [6], but with a different purpose in mind.

7 Quantum Mechanics

Most of the main ideas in the model discussed here have analogues in quantum mechanics. Integration over a space of histories is an important tool in quantum-mechanical calculations, and so our history spaces Ω, Ω_A can be taken over directly into quantum mechanics. The probability distribution $p_A(a)$, however, is more problematical; its nearest analogue is a density matrix, a non-negative definite matrix which will be written $\varrho_A(a, a')$. In analogy with the dynamical condition on the classical probability distribution that dynamically impossible trajectories have zero probability, the density matrix also satisfies a dynamical condition. This condition is derivable from Schrödinger's equation, and has a property of symmetry under time reversal derivable from the fact that the complex conjugate of a wave function satisfies the time-reversed version of the Schrödinger equation. In relativistic quantum mechanics we would also expect it to have a property analogous to (3), enabling us to calculate the density matrix for a region A in terms of the one for a region B which severs the past zone of A. The analogue of (3) would have the form

$$\varrho_A(a, a') = \sum_{b \in \Omega_B} \sum_{b' \in \Omega_B} F_{AB}(a, b) F^{\star}_{AB}(a', b') \varrho_B(b, b') \tag{16}$$

where $F_{AB}(a, b)$ is a propagator and the star denotes a complex conjugate. The propagator may be given the following heuristic interpretation: if the history of the region B should happen to be b, then the wave function of A will be given by $\psi_A(a) = F_{AB}(a, b)$.

It is more difficult to find convincing quantum analogues for properties (ii) and (iii) of the classical probability model. The analogue of the shift invariance condition (4) is easy enough, simply $\varrho_{TA}(Ta, Ta') = \varrho_A(a, a')$. To give a precise formulation of the ergodic property (5), however, requires an understanding of what constitutes an "event" in quantum mechanics, a difficult problem[5] which is far beyond the scope of this article. Nevertheless we do know from experience that events do happen, and so the quantum mechanical model must be considered incomplete if it does not make some kind of provision for events.

Quantum mechanics contains a non-classical concept called *entanglement* which has no exact analogue in classical mechanics: it is formally similar to correlation, but more subtle. Two space-time regions A and B may be said to be entangled if their joint density matrix is not the product of the separate ones, i.e. if

$$\varrho_{A \cup B}(a \times b, a' \times b') \neq \varrho_A(a, b) \varrho_B(a', b'). \tag{17}$$

According to quantum theory, entanglement arises if a bipartite system is prepared in a suitable quantum state and the two parts then move apart from one another, as in the famous EPR paradox[7]. It therefore seems natural to adopt a quantum analogue of Reichenbach's principle, asserting that if two space-time regions are entangled there must exist a common cause. This common cause would be an event in the common past of A and B. If they have no common past, as in Fig. 6, then we would expect A and B to be unentangled.

If we want to formulate a quantum analogue of the law of conditional independence, we need some condition on the common past of the two regions A and B that will preclude the creation within it of any entanglement between A

[5] See, for example, [8,15].

and B. For example, it would presumably be sufficient to require C to be completely empty, both of matter and fields. More leniently, we could allow events to happen inside C provided that any incipient entanglement created thereby was destroyed before it got outside C. A possible way to achive this might be to make the requirement that whatever events take place in C they include something equivalent to making a complete measurement along the entire boundary of C. The measurement would intercept any particles or photons emerging from C and destroy any phase relations between their wave functions that might otherwise go on to generate entanglement between A and B.

8 Conclusion

Now we can look back at the various time-directed processes mentioned in section 2, to see whether their time direction can be derived from Reichenbach's common cause principle or whether it is logically independent from this principle and therefore in a sense accidental. In applying this principle, we shall normally identify the common cause with some interaction between the two correlated systems.

Recording devices, memory. A camera is essentially a closed box which interacts with the outside world for a short time while the shutter is open. The image on the film represents a correlation between the interior of the box and the world outside. According to the common cause principle, such a correlation implies an interaction in the common past; therefore the correlation (the image) can be there only after the shutter was opened, not before. The time direction of other recording devices, such as your memory, can be understood in the same way. For example, if you visit a new place, your memory of it is a correlation between your brain cells and the configuration of the place you went to. This correlation can exist only after the interaction between your body and the place during your visit; so you remember the visit after its occurrence, not before.

Irreversible processes in materials; increase of entropy. Boltzmann's original derivation of his kinetic equation for gases and the consequent H-theorem (the non-decrease of entropy in an isolated gas) [3] depended on his *Stosszahlansatz*, an assumption which says that the velocities of the gas molecules participating in a collision are uncorrelated prior to the collision. After the collision, on the other hand, they will in general be correlated. This is just the time direction we would expect from the the common cause principle applied to these collisions: the correlation comes after the interaction, not before.

In the rigorous derivation of Boltzmann's kinetic equation given by Lanford [9], the collisions are not treated individually but as part of the evolution of an isolated system of many interacting particles. In this case the time direction comes from Lanford's assumption that the particles are uncorrelated at the initial moment of the time evolution he studies (and, because of their subsequent interaction, only at that moment). The time direction given by this assumption can be seen to be consistent with the time direction of the common cause principle if we imagine the interaction to be switched on at this initial moment and then switched off again at some later moment; the common cause principle indicates that the particles are uncorrelated up until the moment when the interaction starts, but are (in general) correlated thereafter, even after the interaction has been switched off. Thus Lanford's assumption need not be seen as an *ad hoc* assumption for fixing the time direction in this particular problem, but instead as a particular case of the general scheme for fixing time directions provided by the common cause principle.

Other rigorous derivations of kinetic or hydrodynamic equations from reversible microscopic dynamics make similar assumption about an uncorrelated initial state, while not assuming anything in particular about the final state, and so their time direction can be understood on the basis the common cause principle in the same way as for Lanford's result. For example, Lebowitz and Spohn [10,11], deriving Fick's law for self-diffusion in a gas consisting of hard spheres of two different colours, assume that the the colours of the particles are initially uncorrelated with the dynamical states of the particles.

Boltzmann also gave a more general argument (for an enthusiastic explanation see [12,13]) for the increase of entropy in an isolated system, which does not use any detailed assumptions about the nature of the interactions, nor does it assume that the particles are uncorrelated at the moment when the evolution process under study begins, which is normally a moment when the system becomes isolated. Instead he assumes that the entropy of the system at this initial time is less than the equilibrium value of the entropy, and one can then argue plausibly that the entropy is likely to move closer to the equilbrium value. i.e. to increase. But what grounds do we have to assume that the lower value of entropy occurs at the moment when the system becomes isolated rather than the later moment when the system ceases to be isolated? Boltzmann's suggestion to base everything on the assumption of a very low initial entropy for the Universe at its initial time $t = t_0$ is very important, but to me it is not a complete answer to the question, since there is no obvious reason why the entropy of a small temporarily isolated part of the universe has to vary with time in the same direction as the entropy of the universe as a whole.

Once again, the common cause principle suggests an answer. The problem is to understand why an isolated system is closer to equilibrium at the end of its period of isolation than at the beginning. When an isolated system is in equilibrium, its correlation with its surroundings is the least possible compatible with its values for the thermodynamic parameters such as energy; whereas if it is not in equilibrium it is more strongly correlated with its surroundings[6]. By the common cause principle, such correlations imply a past interaction between

[6] Such correlations can be regarded as "information" held by the surroundings, which may include human observers, about the system. Indeed there is a quantitative relation, due to L. Szilard, between the amount of this information and the entropy of the system. (For a detailed treatment of this relation, with some references to earlier work, see pp 226-231 of [14].) As time progresses, this information loses its relevance to the current state of the isolated system, the amount of correlation goes down, and the entropy of the system goes up.

system and surroundings; therefore we would expect the correlations to exist at the beginning of the period of isolation when the system has just been interacting with its surroundings, rather than at the end when it has not. (For example, the system might have been prepared in some non-equilibrium state by an experimenter: the correlation between system and experimenter implied by the non-equilibrium state – and the experimenter's knowledge of it – arose from a prior interaction.) In this way the common cause principle provides a rational explanation of why the low-entropy state, in which the system is more strongly correlated with is surroundings, occurs at the beginning of the period of isolation rather than at the end.

Expanding waves. Different points on a spherical wave are correlated. By the common cause principle, this correlation was caused by a previous interaction, in this case a local interaction with an electron in the antenna; therefore the wave comes after the local event that produces it, and must expand rather than contract.

Expanding universe. This is the one item in our list whose time direction clearly does not follow from the common cause principle. It is true that the expansion we see, a correlation between the motions of distant galaxies, implies a past interaction, presumably that which took place at the time of the Big Bang. But a contracting Universe would also be compatible with the common cause principle, since the correlations of the galactic motions which constituted the contraction could still be attributed to a past interaction, namely the long-range gravitational attraction between the galaxies.

Black and white holes. A proper treatment of this subject is beyond the scope of this paper; to produce one we would need to regard the metric of space-time as part of the kinematic description of the Universe instead of regarding it as a fixed background within which the rest of the kinematics takes place. All we can do here is to indicate a few hints that can be obtained by treating black or white holes as part of the background rather than part of the dynamics.

A black hole arises when a very heavy star is no longer hot enough to support itself against its self-gravitation. It is a singularity in curved space-time which begins at the time of collapse and (in classical gravitation theory) remains for ever thereafter. It is called a "black" hole because (again in classical theory) no light, or anything else, comes out of it. Given any point in the non-singular part of space-time, whether inside or outside the event horizon, all the light reaching that point comes from places other than the black hole. The past zones of space-time points near a black hole are bent by the strong graviational field, but topologically they are essentially the same as the ones in Figs. 5 and 7, and so there is no inconsistency with the common cause principle.

The exact time inverse of a black hole would be a white hole that started at the beginning of time and disappeared at a certain moment. Given a point in space-time sufficiently close to the white hole, some or all of the light and other causal influences reaching it come from the white hole rather than from the $t = t_0$ manifold. The past zones of points near the white hole therefore need not extend back to the $t = t_0$ manifold, but may instead end on the white hole itself. The common cause principle applied to this situation would lead us to conclusions about the white hole similar to the ones reached in section 6 about the initial manifold, different pieces of its "surface" (to the extent that a line singularity in space-time can be said to have a surface) being uncorrelated, both with each other and with the $t = t_0$ manifold. It seems, then, that the existence of white holes would be consistent with the common cause principle. The enormous gravitational forces at the surface of a white hole would no doubt have a profound effect on whatever came out of it; indeed one could speculate that the surface of a white hole is at an infinite temperature, or even hotter than that[7].

Acknowledgements

I am much indebted to Ian Percival for many discussions and ideas. I am also indebted to the Societa Italiana di Fondamenti della Fisica, the Istituto Italiano per gli Studi Filosofici and the Royal Society of London for Improving Natural Knowledge for financial support enabling me to attend the conference and talk about this work. I am grateful to Michael Kiessling for translating a difficult passage from Boltzmann's book into English.

References

1. J. S. Bell: 'The theory of local beables', *Speakable and unspeakable in quantum mechanics* (Cambridge University Press 1987), pp 52-62
2. D. Bohm: 'A Suggested Interpretation of the Quantum Theory in terms of "Hidden Variables"', Phys. Rev. **85** 166-179 and 180-193 (1952)
3. L. Boltzmann: *Vorlesungen über Gastheorie* (J. A. Barth, Leipzig, 1896 and 1898), section 5. English translation by S. G. Brush, *Lectures on gas theory* (University of California Press, Berkeley and Los Angeles, 1964)
4. L. Boltzmann: ibid., section 87. English translation by Michael Kiessling (private communication, 2000)
5. L. Boltzmann: ibid., section 89. English translation by S. G. Brush, loc. cit.
6. D. Dürr, S. Goldstein and N. Zanghi: 'Quantum Equilibrium and the Origin of Absolute Uncertainty', J. Stat. Phys. **67** 843-907 (1992)
7. A. Einstein, P. Podolsky and N. Rosen: 'Can Quantum-mechanical Description of Reality be Considered Complete?', Phys. Rev. **47**, 777-780 (1935)
8. R. Haag: 'Fundamental irreversibility and the concept of events', Commun. Math. Phys.**132**, 245-251 (1990)
9. O. Lanford III: 'Time Evolution of Large Classical Systems'. In: *Dynamical Systems: Theory and Applications: Battelle Seattle Rencontres 1974*, ed. by J. Moser (Springer Lecture Notes in Physics **38**, 1975) pp.1-111
10. J. L. Lebowitz and H. Spohn: 'Microscopic Basis for Fick's Law of Self-diffusion', J. Stat. Phys. **28** 539-556 (1982)

[7] The so-called "negative temperatures" are hotter than any positive or even infinite temperature, in the sense that energy flows from any negative temperature to any positive one.

11. J. L. Lebowitz and H. Spohn: 'On the Time Evolution of Macroscopic Systems', Commun. Pure Appl. Math. **36** 595-613 (1983)
12. J. L. Lebowitz: 'Boltzmann's Entropy and Time's Arrow', Physics today **46:9** 32-38 (1993)
13. J. L. Lebowitz: 'Statistical Mechanics: A Selective Review of Two Central Issues', Rev. Mod. Phys. **71**, S346-357 (1999)
14. O. Penrose: *Foundations of statistical mechanics* (Pergamon, Oxford, 1970)
15. O. Penrose: 'Quantum Mechanics and Real Events'. In: *Quantum Chaos – Quantum Measurements*, ed. P. Cvitanović et al. (Kluwer, Netherlands 1992) pp 257-264
16. O. Penrose: 'The "Game of Everything"'. Markov Processes and Related Fields **2**, 167-182 (1996)
17. O. Penrose and I. C. Percival: 'The Direction of Time', Proc. Phys. Soc. **79**, 605-616 (1962)
18. R. Penrose: *The Emperor's New Mind* (Oxford University Press, Oxford 1989), chapter 7
19. H. Reichenbach: *The Direction of Time* (University of California Press, Berkeley 1956), especially section 19 .
20. L. S. Schulman: Opposite Thermodynamic Arrows of Time, Phys. Rev. Letters **83** 5419-5422 (1999)
21. D. W. Sciama: *Modern Cosmology and the Dark Matter Problem* (Cambridge University Press, Cambridge 1993), section 3.2
22. G. Sewell: private communication (1970)

How to Implement Boltzmann's Probabilistic Ideas in a Relativistic World?

Michael K.-H. Kiessling

Rutgers University, Piscataway, NJ 08854, USA

Abstract. This article outlines some of the problems, as well as some recent progress, in the implementation of Boltzmann's probabilistic ideas in a world ruled by relativistic gravity and electromagnetism.

1 Introduction

Boltzmann's probabilistic arguments reduce the explanation of the universal irreversible deterministic behavior of macroscopic systems which, at the microscopic level, are composed of a huge number of 'atoms' whose motion is governed by a deterministic, reversible dynamics, to the postulate of a very 'unlikely' initial state of the universe; see S. Goldstein's contribution to this volume for a very clear exposition of Boltzmann's ideas. Boltzmann's insights, which in my opinion are among the crown jewels of theoretical physics, are simple yet penetrating and subtle. Not surprisingly it is notoriously hard to rigorously demonstrate their validity for concrete dynamical systems of theoretical physics. Beginning with the work of Lanford and King in the mid 70's, we have some definitive results for the classical Newtonian prototype system of many hard microscopic balls (more generally, particles with certain short range repulsive interactions) in the dilute gas phase, though only for very short times; see H. Spohn's article in this proceedings for a very clear explanation of how the irreversible Boltzmann equation emerges with mathematical rigor in the Boltzmann-Grad limit for individual systems.

I am firmly convinced that in the course of time technically more powerful tools will become available for the control of this and for increasingly more realistic systems. However, there is a very large gap between the classical Newtonian world and the relativistic quantum world (whether as quantum fields or superstrings or whatever). In particular, in bridging this gap conceptual problems need to be clarified.

The meaning of the quantum world is already the subject of several sessions at this conference, and I will not add to this discussion except for the remark that I tend to regard the Bohmian mechanics of Dürr, Goldstein, Zanghì and collaborators (see D. Dürr's contribution to this volume) as a very attractive way of rationalizing at least non-relativistic quantum mechanics – to some extent because their rationale in Bohmian mechanics is similar in spirit to Boltzmann's.

A clear discussion of the implications of relativistic causality is given by O. Penrose in his contribution to this proceedings. I want here to draw attention to some other conceptual problems that *classical* relativity imposes on us. I will address two items: a) the entropy of the universe, where gravitation is the main issue, and b) the relativistic dynamics of interacting particle systems, which takes us to the microscopic foundations of electromagnetism.

2 Is the Boltzmann Entropy of the Universe Finite?

In his book "The emperor's new mind," Roger Penrose [33] gives a very clear exposition of Boltzmann's explanation of the 'arrow of time' as being the result of a universe that began in an exceedingly unlikely initial state. Of course, Penrose's discussion is a modern update on Boltzmann, based on the most trusted of our physical theories, viz. general relativity, the big bang theory, and the black hole entropy formula of Bekenstein and Hawking [13]. (For more details on Boltzmann, black holes and the arrow of time, see the contributions by O. Penrose and S. Goldstein in this proceedings). R. Penrose computes the maximum possible entropy of our universe to be a whooping 10^{123}. According to Boltzmann's ideas this now means that the initial condition of the universe was placed in phase space with a precision of one part in $10^{10^{123}}$, an incredibly unlikely feat! Here, the meaning of 'likelihood' is understood not in the sense of relative frequency over an actual ensemble of universes, for we have only one, but in the sense of 'typicality' of a single system, about which we had some discussion at this conference.

Before I explain how one arrives at this estimate for the maximal entropy of our universe, I briefly pause to register the physicist's reductionist judgment of this tiny probability. Clearly, it is entirely acceptable scientifically to postulate that the universe started just where it did (some facts about the initial conditions will have to be postulated anyhow). Furthermore, from the probabilistic view no retroactive conclusions can be drawn from the actual occurrence of a single event, no matter how improbable the event was a priori. Yet, from the traditionally successful reductionist view this extraordinarily small measure of one part in $10^{10^{123}}$ calls for a deeper *explanation* – say in form of a more fundamental physical postulate that would render the initial event an inevitability – rather than just stating it as an axiom itself that the universe started where it did. R. Penrose proposes his "Weyl Curvature Hypothesis" as a sensible possibility. I will not elaborate on this but leave it with the remark that this differential geometric hypothesis amounts to having no black (or white) holes initially. In any event, Penrose's proposal is surely mild in comparison to Einstein's inquiry "whether the Dear Lord actually had a choice in creating the world."

Let us return to Penrose's estimate of the maximum entropy of our universe and see how that number actually emerges. Penrose considers one of the possible space-times of the so-called standard model of cosmology (without the speculative inflation), namely a closed positively curved Friedman-Lemaître universe (a solution of Einstein's field equations). Such a universe has a finite (rest) mass. Penrose next resorts to black hole thermodynamics, which asserts that a black hole has a finite entropy and furthermore that the BH entropy increases when two black holes merge to form a single black hole. Here, BH can stand for either black hole or Bekenstein-Hawking, who provided the formula

$$S_{BH} = M^2 \frac{G}{hc} \tag{1}$$

for the BH entropy (in units of Boltzmann's constant k_B), where G is Newton's constant of universal gravitation, h is Planck's constant, c the speed of light, and M is 'the mass' of the black hole. One has to be a little careful as to what one means by mass, but I will not enter into this and just consider a Schwarzschild black hole (a black hole with no angular momentum and no charge) where mass is read in the traditional way. Assuming that all the mass of the universe has been swallowed up by the black hole, and estimating this mass in terms of the accepted estimate on the number of baryons (essentially protons and neutrons) in it, which is at least 10^{80}, the BH entropy gives Penrose's result. In the computation Penrose neglects the phenomenon of Hawking evaporation for such an enormous black hole, since its time scale would be much larger than the life span of the universe which eventually collapses onto itself in a 'big crunch.' Otherwise, the particular form of the solution of Einstein's equations has not entered.

Where does the BH entropy formula come from? First, there is black hole thermodynamics, invented by Christodoulou and others in the 60's. We stay for simplicity with the Schwarzschild black hole, whose metric in Schwarzschild coordinates reads [32]

$$\mathrm{d}s^2 = -\left(1 - \frac{2GM}{c^2 r}\right)c^2 \mathrm{d}t^2 + \left(1 - \frac{2GM}{c^2 r}\right)^{-1} \mathrm{d}r^2 + r^2\left(\mathrm{d}\theta^2 + \sin^2\theta\, \mathrm{d}\phi^2\right). \tag{2}$$

In this case M is simply the mass visible asymptotically at large distances from the hole. If you merge two such holes of masses M_1 and M_2 you get a new such hole whose mass is simply $M = M_1 + M_2$. For Schwarzschild black holes, this is the first law of black hole thermodynamics. The radius of the black hole's horizon (i.e. the spherical 'surface of no return' surrounding the singularity) is the famous Schwarzschild radius,

$$R_S = \frac{2GM}{c^2}\,. \tag{3}$$

The surface area of the horizon is simply

$$A = 4\pi R_S^2\,, \tag{4}$$

so that $A \propto M^2$. By the simple inequality $M^2 = (M_1 + M_2)^2 > M_1^2 + M_2^2$ (both $M_k > 0$) we have $A > A_1 + A_2$. And while you can merge black holes, you can't separate them, so that the total area can at most increase. For Schwarzschild black holes, this is the second law of black hole thermodynamics. Now, at this stage all that is a purely formal analogy to conventional thermodynamics. At best it suggests that a black hole *might* have a Boltzmann entropy $S \propto A$, viz. $S \propto M^2$. The contribution of Bekenstein [4] is to have suggested that the black hole entropy actually *is* a Boltzmann entropy, while Hawking has substantiated this analogy by relativistic quantum field theoretical considerations, which include the quantitative derivation of the constant of proportionality between entropy and area, the Hawking temperature of a black hole, and the evaporation phenomenon [13]. Relativistic quantum field theory of course should not be misread here as the elusive quantum gravity but as some ingenious application of a massless Hermitian field theory in the curved space-time of a black hole.

Granted that the BH entropy may actually be *a* Boltzmann entropy, I will now address the question of whether the finite maximum BH entropy is really *the* Boltzmann entropy of a closed general relativistic universe with 10^{80} Fermions. I take a stepwise approach to the Boltzmann entropy of a gravitating universe, gradually adding more realism. Interestingly, this approach suggests that the Boltzmann entropy of the universe may actually be unbounded above! As for the unlikelihood of the initial state of the universe, this only underscores Penrose's point about the need for a deeper postulate (physically, $10^{-10^{123}}$ or 0 don't make much of a difference for this argument). However, it puts an interesting spin on this problem, for on an infinite entropy scale *any* finite entropy configuration qualifies as such an unlikely initial state, even that of the biggest black hole.

Following Penrose I take a closed, positively curved model universe to have a finite 'space' in which there are finitely many particles, having a finite energy at a certain cosmic time. For the ensuing considerations a snapshot suffices, whence I simply consider a finite number N of identical particles with Newtonian gravitational interactions on $\mathbb{S}^3$, having a total energy E at a fixed time t.

To depart from familiar grounds, I first consider classical Newtonian point particles. Then Boltzmann's entropy (in units of k_B) is formally given by the conventional formula

$$S_{Cl}(E,N) = \ln\left(\frac{E_*}{N!h^{3N}}\int_{(\mathbb{S}^3)^N}\int_{\mathbb{R}^{3N}}\delta(H^{(N)}-E)\mathrm{d}x_1...\mathrm{d}p_N\right), \tag{5}$$

where

$$H^{(N)} = \sum_{1\le k\le N}\frac{p_k^2}{2m} - \sum_{1\le k<l\le N}\frac{Gm^2}{\|x_k - x_l\|} \tag{6}$$

is the Hamiltonian. Here, points on $\mathbb{S}^3$ have been identified with vectors $x_k \in \mathbb{R}^4$ of Euclidean length $|x_k| = 1$, and the distance function $\|x_k - x_l\|$ between points x_k and x_l is a monotonic function of the chordal distance on $\mathbb{S}^3$ which reduces to the latter when x_k is near x_l. E_* is an energy unit. It is easy to perform the integrations over the momenta p, leaving us with a $3N$ fold spatial integration

$$S_{Cl}(E,N) = \ln\left(C\int_{(\mathbb{S}^3)^N}\left(E + \sum_{1\le k<l\le N}\frac{Gm^2}{\|x_k - x_l\|}\right)_+^{\frac{3N-2}{2}}\mathrm{d}x_1...\mathrm{d}x_N\right), \tag{7}$$

where C is some constant, and the subscript $_+$ means the positive part. By direct inspection one sees that this remaining integral exists only for $N \le 2$ ($N = 1$ is trivial, of course), while it diverges for $N \ge 3$. So the entropy of such a finite classical universe of Newtonian point particles is infinite for finite E and $N > 2$.

It is instructive to know how the unbounded Boltzmann entropy shows itself in the dynamical evolution of N Newtonian point particles with Newtonian gravity on $\mathbb{S}^3$ with a spatially uniform initial distribution. A very detailed, though not entirely rigorous study of the dynamical scenario, has been carried out in a monumental paper by Heggie [14]. Heggie's work indicates that the system 'splits' into a tightly bound binary system and an expanded, hot halo containing all other particles. The halo is heated at the expense of the binary, which gets bound together ever tighter, liberating an unlimited amount of gravitational energy. The halo carries the entropy. Notice that the infinity comes from the local singularity of Newtonian gravity, not from its long range character (our space is finite).

Next, I add more realism and consider non-relativistic quantum mechanics for Fermions interacting via Newtonian gravity. This will definitely stabilize the local singularity of the Newtonian interaction, first because of the uncertainty principle (in form of a Sobolev inequality; Heisenberg's principle won't do. See [23]), and second because of the Pauli exclusion principle for Fermions. The Hamiltonian is the same, except that now $p_k = (\hbar/i)\nabla_k$, where $\hbar$ is Planck's constant divided by 2π, and it now acts on the antisymmetric N particle wave functions. By following verbatim a proof of Lévy-Leblond [22] for the Euclidean case, we find that the Hamiltonian is bounded below as inf spec $H \propto -N^{7/3}$, and since our space is compact, filled with N Fermions, and the operator self-adjoint, we have a discrete spectrum for which Weyl's asymptotic law for the counting of eigenvalues holds. Thus, for finite N and $E > E_0$, where E_0 is the ground state energy, the non-relativistic quantum mechanical Boltzmann entropy is finite,

$$S^{nr}_{QM}(E,N) = \ln \mathrm{Tr} P_{E,\triangle E} < \infty\,. \tag{8}$$

(Here, $P_{E,\triangle E}$ is the projector onto a small interval $\triangle E$ around E; projecting onto E would not work because of the discrete spectrum).

Next, let us become one step more realistic and take special relativity into account while still keeping Newtonian gravity, as in Chandrasekhar's theory of white dwarf stars. At this stage we already need some compromise, for a relativistic many body Hamiltonian is a delicate problem in itself. A rigorous analysis can be carried out (and has been in the Euclidean setting, by Lieb and Thirring [24] and Lieb and Yau [25]; see also [23]) for the ground state problem of the pseudo-relativistic Hamiltonian

$$H^{(N)} = \sum_{1\leq k\leq N} \sqrt{m^2c^4 + p_k^2c^2} - \sum_{1\leq k<l\leq N} \frac{Gm^2}{|x_k - x_l|} \tag{9}$$

for which the properly special relativistic equations of Chandrasekhar, here of a one species Fermion system, emerge in a large N scaling limit. Most importantly, the Hamiltonian is bounded below now only if $N < N^*$, with (neglecting spin counting)

$$N^* = C\frac{9}{16\sqrt{\pi}}\Big(\frac{hc}{Gm^2}\Big)^{3/2} \tag{10}$$

where $C = 0.87..$ is determined by solving a nonlinear PDE numerically. If m is the neutron mass, then $N^* \approx 10^{58}$. Systems which are more massive than about one solar mass collapse. What does this imply for the entropy? For $N < N^*$ the same reasoning as in the non-relativistic case leads again to a finite Boltzmann entropy for $E > E_0$, where E_0 is the ground state energy. However, this argument fails in the supercritical case $N > N^*$ relevant to our universe as a whole. We need to look for a different argument.

Since we are interested in large N, an educated guess can be made using the semi-classical theory. The semi-classical theory is a large N limit, whenever that limit exists, of the microscopic quantum model governed by $H^{(N)}$ in which particle correlations are completely dominated by the mean one-particle behavior. Thus, in the semi-classical theory for N relativistic, gravitating Fermions of mass m one treats the Fermions as distributed independently and identically according to a probability density function $f(x,p)$ on $\mathbb{S}^3 \times \mathbb{R}^3$ satisfying the bound $h^3 f < 1$. Asymptotically as $N \nearrow \infty$, the quantum mechanical Boltzmann entropy for such a large special relativistic system of gravitationally interacting Fermions is given by

$$S_{QM}^{sr}(E,N) = \sup_f S_{sCl}^{sr}(f,N)\,, \tag{11}$$

where

$$S_{sCl}^{sr}(f,N) = -N \int_{\mathbb{R}^3}\int_{\mathbb{S}^3} \Big(f \ln(h^3 f) + (h^{-3} - f)\ln(1 - h^3 f)\Big) \mathrm{d}x\mathrm{d}p \tag{12}$$

is the semi-classical entropy functional (I neglect irrelevant factors of 2 for counting neutron spin), and the sup is taken over all probability densities f for which $h^3 f < 1$ and $E_{sCl}^{sr}(f,N) = E$, where

$$\begin{aligned} E_{sCl}^{sr}(f,N) = N^{\frac{4}{3}} \int_{\mathbb{R}^3}\int_{\mathbb{S}^3} \sqrt{m^2c^4 + p^2c^2}\, f(x,p)\mathrm{d}x\mathrm{d}p \\ -N^2 \frac{1}{2} \int_{\mathbb{S}^3}\int_{\mathbb{S}^3} \frac{Gm^2}{|x-y|} \rho(x)\rho(y)\mathrm{d}x\mathrm{d}y \end{aligned} \tag{13}$$

is the (pseudo) special relativistic semi-classical energy functional, with

$$\rho(x) = \int_{\mathbb{R}^3} f(x,p)\mathrm{d}p\,. \tag{14}$$

The extra $N^{1/3}$ factor at the kinetic energy is all that remains from Pauli's exclusion principle for N Fermions, which at the microscopic level rules out that two or more Fermions are in the same single-particle state.

Now, we want to use this theory in the regime where the scaling limit would not exist, i.e. $N > N^*$, and this has to be justified. A rigorous justification is not so easy to come by, but note that the semi-classical theory is physically very plausible and used with impunity in stellar structure calculations for hot stars whose mass may exceed the Chandrasekhar limit by a factor 10 or more. In this spirit, I simply use this semi-classical relativistic theory to show that the entropy of such a special relativistic universe does not have an upper bound if $N > 1.2N^*$, for any given E. The non-optimal factor 1.2 is an artifact of the method of proof, which surely can be improved.

My proof uses a suitable maximizing sequence of trial functions $f(x,p)$ along which the entropy surpasses any prescribed value. The sequence of trial densities itself does not converge. A similar proof was designed by Antonov for the corresponding non-relativistic classical theory [2]; however, Antonov's trial functions cannot be used for proving the unboundedness of entropy in our semi-classical, special relativistic theory. While the strategy is the same, namely to break up the system into a core which collapses, liberating gravitational energy, and a halo which picks up that energy, thus heating up and in the course boosting its entropy beyond any bound, the breakup into core and halo is quite different in both theories.

In fact, to simplify the calculation I choose a slightly less optimal breakup of the system which suffices for the present purposes. Thus, we take our $N > N^*$ Fermions and split them into two subsystems, one with N_1 and one with N_2 particles, such that $N_1 + N_2 = N$. Both subsystems are placed far apart so that they interact only by a mutual, essentially constant gravitational energy which can be neglected. As a consequence of this separation, the total energy of the full system is simply the sum of the two energies of the two individual subsystems (up to the discarded irrelevant additive constant). Likewise, the entropy of the full system is simply the sum of the two entropies of the two individual subsystems. Provided that N_1 and N_2 are both sufficiently large, semi-classical asymptotics will hold for each subsystem individually. In order to collapse one subsystem, say number 1, we need to choose $N_1 > N^*$ anyhow. While N_2 need not exceed N^*, it cannot be arbitrarily small for semi-classical asymptotics to hold. Something like $N_2 > 10^{-20}N^*$ will surely do, however.

I have yet to specify the probability densities f_1 and f_2 for the particles of the two subsystems. Both f_1 and f_2 are of the type $(|B_R||B_P|)^{-1}\chi_{B_R(x_0)}(x)\chi_{B_P}(p)$, where $B_R(x_0)$ is a ball in $\mathbb{S}^3$ of geodesic radius R centered at x_0, B_P is a ball in $\mathbb{R}^3$ of radius P, centered at 0, and $|B|$ denotes the corresponding volume of a ball. Moreover, χ denotes characteristic function, taking the value 1 on the domain indicated as suffix and vanishing outside of it. There is enough freedom left in choosing the radii in space and momentum space, as well as the locations in space. Our goal is to show that the R's and P's can be chosen such that the total semi-classical entropy $S^{sr}_{sCl} = S^{sr}_{sCl}(f_1, N_1) + S^{sr}_{sCl}(f_2, N_2)$ surpasses any prescribed value without violating the energy constraint $E^{sr}_{sCl}(f_1, N_1) + E^{sr}_{sCl}(f_2, N_2) = E$ (= *const.*) nor the exclusion principle $h^3 f_n < 1$ for $n = 1, 2$.

Explicitly, the semi-classical entropy for our total system reads

$$S^{sr}_{sCl} = \sum_{n=1,2} N_n \left[\ln \frac{|B_{R_n}||B_{P_n}|}{h^3} - \left(\frac{|B_{R_n}||B_{P_n}|}{h^3} - 1 \right) \ln \left(1 - \frac{h^3}{|B_{R_n}||B_{P_n}|} \right) \right], \quad (15)$$

and the total energy constraint reads (in good approximation)

$$E = \sum_{n=1,2} \left(N_n^{4/3} \frac{3}{4} P_n c - N_n^2 \frac{3}{5} \frac{Gm^2}{R_n} \right). \tag{16}$$

Here I have made use of the fact that we will shrink R_1 arbitrarily so that the gravitational energy of subsystem 1 is essentially the same as in $\mathbb{R}^3$, and similarly I have chosen R_2 sufficiently small (though fixed) to allow this replacement. Furthermore I made use of the fact that the shrinking of R_1 will be compensated by an expansion of P_1 and P_2 in such a way that, the exclusion principle is obeyed, yet also the energy of subsystem 1 becomes arbitrarily negative. Since the P's will become arbitrarily large, we can operate with the asymptotic form for large P_n of the exact formula for the kinetic (plus rest) energy $E_{kin} = (3/8)\sum_{n=1,2} N_n^{4/3} mc^2((1 + 2X_n^{-2})\sqrt{1+X_n^2} - X_n^{-3} \operatorname{arsinh} X_n)$, where $X_n = P_n/mc$.

We determine P_1 in terms of R_1 by demanding that the energy E_1 of subsystem 1 decreases to $-\infty$ as R_1 shrinks to zero, given fixed N_1, more specifically like $E_1 = -\varepsilon/R_1$ with ε an arbitrarily small, strictly positive constant. Notice moreover that we have to satisfy the condition $h^3 f_n < 1$ as well. This constraint for f_1 gives us a lower bound on N_1, which up to corrections of $O(\varepsilon)$ is given by

$$N_1 > \frac{15\sqrt{5}}{32\pi} \left(\frac{hc}{Gm^2} \right)^{3/2} \approx 1.199 N^* , \tag{17}$$

where the same mass m is to be used as in the formula for N^*. This lower bound on N_1 that allows us to shrink R_1 arbitrarily in our special relativistic semi-classical functional is essentially the critical N^* which obtains also from the boundedness properties of the pseudo-relativistic Hamiltonian (as in fact it should), the factor 1.199.. coming from the non-optimal trial densities. Thus, if we choose $N > 1.2N^*$, we can pick $N_1 > 1.199N^*$ and $N_2 > 10^{-20}N^*$ and then let $R_1 \to 0$, obtaining

$$P_1 = \left(N_1^{2/3} \frac{4}{5} \frac{Gm^2}{c} - N_1^{-4/3} \frac{4}{3} \frac{\varepsilon}{c} \right) \frac{1}{R_1} \tag{18}$$

from our postulated behavior of E_1. Total energy conservation and the assumed constancy of R_2 now give us

$$P_2 = N_2^{2/3} \frac{4}{5} \frac{Gm^2}{R_2 c} + N_2^{-4/3} \frac{4}{3c} \left(E + \frac{\varepsilon}{R_1} \right), \tag{19}$$

which has the asymptotic behavior

$$P_2 \sim N_2^{-4/3} \frac{4\varepsilon}{3c} \frac{1}{R_1} \tag{20}$$

as $R_1 \to 0$. With R_2 constant and P_2 increasing beyond all bounds, it now follows that the constraint $h^3 f_2 < 1$ holds as well.

Finally, given arbitrary fixed E, and $N > 1.2N^*$, for the special relativistic quantum Boltzmann entropy we find from the semi-classical analysis that

$$S^{sr}_{QM}(E,N) > \textit{const.} + 3N_2 \ln \frac{R_2}{R_1}, \tag{21}$$

for arbitrary R_1. Whence, letting $R_1 \searrow 0$, we find that the right hand side $\nearrow \infty$. The proof that the quantum Boltzmann entropy is infinite for supercritical special relativistic masses is complete.

We summarize these findings as follows. While quantum mechanics stabilizes the short range singularity of the Newtonian gravity, special relativity counteracts this to some extent, spoiling the lower bound for the Hamiltonian, and as result of this also the finiteness of the Boltzmann entropy, if a critical N is surpassed.

In our last step we finally take general relativity into account. General relativity will enhance the destabilization of the system towards collapse, and so one should naively expect that the Boltzmann entropy is infinite also now. However, the BH entropy is finite. How can we understand this? Since radiation has not been taken into account at all, neither in the special relativistic, Newtonian gravity model, nor in the BH entropy formula, the resolution of this putative paradox should not be sought in the Hawking evaporation of black holes. A possible hint comes from the observation that, if in the semi-classical, special relativistic theory we use a trial density that describes the collapse of all the matter onto itself (the closest analog in that model to a universe whose matter has formed a single black hole), then we get the finite entropy

$$S^{sr}_{sCl}(M) = 2N \ln N + O(N). \tag{22}$$

This seems to suggest that also in general relativity one might get higher entropies by splitting the system into a collapsing core and a halo which receives the liberated energy from the core and carries the entropy to infinity. The problem is that black holes are surrounded by a horizon. This would make it impossible to transfer the gravitational energy liberated in the collapse of part of the system to the remaining halo, in this way cutting off the supply line that would boost the entropy beyond all bound. However, all known black hole solutions are axisymmetric, which is hardly an exact symmetry in Nature. It is not clear whether a closed, nonsymmetric horizon can form to give a nonsymmetric black hole. (I owe this thought to Christodoulou.) This would imply that black holes do not strictly exist in Nature, but naked singularities would, in violation of the cosmic censorship hypothesis. In that case a transfer of energy to a halo would become possible so that the Boltzmann entropy could be infinite also in the general relativistic case. This would imply that the maximum BH entropy is not the Boltzmann entropy of that universe.

Thus, while BH entropy may actually be *a* Boltzmann entropy (superstring theorist claim they can do the counting of states), the BH entropy of a black hole containing all the mass of the universe may not give the Boltzmann entropy of that universe. It surely is an interesting problem to investigate, but a definitive answer presumably requires a consistent theory of quantum gravity. Presently the jury is still out on whether this is possible.

Meanwhile, it is legitimate to contemplate that the Boltzmann entropy of the (closed and finite) universe may actually be infinite. This pushes Boltzmann's ideas to the extreme. Strictly speaking, the initial state of the universe must have been not just exceedingly unlikely but 'sheer impossible.' Moreover, measured against this entropy scale *any* finite entropy state would qualify as 'impossible initial state.' Clearly that would even allow many states which are not compatible at all with the actual dynamics of the universe, which from a different perspective underscores Penrose's call for a deeper principle that selects the initial state.

3 Relativistic Dynamics of Interacting Particle Systems

I now switch gears and consider the foundations of a microscopic relativistic dynamics of interacting particles. This problem lies at the heart of both the correct formulation of relativistic equilibrium statistical mechanics and of the probabilistic derivation of the relativistic analogs of Boltzmann's kinetic equation for the prototypical relativistic interactions, gravity and electromagnetism. Pauli discusses the relativistic equilibrium statistical mechanics for noninteracting ideal gases [32] but is silent as to how to formulate it for interacting particles. In kinetic theory, very reasonable and beautiful relativistic kinetic equations can readily be written down both for gravity and for electromagnetism, but it is not at all clear how to consistently formulate the underlying microscopic models of interacting relativistic particles that these kinetic equations are supposed to describe.

Let me explain this situation in more detail at hand of the kinetic theory. First, how do we know that the relativistic kinetic equations are indeed reasonable? The answer is that we do have a secure point of departure in the non-relativistic kinetic theory for systems with gravitational (Newton) or electrostatic (Coulomb) interactions. Their kinetic theory differs significantly from that of systems of particles with very short range repulsive interactions, e.g. hard spheres. In the case of systems with short range interactions, the kinetic regime is that of a dilute gas, in which the leading dynamics is just reversible 'free streaming' of noninteracting particles; small corrections due to occasional encounters of two particles (pair collisions) are responsible for the irreversible behavior over much longer time scales; triple or higher encounters are entirely negligible. The appropriate equation for these systems with short range repulsive interaction is Boltzmann's kinetic equation, and its derivation from a microscopic model has so far been possible only for short times; cf. the contribution to this volume by H. Spohn. In contrast, the truly long range character of both Newton and Coulomb interactions guarantees that each particle always interacts with all the other particles of the system. The kinetic regime here emerges at the other extreme, where so many particles (in the limit, infinitely many) interact simultaneously that the leading order dynamics is a reversible mean-field dynamics generated self-consistently by the common single particle distribution of the particles. Irreversibility enters as correction due to many small encounters of a particle with the distortions in the smooth mean-field that are caused by the nearby presence of other individual particles. For Newton and Coulomb systems, the leading order, reversible mean-field dynamics is governed by the Vlasov equations, the irreversible dynamics by the Balescu-Lenard equations. For the non-relativistic Vlasov equation with regularized Newton or Coulomb interactions a rigorous derivation valid for all times has been given by Braun and Hepp [6]. For the non-relativistic Balescu-Lenard equation only formal derivations exist. In any case, it is straightforward to write down relativistic generalizations for gravitating and for electromagnetic systems that reduce to these non-relativistic models in the limit $c \to \infty$.

To have a concrete example, I now list the equations of the reversible, relativistic Vlasov theory. The electromagnetic systems in flat Minkowski space are governed by what is commonly called relativistic Vlasov-Maxwell equations, [40], which read (k indexes the particle species)

$$\frac{\partial f_k(x,p,t)}{\partial t} + v_k \cdot \frac{\partial f_k(x,p,t)}{\partial x} + e_k\Big(E + \frac{v_k}{c} \times B\Big) \cdot \frac{\partial f_k(x,p,t)}{\partial p} = 0 \tag{23}$$

for the phase space probability densities f_k, and

$$\frac{1}{c}\frac{\partial}{\partial t}B(x,t) + \nabla \times E(x,t) = 0\,, \tag{24}$$

$$-\frac{1}{c}\frac{\partial}{\partial t}E(x,t) + \nabla \times B(x,t) = 4\pi\frac{1}{c}\sum_k N_k e_k \int_{\mathbb{R}^3} v_k f_k(x,p,t)\mathrm{d}p\,, \tag{25}$$

$$\nabla \cdot B(x,t) = 0\,, \tag{26}$$

$$\nabla \cdot E(x,t) = 4\pi \sum_k N_k e_k \int_{\mathbb{R}^3} f_k(x,p,t)\mathrm{d}p\,, \tag{27}$$

for the electric field E and magnetic field B at the point $x \in \mathbb{R}^3$ at time $t \in \mathbb{R}$. Here,

$$v_k = \frac{p/m_k}{\sqrt{1 + p^2/m_k^2 c^2}} \tag{28}$$

is the velocity of a particle with momentum p, rest mass m_k, charge e_k. The relativistic Vlasov-Maxwell equations are Lorentz covariant, though here not written in a manifestly covariant format. They preserve the conventional integrals of mass, energy, momentum, angular momentum, Boltzmann's H function, and infinitely many other integrals. Global well-posedness of the Cauchy problem for small initial data is known, and for large data under the additional assumption that no singularities occur near the light cone, see [40,10–12].

The corresponding kinetic equations for a general relativistic gravitating system are the Vlasov-Einstein equations. I skip their presentation, but remark that they are precisely what you expect them to be, once you have seen the above Vlasov-Maxwell equations written in manifestly covariant form. Excellent discussions can be found in the recent works of Rein [34] and the earlier article by Ehlers [8].

The ease with which one can write down the Vlasov-Maxwell and Vlasov-Einstein equations contrasts sharply with the enormous difficulties one faces in writing down the corresponding consistent microscopic equations which these kinetic equations are supposed to approximate in the kinetic regime. The Vlasov-Maxwell respectively -Einstein equations are purely classical equations, whence it should be possible to derive them from a classical microscopic model. However, we don't seem to have any consistent relativistic microscopic model of classical particles interacting with (their) radiation fields, be it electromagnetic or gravitational. This is the notorious radiation reaction problem. For recent work on the electromagnetic point particle problem, see [9,31].

The classical electron theory of the late 19th and early 20th century was an early, pre-relativistic and pre-quantum attempt to construct a consistent microscopic dynamical theory of electrons and their electromagnetic field. To overcome the problem of infinite radiation reactions and self energies of point charges, that theory postulates, stable extended 'atoms of electricity' which generate the electromagnetic fields, the Maxwell-Lorentz equations for the dynamics of these fields, and Newton's equation of motion for the dynamics of these electrons, equipped with the Lorentz force averaged over an electron. Very recently now, W. Appel from the ENS Lyon and I succeeded [3] in constructing a working relativistic Lorentz electrodynamics (LED), following an important lead taken by Nodvik [30]. This may open up the door for the first serious attempt to give a microscopic foundation of relativistic Vlasov-Maxwell theory. But it also sheds new light on the physical viability of LED as a potential classical limit of QED.

In the remaining pages I outline the LED of Appel and myself. A comprehensive discussion will appear elsewhere [3]. I pave the way to our relativistic LED with a brief review of the semi-relativistic theory.

In the semi-relativistic electrodynamics of Abraham-Lorentz the charge ($-e$ for an electron) is distributed around the instantaneous location $q(t) \in \mathbb{R}^3$ of the particle by a charge density f_{e} with $SO(3)$ symmetry and compact support, satisfying $\int_{\mathbb{R}^3} f_{\mathrm{e}}(x)\mathrm{d}^3x = -e$. The charge density f_{e} is rigidly carried along by the particle with linear velocity $\dot{q}(t)$, a point advocated in particular by Abraham [1], and rotating rigidly with angular velocity $\omega(t)$. Hence, charge and current densities, $\rho(x,t)$ and $\jmath(x,t)$, of a single particle are given by the Abraham-Lorentz expressions

$$\rho(x,t) = f_{\mathrm{e}}(x - q(t))\,, \tag{29}$$

$$\jmath(x,t) = f_{\mathrm{e}}(x - q(t))\,v(x,t)\,, \quad \text{with} \quad v(x,t) = \dot{q}(t) + \omega(t) \times (x - q(t))\,. \tag{30}$$

Considering only a single particle interacting with the electromagnetic field, (29) and (30) are the source terms in the classical Maxwell-Lorentz equations

$$\frac{1}{c}\frac{\partial}{\partial t}B(x,t) + \nabla \times E(x,t) = 0\,, \tag{31}$$

$$-\frac{1}{c}\frac{\partial}{\partial t}E(x,t) + \nabla \times B(x,t) = 4\pi\frac{1}{c}\jmath(x,t)\,, \tag{32}$$

$$\nabla \cdot B(x,t) = 0\,, \tag{33}$$

$$\nabla \cdot E(x,t) = 4\pi\rho(x,t)\,, \tag{34}$$

where $E(x,t) \in \mathbb{R}^3$ is the electric field and $B(x,t) \in \mathbb{R}^3$ the magnetic field at the point $x \in \mathbb{R}^3$ at time $t \in \mathbb{R}$. The evolution equations for the dynamical variables of the particle, i.e. position $q(t)$, linear velocity $\dot{q}(t)$, and angular velocity $\omega(t)$, are Newton's and Euler's equations of motion, equipped with the Abraham-Lorentz expressions for the volume-averaged Lorentz force and torque [26] felt by the 'electron.' Newton's equation here reads

$$\frac{\mathrm{d}p_{\mathrm{b}}}{\mathrm{d}t} = \int_{\mathbb{R}^3} \left[E(x,t) + \frac{1}{c}v(x,t) \times B(x,t)\right] f_{\mathrm{e}}(x - q(t))\,\mathrm{d}x\,, \tag{35}$$

where

$$p_{\mathrm{b}} = m_{\mathrm{b}}\dot{q} \tag{36}$$

is the linear momentum of the bare particle, with m_{b} its bare mass ('material mass' in [28]). Euler's equation here reads

$$\frac{\mathrm{d}s_{\mathrm{b}}}{\mathrm{d}t} = \int_{\mathbb{R}^3} (x - q(t)) \times \left[E(x,t) + \frac{1}{c}v(x,t) \times B(x,t)\right] f_{\mathrm{e}}(x - q(t))\,\mathrm{d}x\,, \tag{37}$$

where

$$s_{\mathrm{b}} = I_{\mathrm{b}}\omega \tag{38}$$

is the classical particle spin of the bare particle associated with the bare moment of inertia I_{b}. In [16] it was shown that this model satisfies the conventional conservation laws for charge, energy, linear and angular momentum. When the bare inertias $m_{\mathrm{b}} \neq 0$ and $I_{\mathrm{b}} \neq 0$, the semi-relativistic equations listed above pose a Cauchy problem for the following initial data, posed at time $t = t_0$: for the mechanical variables of the particles, the data are $q(t_0)$, $\dot{q}(t_0)$, and $\omega(t_0)$; and for the fields, $B(x,t_0)$ satisfying (33), and $E(x,t_0)$ satisfying (34) at $t = t_0$. Notice that (33) and (34) are merely initial constraints, which then remain satisfied by the fields $B(x,t)$ and $E(x,t)$ in the ensuing evolution. For (33) this is seen by taking the divergence of (31). For (34) this is seen by taking the divergence of (32) and the time-derivative of (34), then using the continuity equation for the charge, which is a consequence of (29) and (30) alone. Abraham [1] actually insisted on setting $m_{\mathrm{b}} = 0 = I_{\mathrm{b}}$, but note that this is in serious conflict with the initial value problem just explained.

Interestingly, most works after Abraham [1], e.g. [27,28,36,15,35,41], consider a model 'without spin,' obtained by discarding (37) and setting $\omega \equiv 0$ identically in the remaining equations. (To just discard (37) is not entirely consistent with the conservation of angular momentum understood in the conventional physical sense [16]. A consistent procedure is to take $\omega \to 0$ jointly with $I_b \to \infty$ such that $s_b \to s_b^*$. The particle motion is then the same as in the model without spin, but now spin is still part of the model, following passively along in such a way that angular momentum is conserved [3].) This simplified model without spin was recently treated rigorously in the papers [5,19–21], where for technical reasons the Einsteinian momentum

$$p_{\rm b} = \frac{m_{\rm b}\dot{q}}{\sqrt{1-|\dot{q}|^2/c^2}} \tag{39}$$

with $m_{\rm b} \neq 0$ is used instead of the Newtonian (36) (see also [17,18] for a simpler scalar theory). In [5,19], the global existence and uniqueness for the Cauchy problem of a particle without spin and $m_{\rm b} \neq 0$ was proven. The papers [19–21] address the long time asymptotics of the dynamics with $m_{\rm b} > 0$ and derive effective equations of motion in slowly varying external fields. Moreover, in [5] it was shown that the motion is stable if $m_{\rm b} > 0$ and unstable if $m_{\rm b} < 0$. Their work shows that the bare mass should not be negative in a theory without spin, contrary to what was advocated by Dirac [7]. Comparable results for the semi-relativistic equations including spin are not yet available.

I now turn to the relativistic LED. Our relativistic dynamical equations can be written in close analogy to the semi-relativistic equations by following the conventions of C.W. Misner, K. Thorne and J.A. Wheeler [29] and using component-free notation to facilitate the comparison with the semi-relativistic theory. A Lorentz frame $\{\mathbf{e}_0, \mathbf{e}_1, \mathbf{e}_2, \mathbf{e}_3\}$ satisfies the elementary inner product rules

$$\mathbf{e}_\mu \cdot \mathbf{e}_\nu = \begin{cases} -1 \text{ for } \mu = \nu = 0 \\ 1 \text{ for } \mu = \nu > 0 \\ 0 \text{ for } \mu \neq \nu \end{cases} . \tag{40}$$

We abbreviate $\mathbf{x} \cdot \mathbf{x} = \|\mathbf{x}\|^2$, where $\|\mathbf{x}\|$ is the Minkowski norm of $\mathbf{x}$, defined as the principal value of $(\mathbf{x} \cdot \mathbf{x})^{1/2}$. Notice that $\|\mathbf{x}\|^2$ is negative for time-like vectors. The tensor product between any two four-vectors $\mathbf{a}$ and $\mathbf{b}$ is a tensor of rank two, denoted by $\mathbf{a} \otimes \mathbf{b}$, and defined by its inner-product action on four-vectors thus, $(\mathbf{a} \otimes \mathbf{b}) \cdot \mathbf{x} = \mathbf{a}(\mathbf{b} \cdot \mathbf{x})$ and $\mathbf{x} \cdot (\mathbf{a} \otimes \mathbf{b}) = (\mathbf{a} \cdot \mathbf{x})\mathbf{b}$. Any tensor of rank two, T, can be uniquely written as $\mathsf{T} = T^{\mu\nu}\mathbf{e}_\mu \otimes \mathbf{e}_\nu$. In particular, the metric tensor reads $\mathbf{g} = g^{\mu\nu}\mathbf{e}_\mu \otimes \mathbf{e}_\nu$ where $g^{\mu\nu} = \mathbf{e}_\mu \cdot \mathbf{e}_\nu$ as given in (40). Notice that $\mathbf{g} \cdot \mathbf{x} = \mathbf{x}$, i.e. $\mathbf{g}$ acts as identity on four-vectors. The exterior product between two four-vectors defines an anti-symmetric tensor of rank two, denoted by a wedge product $\mathbf{a} \wedge \mathbf{b}$, and given by $\mathbf{a} \wedge \mathbf{b} = \mathbf{a} \otimes \mathbf{b} - \mathbf{b} \otimes \mathbf{a}$. Notice that $(\mathbf{a} \wedge \mathbf{b}) \cdot \mathbf{x} = -\mathbf{x} \cdot (\mathbf{a} \wedge \mathbf{b})$. The four-trace, or contraction, of a tensor of rank two is given by sum over the diagonal after multiplying the time-component by negative one. Thus, the four-trace of an antisymmetric tensor vanishes, and that $\mathrm{Tr}\,\mathbf{g} = 4$. For a differentiable function $f(\mathbf{x})$, its four-gradient is denoted by $\partial_{\mathbf{x}} f$. The four-Laplacian is simply the (negative of the) d'Alembertian, or wave operator, and given by $\partial_{\mathbf{x}} \cdot \partial_{\mathbf{x}} = -\Box$. This fixes the general notation.

Kinematically, the motion of the particle will be described by one time-like oriented space-time curve $\tau \mapsto \mathbf{q}(\tau) \in \mathbb{R}^{1,3}$ for the translations, and one space-like oriented space-time curve $\tau \mapsto \mathbf{w}_{\rm E}(\tau) \in \mathbb{R}^{1,3}$ for the inertial gyrations, where $\tau \in \mathbb{R}$ is the proper time of the particle, i.e. $c^2(\mathrm{d}\tau)^2 = -\mathrm{d}\mathbf{q} \cdot \mathrm{d}\mathbf{q}$. I shall switch to units such that $c = 1$. By $\mathbf{u}(\tau) = \mathrm{d}\mathbf{q}/\mathrm{d}\tau$ we denote the four-velocity of the particle. We have $\mathbf{u} \cdot \mathbf{u} = -1$. We will sometimes write $\mathbf{u} = \mathring{\mathbf{q}}$. By $\boldsymbol{\Omega}_{\rm T} = \mathring{\mathbf{u}} \wedge \mathbf{u}$

we denote the antisymmetric tensor of *Thomas precession* [37] in Minkowski space-time ($\boldsymbol{\Omega}_{\mathrm{FW}}$ in [29]). By $\boldsymbol{\Omega}_{\mathrm{E}}$ we denote the antisymmetric tensor of the inertial *Euler rotation* w.r.t. that co-moving particle frame $\{\bar{\mathbf{e}}_\mu\}_{\mu=0,\ldots,3}$, with $\bar{\mathbf{e}}_0 = \boldsymbol{u}$, which satisfies the law of *Fermi-Walker transport* [29]

$$\frac{\mathrm{d}}{\mathrm{d}\tau}\bar{\mathbf{e}}_\mu = -\boldsymbol{\Omega}_{\mathrm{T}} \cdot \bar{\mathbf{e}}_\mu \,. \tag{41}$$

Our four-vector $\mathbf{w}_{\mathrm{E}}$ that describes the particle's inertial rotation is the dual to the tensor $\boldsymbol{\Omega}_{\mathrm{E}}$, whence $\boldsymbol{\Omega}_{\mathrm{E}} \cdot \mathbf{w}_{\mathrm{E}} = \boldsymbol{0}$, and since $\boldsymbol{\Omega}_{\mathrm{E}} \cdot \boldsymbol{u} = \boldsymbol{0}$, also $\mathbf{w}_{\mathrm{E}} \cdot \boldsymbol{u} = \boldsymbol{0}$, i.e. $\mathbf{w}_{\mathrm{E}}$ is space-like.

The 'classical electron' carries a negative unit of the elementary charge e, distributed in the particle's rest frames by a spherically symmetric electrical charge density $f_{\mathrm{e}}(|\ .\ |) : \mathbb{R}^3 \to \mathbb{R}^-$, satisfying

$$\int_{\mathbb{R}^3} f_{\mathrm{e}}(|x|)\,\mathrm{d}x = -e. \tag{42}$$

In general, $f_{\mathrm{e}} : \mathbb{R}^+ \to \mathbb{R}^-$ is a measure. The relativistic analog of (29) and (30) for such a charge distribution is the four-current density computed by Nodvik [30],

$$\mathsf{j}(\mathbf{x}) = \int_{-\infty}^{+\infty} \Big(\boldsymbol{u} - \boldsymbol{\Omega}\cdot(\mathbf{x}-\boldsymbol{q})\Big)\, f_{\mathrm{e}}\big(\,\|\mathbf{x}-\boldsymbol{q}\|\,\big)\, \delta\big(\boldsymbol{u}\cdot(\mathbf{x}-\boldsymbol{q})\big)\,\mathrm{d}\tau, \tag{43}$$

where

$$\boldsymbol{\Omega} = \boldsymbol{\Omega}_{\mathrm{T}} + \boldsymbol{\Omega}_{\mathrm{E}}\,. \tag{44}$$

The four-current density (43) is the source term in the inhomogeneous Maxwell-Lorentz equations

$$\partial_{\mathbf{x}} \cdot \mathsf{F}(\mathbf{x}) = 4\pi \mathsf{j}(\mathbf{x})\,, \tag{45}$$

supplemented by the homogeneous Maxwell equations

$$\partial_{\mathbf{x}} \cdot {}^{*}\mathsf{F}(\mathbf{x}) = \boldsymbol{0}\,. \tag{46}$$

Here, F is the standard anti-symmetric, rank-two, electromagnetic field strength tensor, and ${}^{*}\mathsf{F}$ is the star dual to F.

Different from the semi-relativistic theory, we now also need a compactly supported spherical mass density $f_{\mathrm{m}}(|\ .\ |) : \mathbb{R}^3 \to \mathbb{R}^+$, where $f_{\mathrm{m}} : \mathbb{R}^+ \to \mathbb{R}^+$ is in general a measure giving the particle a positive bare rest mass

$$m_{\mathrm{b}} = \int_{\mathbb{R}^3} f_{\mathrm{m}}(|x|)\,\mathrm{d}x \tag{47}$$

and positive non-relativistic moment of inertia

$$I_{\mathrm{b}} = \frac{2}{3}\int_{\mathbb{R}^3} |x|^2 f_{\mathrm{m}}(|x|)\,\mathrm{d}x\,. \tag{48}$$

We are now ready for the dynamical equations for the translational and gyrational degrees of freedom. Translational motions are ruled by

$$\frac{\mathrm{d}}{\mathrm{d}\tau}\boldsymbol{p} = \int_{\mathbb{R}^{1,3}} \mathsf{F}(\boldsymbol{q}+\mathbf{x})\cdot\Big(\boldsymbol{u} - \boldsymbol{\Omega}\cdot\mathbf{x}\Big) f_{\mathrm{e}}(\|\mathbf{x}\|)\,\delta(\boldsymbol{u}\cdot\mathbf{x})\,\mathrm{d}\mathbf{x}, \tag{49}$$

where

$$\boldsymbol{p}(\tau) = \mathsf{M}(\tau)\cdot\boldsymbol{u}(\tau) \tag{50}$$

is the *Minkowski four-momentum*, and

$$\begin{aligned}\mathsf{M} = &\int_{\mathbb{R}^{1,3}} \frac{1}{\sqrt{1-\|\boldsymbol{\Omega}_{\mathrm{E}}\cdot\mathbf{x}\|^2}} f_{\mathrm{m}}(\,\|\mathbf{x}\|\,)\,\delta(\boldsymbol{u}\cdot\mathbf{x})\,\mathrm{d}\mathbf{x}\,\mathbf{g}\\ &-\int_{\mathbb{R}^{1,3}} \big[\mathbf{x}\otimes\mathbf{x}, [\mathsf{F}(\boldsymbol{q}+\mathbf{x}), \boldsymbol{\Omega}_{\mathrm{E}}]_+\big]_+ f_{\mathrm{e}}(\,\|\mathbf{x}\|\,)\,\delta(\boldsymbol{u}\cdot\mathbf{x})\,\mathrm{d}\mathbf{x}\end{aligned} \tag{51}$$

the *symmetric Minkowski mass tensor.* The first term is simply the rotational bare mass of the particle. The second term originates in an electromagnetic spin-orbit coupling and describes a generally anisotropic translational inertia of the spinning charged particle. The gyrational motions are ruled by

$$\begin{aligned}&\frac{\mathrm{d}}{\mathrm{d}\tau}\mathsf{S}_{\mathrm{b}} + \big[\boldsymbol{\Omega}_{\mathrm{T}}, \mathsf{S}_{\mathrm{b}}\big]_- =\\ &\int_{\mathbb{R}^{1,3}} \mathbf{x}\wedge(\mathbf{g}+\boldsymbol{u}\otimes\boldsymbol{u})\cdot\mathsf{F}(\boldsymbol{q}+\mathbf{x})\cdot\Big(\boldsymbol{u}-\boldsymbol{\Omega}\cdot\mathbf{x}\Big) f_{\mathrm{e}}(\|\mathbf{x}\|)\,\delta(\boldsymbol{u}\cdot\mathbf{x})\,\mathrm{d}\mathbf{x},\end{aligned} \tag{52}$$

where

$$\mathsf{S}_{\mathrm{b}}(\tau) = -\int_{\mathbb{R}^{1,3}} \mathbf{x}\wedge\frac{\boldsymbol{\Omega}_{\mathrm{E}}(\tau)\cdot\mathbf{x}}{\sqrt{1-\|\boldsymbol{\Omega}_{\mathrm{E}}(\tau)\cdot\mathbf{x}\|^2}} f_{\mathrm{m}}(\,\|\mathbf{x}\|\,)\,\delta\big(\boldsymbol{u}(\tau)\cdot\mathbf{x}\big)\,\mathrm{d}\mathbf{x} \tag{53}$$

is the spin angular momentum (about $\boldsymbol{q}$) of the bare particle, satisfying $\mathsf{S}_{\mathrm{b}}\cdot\boldsymbol{u} = 0$.

These are our equations of relativistic LED with bare inertias. They have to be supplemented by the Cauchy data for $\boldsymbol{q}$, $\boldsymbol{u}$, with $\boldsymbol{u}$ satisfying $\|\boldsymbol{u}\|^2 = -1$, furthermore for $\boldsymbol{\Omega}_{\mathrm{E}}$ and F.

For the special case that only the spin degree of freedom is coupled to the field in the dynamics, we proved global existence and uniqueness for small 'coupling constant.' For the general case, only local well-posedness under certain conditions is known. Assuming global regularity, our equations satisfy the conventional conservation laws analogous to the semi-relativistic theory with spin. In addition, there is a further conservation law for the magnitude of the renormalized spin. Most notably: the renormalized electron is a soliton. It is this feature which guarantees that 'an electron remains an electron.'

I end with a note on the physical viability of a purely electromagnetic LED. When Uhlenbeck and Goudsmit [39] proposed the idea of an electron spin magnetic moment, Lorentz objected that the postulated magnitude of about one

Bohr magneton would require equatorial rotation speeds of the purely electromagnetic electron of about 10 times the speed of light ([38], p. 35). The acceptance of electron spin as some form of electron-intrinsic angular momentum turned the Lorentz electron's superluminal equatorial speed into a weighty argument against the physical viability of LED as a classical limit of QED, which has persisted ever since. However, judged from a modern perspective on mass renormalization the purely electromagnetic calculations by Lorentz are not proper. We have carried out a renormalization flow analysis for a stationary rotating charged particle with its charge and positive bare rest mass distributed uniformly over a sphere. For matched physical electron data (mass, charge, and magnetic moment), the model electron reaches equatorial speed of light *precisely* when its bare rest mass vanishes, which happens when its radial size is of the order of the Compton radius of the physical electron, *not* the classical electron radius! The renormalized mass of this purely electromagnetic limit particle consists of the traditional electromagnetic contribution plus a 'photonic' contribution associated with its luminal rotational motion while its bare rest mass is zero. Its spin is $3\hbar/2$ plus corrections of order α.

Acknowledgments

My gratitude goes to the organizers of this remarkable conference. My thanks go also to the Societa Italiana di Fondamenti della Fisica and to the Instituto Italiano per gli Studi Filosifici for their generous support of the conference attendees. Part of my work was funded by the NSF through Grant # DMS 9623220, which is gratefully acknowledged. Finally, I am indebted to Sheldon Goldstein for valuable comments on an earlier draft of this paper.

References

1. M. Abraham, Ann. Phys. **10**, 105 (1903)
2. V.A. Antonov, Vest. Leningrad Gas. Univ. **7**, 135 (1962) (English transl.: 'Most probable phase distribution in spherical star systems and conditions for its existence.' In: *Dynamics of Star Clusters,* ed. by J. Goodman and P. Hut, IAU 1985), pp. 525–540)
3. W. Appel, M. K.-H. Kiessling, 'Mass and spin renormalization in Lorentz electrodynamics,' Annals Phys. (NY) (in press 2000)
4. J.D. Bekenstein, Phys. Rev. D. **7**, 2333 (1973)
5. G. Bauer, D. Dürr, 'The Maxwell-Lorentz system of a rigid charge distribution,' Preprint, Ludwig Maximilian Universität München (1999)
6. W. Braun, K. Hepp, Commun. Math. Phys. **56**, 101 (1977)
7. P.A.M. Dirac, Proc. Roy. Soc. A **167**, 148 (1938)
8. J. Ehlers, 'General relativity and kinetic theory.' In: *General relativity and cosmology, Proceedings of the International School of Physics Enrico Fermi vol. 47*, ed. by R. K. Sachs (Academic Press, New York 1971), pp. 1–70
9. H.-P. Gittel, J. Kijowski, E. Zeidler, Commun. Math. Phys. **198**, 711 (1998)

10. R. Glassey, J. Schaeffer, Comm. Math. Phys. **119**, 353 (1988)
11. R. Glassey, W. Strauss, Comm. Math. Phys. **113** 191 (1987)
12. R. Glassey, W. Strauss, Arch. Rat. Mech. Analysis **92**, 59 (1986)
13. S. Hawking, Comm. Math. Phys. **43**, 199 (1975)
14. D.C. Heggie, Mon. Not. R. astr. Soc. **173**, 729 (1975)
15. J.D. Jackson, *Classical Electrodynamics*, 3rd ed., (Wiley, New York 1999)
16. M. K.-H. Kiessling, Phys. Lett. A **258**, 197 (1999)
17. A. Komech H. Spohn, M. Kunze, Commun. PDE **22**, 307 (1997)
18. A. Komech M. Kunze, H. Spohn, Commun. Math. Phys. **203**, 1 (1999)
19. A. Komech H. Spohn, 'Long-time asymptotics for the coupled Maxwell-Lorentz Equations,' J. Diff. Equations (in press)
20. M. Kunze, H. Spohn, 'Adiabatic limit of the Maxwell-Lorentz equations,' Ann. Inst. H. Poincaré, Phys. Theor. (in press)
21. M. Kunze, H. Spohn, 'Radiation reaction and center manifolds,' SIAM J. Math. Anal. (in press)
22. J.M. Lévy-Leblond, J. Math. Phys. **10**, 806 (1969)
23. E.H. Lieb, Bull. Am. Math. Soc. **22**, 1 (1990)
24. E.H. Lieb, W. Thirring, Ann. Phys. **155**, 494 (1984)
25. E.H. Lieb, H.T. Yau, Commun. Math. Phys. **112**, 147 (1987)
26. H.A. Lorentz, Arch. Néerl. Sci. Exactes Nat. **25**, 363 (1892)
27. H.A. Lorentz, Proc. Acad. Wet. Amsterdam, **6** (1904)
28. H.A. Lorentz, *The Theory of electrons and its applications to the phenomena of light and radiant heat*, (Dover, New York 1952)
29. C.W. Misner, K.S. Thorne, J.A. Wheeler, *Gravitation* (W.H. Freeman Co., New York 1973)
30. J.S. Nodvik, Ann. Phys. **28**, 225 (1964)
31. D. Noja, A. Posilicano, *Delta interactions and electrodynamics of point particles*, LANL e-print math-ph/9907009 (1999)
32. W. Pauli, *Theory of relativity*, Dover (1958)
33. R. Penrose, *The emperors new mind*, (Oxford Univ. Press, Oxford 1989)
34. G. Rein, *The Vlasov-Einstein system with surface symmetry*, Habilitationsschrift, Ludwig Maximilian Universität München (1995)
35. F. Rohrlich, *Classical charged particles*, (Addison Wesley, Redwood City, CA 1990)
36. A. Sommerfeld, *Electrodynamics*, (Academic Press, New York 1952)
37. L.H. Thomas, Nature **117**, 514 (1926); Phil. Mag. **3**, 1 (1927)
38. S. Tomonaga, *The story of spin*, (Univ. Chicago Press, Chicago 1997)
39. G.E. Uhlenbeck, S.A. Goudsmit, Nature **117**, 264 (1926)
40. A.A. Vlasov, *Many-particle theory and its application to plasma*, (Gordon and Breach, New York, 1961)
41. A.D. Yaghjian, *Relativistic dynamics of a charged sphere*, Lect. Notes Phys. **m11**, (Springer, Berlin, 1992)

Part III

Quantum Mechanics

Probability in Orthodox Quantum Mechanics: Probability as a Postulate Versus Probability as an Emergent Phenomenon

Stephen L. Adler

Institute for Advanced Study, Einstein Drive, Princeton NJ 08540, USA

Abstract. The role of probability in quantum mechanics is reviewed, with a discussion of the "orthodox" versus the statistical interpretative frameworks, and of a number of related issues. After a brief summary of sources of unease with quantum mechanics, a survey is given of attempts either to give a new interpretive framework assuming quantum mechanics is exact, or to modify quantum mechanics assuming it is a very accurate approximation to a more fundamental theory. This survey focuses particularly on the issue of whether probabilites in quantum mechanics are postulated or emergent.

1 Orthodox Quantum Mechanics and Issues It Raises

Quantum mechanics (QM) is our most successful physical theory, encompassing phenomena as diverse as chemical bonding, the band structure of solids, and the standard model of particle physics. But the probabilistic aspect of quantum mechanics has been a source of unease from the outset. As surveyed by Home [1], the Founding Fathers were divided over this aspect of the theory that they had created: Bohr, Born, and Heisenberg were comfortable with the probabilistic structure of quantum theory, whereas Einstein and Schrödinger had profound reservations. This unease, and division, have continued to the present.

1.1 Postulates of Quantum Mechanics

Let us begin with a review of the postulates of QM, in the arena of a complex Hilbert space, following for the larger part the presentation of Ballentine [2].

- Observables are associated with self-adjoint operators. Thus we have

$$S = \sum_n s_n P_n \quad , \tag{1}$$

$$P_n = \sum_a |a, s_n\rangle\langle a, s_n| \quad , \tag{2}$$

  with S an operator, s_n its eigenvalues, P_n the corresponding orthogonal projectors, and a a label that distinguishes degenerate eigenvectors.
- Each state is associated with a density matrix ρ, which is self-adjoint, non-negative, and has trace unity,

$$\rho = \rho^\dagger \ , \tag{3}$$

$$\rho \geq 0 \ , \quad \mathrm{Tr}\rho = 1 \ , \tag{4}$$

so that the spectral decomposition of ρ takes the form

$$\rho = \sum_n \rho_n |\phi_n\rangle\langle\phi_n| \ , \tag{5}$$

$$0 \le \rho_n \le 1 \ , \quad \sum_n \rho_n = 1 \quad . \tag{6}$$

- A pure state is asssociated with an idempotent density matrix, satisfying $\rho^2 = \rho$. This condition, together with the condition of unit trace, implies that there is exactly one nonzero eigenvalue $\rho_n = 1$, with all other eigenvalues $\rho_{n'}$, $n' \neq n$ vanishing, and so for a pure state the spectral decomposition or the density matrix reduces to

$$\rho = |\phi_n\rangle\langle\phi_n| \quad . \tag{7}$$

- The average of an observable S in a general state ρ is given by

$$\langle S\rangle = \mathrm{Tr}\rho S \quad . \tag{8}$$

In particular, for a pure state with $\rho = |\psi\rangle\langle\psi|$, $\langle\psi|\psi\rangle = 1$, we have

$$\langle S\rangle = \langle\psi|S|\psi\rangle \quad . \tag{9}$$

- An observable S can only take the values s_n of its eigenvalues. The *probability* of finding the eigenvalue s_n in a normalized pure state $|\psi_n\rangle$ is

$$p_n = \sum_a |\langle\psi|a, s_n\rangle|^2 \quad . \tag{10}$$

- Within coherent sectors of Hilbert space, superpositions of pure states are pure states, and self-adjoint compositions of observables are observables.
- The dynamics of the density matrix ρ and of a state ψ in Hilbert space is specified by

$$\rho(t) = U\rho(t_0)U^{-1} \quad , \tag{11}$$

$$|\psi(t)\rangle = U|\psi(t_0)\rangle \quad . \tag{12}$$

Here $U(t, t_0)$ is a unitary operator, which for small $t - t_0 = \delta t$ takes the form

$$U = \exp\left(-\frac{\mathrm{i}}{\hbar}\mathrm{H(t)}\delta\mathrm{t}\right) \quad , \tag{13}$$

defining the (possibly time dependent) system Hamiltonian. This dynamics is termed the "U operation" by Penrose [3].
- We finally come to the effect of a measurement. After the measurement of the eigenvalue s_n in a pure state, the new system state is

$$|\psi'\rangle = \frac{\sum_a |a, s_n\rangle\langle a, s_n|\psi\rangle}{[\sum_a |\langle a, s_n|\psi\rangle|^2]^{\frac{1}{2}}} \quad . \tag{14}$$

This equation summarizes what is termed the "R operation" by Penrose.

1.2 Interpretive Framework

While everyone agrees that the above postulates provide a practical set of rules for making predictions in quantum mechanics, and that these predictions to date have always agreed with experiment, there is a dichotomy when it comes to giving an interpretive framework for the rules.

- On the one hand, we have the "orthodox" interpretation, as given, for example, in the text of Mandl [4]. This asserts that the state $|\psi\rangle$ gives a complete description of an *individual* system, and that (14) corresponds to "reduction" of the individual's state vector.
- On the other hand, we have the "statistical" interpretation, as discuused, for example, in the review of Ballentine [2], according to which the state $|\psi\rangle$ describes certain statistical properties of an *ensemble* of similarly prepared systems. According to this interpretation, (14) corresponds to the preparation of a new ensemble by a measurement. There *may* be, or there *may not* be, hidden variables that specify a complete, nonstatistical interpretation of individual systems: the statistical interpretation is agnostic with respct to this issue.

1.3 Why the "R" Operation Is Needed

To see why the "R" operation is needed, let us demonstrate that the measurement process cannot be represented by a *deterministic*, *unitary* evolution on a closed system. Let us consider a Stern-Gerlach experiment, in which an initial spin eigenstate $|\psi\rangle$ with eigenvalue 1/2 along the x axis is separated, by means of a magnetic field that is inhomogeneous along the z axis, into orthonormal states $|\psi_\uparrow\rangle$, $|\psi_\downarrow\rangle$ that have respective spin eigenvalues $1/2, -1/2$ along the z axis. Thus, at the detector one sees either $|\psi_\uparrow\rangle$ or $|\psi_\downarrow\rangle$, with

$$\langle\psi_\downarrow|\psi_\uparrow\rangle = 0 \quad . \tag{15}$$

Let us suppose that these final states evolved from the initial state by the deterministic unitary "U"process, which would imply that

$$|\psi_\uparrow\rangle = U|\psi\rangle \ , \tag{16}$$

$$|\psi_\downarrow\rangle = U|\psi\rangle \quad . \tag{17}$$

This would imply

$$0 = \langle\psi_\downarrow|\psi_\uparrow\rangle = \langle\psi|U^\dagger U|\psi\rangle = \langle\psi|\psi\rangle = 1 \quad , \tag{18}$$

which is a contradiction. Hence, the measurement process, in which an initial coherent superposition of states leads to a definite but unpredictable outcome, cannot be described by a deterministic unitary time evolution. Thus measurement involves a physical process, which we have called the "R" process, that is distinct from the deterministic unitary "U" process that governs the unobserved evolution of the quantum system.

However, the "R" process does not have to be nonunitary; one can have for the ith atom going through the apparatus the evolution

$$|\psi_i\rangle = U_i|\psi\rangle \quad , \tag{19}$$

with U_i a unitary evolution that is different for each i. This is possible because any path through Hilbert space connecting two normalized pure states can be described by a succession of infinitesimal unitary transformations. To prove this, it suffices to consider the infinitesimal segment $|\psi\rangle \to |\psi\rangle + |\mathrm{d}\psi\rangle$, with $\langle\psi|\mathrm{d}\psi\rangle = 0$. If we take

$$U = 1 + |\mathrm{d}\psi\rangle\langle\psi| - |\psi\rangle\langle\mathrm{d}\psi| \quad , \tag{20}$$

$$U^\dagger = 1 - |\mathrm{d}\psi\rangle\langle\psi| + |\psi\rangle\langle\mathrm{d}\psi| \quad , \tag{21}$$

then we have $U^\dagger U = UU^\dagger = 1$ up to an error of second order, and $U|\psi\rangle = |\psi\rangle + |\mathrm{d}\psi\rangle$, as needed. So it is perfectly possible for the "R" process to be described by a *stochastic* unitary process, constructed from a sequence of random or partially random infinitesimal unitary transformations, and we will see in Sect. 3.1 examples of how this can be accomplished.

1.4 Micro Versus Macro

In QM, the probability is the squared modulus of the probability amplitude. Probability amplitudes superimpose coherently, and between measurements they evolve in time by the "U" process. Thus, in the microscopic realm:

- One sees coherent superpositions.
- Amplitudes evolve through deterministic, unitary evolution.

On the other hand, in the macroscopic realm:

- One does *not* see coherent superpositions, e.g., of dead and alive cats.
- Measurements involve the "R" process, which is not deterministic unitary.

This dichotomy leaves us with the following questions (for more detailed discussions, see Penrose [3] and Leggett [5]):

- Where is the dividing line between "micro" and "macro"?
- What is responsible for it?

1.5 Postulated Versus Emergent Probability

A unique feature of orthodox QM is that it is the only probabilistic theory where the probabilities are *postulated* ab initio, and are not *emergent* from unobserved, deterministic phenomena at a deeper level. A typical theory where probabilites are emergent is statistical mechanics. In statistical mechanics one starts from a probability postulate of uniform phase space occupation. This assumption, and the related concept of an equilibrium ensemble (which reflects the implications of conserved quantities for the phase space occupation), is consistent because of Liouville's theorem, which implies that a uniform distribution is preserved in time by the dynamics.

However, these probabilistic statements are not the end of the story in statistical mechanics. There are underlying laws – the equations of classical molecular dynamics – which are deterministic; no probabilities enter into their formulation. These laws lead, by a process that is still not completely understood (as reflected in discussions of ergodicity at this Conference), to an effectively uniform phase space distribution for systems that are sufficiently complex. Thus, the probabilistic theory of statistical mechanics is *emergent* from the deterministic theory of classical mechanics.

1.6 Recapitulation

To sum up, there are a number of sources of unease about QM:

- There is no predictive description of individuals.
- There is a micro-macro divide of unclear origin.
- There is a probabilistic structure that is postulated rather than emergent.

But QM works! Many subtle and remarkable predictions of QM are experimentally verified in many different physical arenas. Thus either

- (A) QM is exact, but to deal with the sources of unease it needs reinterpretation at the conceptual level (although no modification of the standard postulates is needed to use QM as a practical computational and predictive tool).
- (B) QM is not exact, but rather is a very accurate asymptotic approximation to a deeper level theory.

I do not believe that it is just a matter of taste which of these possibilites is chosen, because the distinction between (A) and (B) is relevant to the issue of Planck scale unification of the forces and the particle spectrum. If QM changes, it may profoundly influence the ground rules for unifying gravity with the other interactions.

2 Reinterpretations of Quantum Mechanics Assuming It Is Exact

Let us now review four differing approaches based on premise (A), that QM is exactly correct but in need of reinterpretation. Our focus in each case will be on the extent to which the probabilistic structure is postulated or is emergent.

2.1 Everett's "Many Worlds" Interpretation

In the "Many Worlds" interpretation introduced by Everett [6] and discussed in further detail in the articles collected in [7], there is no state vector reduction; instead, there is only Schrödinger evolution of the wave function of the entire universe. To describe N successive measurements in this interpretation requires an N-fold tensor product wave function.

Probability is not emergent, but rather is postulated in the Everett picture. Everett introduces a measure on the coefficients of the final superposition resulting from N successive measurements, which as $N \to \infty$ behaves mathematically like the usual QM probability rule. There is a logical jump (or an implicit assumption – this is still a matter of debate) in going from the Everett measure on tensor product coefficients to statements about the relative frequencies of experimental outcomes.

2.2 The Histories Approach

The so-called "Histories" approach has been extensively developed recently by Griffiths, Omnès, and Gell-Mann and Hartle (for a review and references see [8], and for a semipopular account see [9]). The histories approach takes as a given that QM is a stochastic theory; probability is introduced as a postulate, and is not emergent. The basic objects in the histories approach are time-dependent projectors $E_k(t_k)$ associated with properties occurring in a history. The probability of a history is then postulated to be given by

$$p = \mathrm{Tr}[E_n(t_n)...E_1(t_1)\rho E_1(t_1)...E_n(t_n)] \quad , \tag{22}$$

with ρ the density matrix at the intial time. This definition can be shown to lead, under appropriate circumstances, to all of the expected properties of probabilities. In this interpretation, state vector reduction appears only in the statistical interpretation sense discussed above, as a rule for relating the density matrix after a measurement to the density matrix before the measurement.

In both the "Many Worlds" and the "Histories" interpretations, there is by definition no concept of the "individual". We shall now discuss two other currently studied interpretations of QM that enlarge the mathematical framework to give an "individual".

2.3 Bohmian Mechanics

There has been a recent revival of interest in Bohmian mechanics (see [10] for a technical account and references, and [11] for a semipopular account). In Bohmian mechanics, in addition to the Schrödinger equation for the N-body wave function $\psi(\boldsymbol{q}_1, ..., \boldsymbol{q}_N, t)$,

$$i\hbar \frac{\partial \psi}{\partial t} = -\sum_{k=1}^{N} \frac{\hbar^2}{2m_k} \nabla^2_{\boldsymbol{q}_k} \psi + V\psi \quad , \tag{23}$$

the mathematical framework is enlarged by introducing hidden "particles" moving in configuration space with coordinates $\boldsymbol{Q}_k$ and velocities

$$\boldsymbol{v}_k = \frac{\mathrm{d}\boldsymbol{Q}_k}{\mathrm{d}t} = \frac{\hbar}{m_k}\mathrm{Im}\nabla_{\boldsymbol{Q}_k}\log\psi(\boldsymbol{Q}_1, ..., \boldsymbol{Q}_k, t) \quad . \tag{24}$$

The state of the "individual" is then specified by giving both the wave function *and* the coordinates $\boldsymbol{Q}_k$ of the hidden particles. A probability postulate is introduced, that the probability distribution on configuration space obeys $p = |\psi|^2$ at some initial time t_0. The Bohmian equations given above then imply that this remains true for all times subsequent to t_0; the logic here resembles the use of the Liouville theorem in statistical mechanics. Unlike statistical mechanics, Bohmian mechanics has no underlying molecular dynamics-like layer, so the probabilites are not prima facie emergent. We note, however, that in [10] arguments are given (and are further discussed at this Conference) that the Bohmian probability postulate follows from considerations of "typicality" of initial configurations (in distinction to the ergodicity arguments used in attempts to derive the postulates of statistical mechanics from the equations of molecular dynamics).

2.4 The Ax–Kochen Proposal

Recently Ax and Kochen [12] have extended the mathematical framework of QM in a different way to encompass the "individual". They identify the *ray* with the ensemble, and the *ray representative*, i.e. the $U(1)$ phase associated with a particular state vector, with the individual. They then give a mathematical construction to specify a unique physical state from knowledge of the toroid of phases. They introduce a probability assumption, that the a priori distribution of phases is uniform, and then show that, by their construction, this implies that the probabilities of outcomes obey the usual QM rule. Thus, probability in the Ax-Kochen interpretation is not emergent, but their probabilistic postulate is arguably weaker than that in standard QM or in Bohmian QM.

2.5 Are Interpretations of Quantum Mechanics Falsifiable?

We conclude this brief survey of alternative interpretations of an assumed exact quantum mechanics by posing the question, can the interpretations given above be falsified? By construction, the four interpretations described above are designed to agree with the predictions of standard QM. Clearly, if an interpretation could be shown to *differ* in some prediction from that of QM, and if this difference in predictions were resolved experimentally in favor of standard QM (in the way that the Bell inequalities have been tested and favor QM), then the interpretation would be falsified. But suppose that an interpretation makes empirical predictions that, within the domain in which the rules of QM apply, are without exception indistinguishable from the predictions of QM. Then is it possible, in principle, to falsify that interpretation?

The answer, I believe, may be "yes", because none of the interpretations described above gives a quantitative account of the micro-macro divide – that is, when do we, and when don't we, expect to see coherent superpositions? To the extent that this becomes an experimentally answerable question, and to the extent that one can get corresponding predictions from the interpretations sketched above, one might be able to distinguish between different interpretive frameworks for an exact QM.

3 Theories Where Quantum Mechanics Is Modified

Let us turn now to approaches based on premise (B), that QM is a very accurate approximation to a deeper level theory. We will first discuss phenomenological approaches based on this premise, and then turn to attempts at a more fundamental theory.

3.1 Phenomenological Modifications: Stochastic Models

As we discussed in Sect. 1.3, although the "R" process cannot be described by a deterministic unitary evolution, it is perfectly admissible for it to be described by a unitary evolution that differs for each individual measurement act, and in particular by a stochastic unitary evolution. Considerable effort has gone over the past two decades into attempts to unify the "U" process and the "R" process into a single dynamical rule, by formulating phenomenological modifications of the Schrödinger equation in which the "individual" is described by a stochastic unitary evolution of a pure state. The physical motivation for such modifications is that if quantum theory is an approximation to physics at a deeper level, there could be small fluctuation or "Brownian motion" corrections to this physics, which determine the outcomes for individual systems.

The natural mathematical language for formulating stochastic modifications of the Schrödinger equation is the Itô stochastic calculus, which is basically a differential calculus version of the theory of Gaussian random variables. (For a clear exposition of the Itô rules, see Gardiner [13].) One introduces the stochastic Itô differential $\mathrm{d}W_t$, which obeys the rules

$$(\mathrm{d}W_t)^2 = \mathrm{d}t \ , \quad \mathrm{d}W_t \mathrm{d}t = 0 \ ; \tag{25}$$

thus $\mathrm{d}W_t$ is a fluctuating variable with magnitude $(\mathrm{d}t)^{\frac{1}{2}}$, and as is familiar from the theory of path integrals, quantities of order $\mathrm{d}t$ are retained while those of order $(\mathrm{d}t)^{\frac{3}{2}}$ are dropped. Let us now consider the following equivalent stochastic evolutions (introduced at various times, and in various forms, by Diósi; Ghirardi, Rimini, and Weber; Gisin; Hughston; Pearle; and Percival – for references, see [14] and [15]). Letting $|z\rangle$ be a pure state, and $\rho = |z\rangle\langle z|/\langle z|z\rangle$ be the corresponding density matrix, we can write a stochastic pure state evolution

$$d|z\rangle = [\alpha dt + \beta dW_t]|z\rangle \ , \tag{26}$$

$$\alpha = -iH - \frac{1}{8}\sigma^2[A - \langle A\rangle]^2 \ , \quad \beta = \frac{1}{2}\sigma[A - \langle A\rangle] \ , \tag{27}$$

$$A = A^\dagger \ , \quad \langle A\rangle = \langle z|A|z\rangle/\langle z|z\rangle \quad , \tag{28}$$

or the equivalent [15] density matrix evolution

$$d\rho = -i[H,\rho]dt - \frac{1}{8}\sigma^2[A,[A,\rho]]dt + \frac{1}{2}\sigma[\rho,[\rho,A]]dW_t \quad . \tag{29}$$

Here we have taken units with $\hbar = 1$, σ is a numerical parameter which governs the strength of the stochastic process that modifies the standard Schrödinger dynamics, and one can generalize the above equations by replacing $A \to A^j$, $dW_t \to dW_t^j$ and including a sum over j in each term. Letting $E[\rho]$ denote the stochastic expectation of ρ with respect to dW_t (*not* the same as the quantum expectation $\langle\rho\rangle$), the evolution of ρ implies the following Lindblad type evolution of $E[\rho]$,

$$\frac{dE[\rho]}{dt} = -i[H,E[\rho]] - \frac{1}{8}\sigma^2[A,[A,E[\rho]]] \quad . \tag{30}$$

Let us now ask [15], when does this equation admit stationary solutions $E[\rho]_S$, for which

$$\frac{dE[\rho]_S}{dt} = 0 = -i[H,E[\rho]_S] - \frac{1}{8}\sigma^2[A,[A,E[\rho]_S]] \quad ? \tag{31}$$

Multiplying by $E[\rho]_S$, taking the trace, and using cyclic permutation under the trace, which implies that

$$\mathrm{Tr}E[\rho]_S[H,E[\rho]_S] = \mathrm{Tr}H[E[\rho]_S,E[\rho]_S] = 0 \quad , \tag{32}$$

we get the condition

$$0 = -\frac{1}{8}\sigma^2\mathrm{Tr}E[\rho]_S[A,[A,E[\rho]_S]] \tag{33}$$

$$= -\frac{1}{8}\sigma^2\mathrm{Tr}[A,E[\rho]_S][A,E[\rho]_S]^\dagger \quad . \tag{34}$$

Since the argument of the final trace is positive semidefinite, it must vanish, and so we learn by reference to the evolution equation for $E[\rho]_S$ that we must have

$$[A,E[\rho]_S] = 0 \ , \quad [H,E[\rho]_S] = 0 \quad , \tag{35}$$

in other words, a stationary value $E[\rho]_S$ must commute with both the Hamiltonian H and with the operator A which drives the dissipative process.

Various cases are possible, depending on the choice of A:

- For an *energy driven process*, with $A = H$, the stationary value $E[\rho]_S$ can be any function of H. One can then prove [15] *with no approximations* that, when all energy eigenstates are nondegenerate, in the limit of large times, ρ approaches an energy eigenstate projector $|e\rangle\langle e|$, with each such projector occuring as the outcome of the stochastic process with the corresponding probability $P_e = \mathrm{Tr}\rho(0)|e\rangle\langle e|$, with $\rho(0)$ the initial time density matrix. Correspondingly, the stochastic expectation of ρ, which is what we customarily term the density matrix, evolves from a pure state density matrix $E[\rho(0)] = \rho(0)$ to the mixed state density matrix $E[\rho]_S = \sum_e P_e|e\rangle\langle e|$. Thus, for an energy driven process (or more generally, processes in which there are several A^j which all commute with the Hamiltonian), the QM probability rule is *emergent* from the phenomenological stochastic dynamics.

  As discussed in [14], if one assumes that the parameter σ and the corresponding stochastic process originate from Planck scale physics, one gets the estimate (again in units with $\hbar$=1) $\sigma \sim M_{\mathrm{Planck}}^{-\frac{1}{2}}$, which implies a characterstic state vector reduction time scale

$$t_R \sim \left(\frac{2.8\mathrm{MeV}}{\Delta E}\right)^2 \mathrm{sec} \quad , \tag{36}$$

  with ΔE the energy dispersion of the initial state. An important question that remains to be answered is whether this estimate gives a satisfactory phenomenology for state vector reduction in all cases, when the characteristic ΔE arising from environmental interaction effects is assumed. That is,does the predicted micro-macro divide always occur in the right place?
- For a *localization process* (the Ghirardi, Pearle, Rimini form of the original Ghirardi, Rimini, Weber idea; for references see [14], [15]), one takes A to be an operator that produces a Gaussian localization, or one uses multiple A^j corresponding to many such localizations. Since the kinetic term $\boldsymbol{p}^2/(2M)$ in the Hamiltonian H does not commute with $\boldsymbol{x}$, and thus does not commute with a localizing operator A, there now is *no* stationary limit unless the usual Schrödinger evolution term in the stochastic Schrödinger equation is neglected. In this approximation, in which only the stochastic terms are kept, one gets similar results to those found in the energy driven case, with H eigenstates replaced now by A eigenstates.

  For the localizing case, an important issue that remains to be addressed is whether the phenomenological theory can be made relativistic. This may be a more severe problem than in the energy driven case because, while the Hamiltonian operator H of Schrödinger dynamics appears in a similar role in quantum field theory, the coordinate $\boldsymbol{x}$ appears in quantum field theory as a label for degrees of freedom, rather than as an operator.

3.2 Fundamental Modifications

Of course, even though the phenomenological stochastic Schrödinger equations discussed above give exactly (in the energy driven case) or approximately (in the localizing case) an emergent QM probability rule, there is still a probabilistic postulate in the form of the appearance of the Itô differential dW_t. Since these equations have the characteristic form expected for the dynamics of open quantum systems, it is natural to ask whether they are simply the Brownian motion description of some underlying dynamics. Specifically, can one achieve a fully emergent probabilistic structure at the QM level from a pre-quantum dynamics that is not probabilistic?

Two approaches of this type have been discussed in the literature:

- In [16] 't Hooft has proposed that quantum states are the equilibrium limit orbits or Poincarè cycles of an underlying chaotic, dissipative, deterministic theory.
- In [17] we have proposed as a possible pre-quantum dynamics a "generalized quantum" or "trace" dynamics, obtained by setting up a generalized classical dynamics of noncommuting phase space variables $\{q_r\}$, $\{p_r\}$ with no a priori commutativity properties beyond cyclic permutation inside a trace. One can show that, with an approximation similar to assuming a large hierarchy between the pre-quantum and the QM energy scales, that by an equipartition argument the canonical commutation relations of QM are an emergent property of the statistical mechanics of this system.

Both of these proposed approaches to an emergent probability structure in QM are at present programmatic, and significant open questions remain: Can one construct an effective wave function and Schrödinger equation from the pre-quantum dynamics? What do the leading fluctuation corrections look like, and are they the mechanism responsible for state vector reduction? Can one use them to make contact with the phenomenological stochastic extensions of the Schrödinger equation discussed above? Affirmative answers to these questions would yield the probabilistic structure of QM as an emergent phenomenon, in close analogy with the origins, in the underlying deterministic layer of molecular dynamics, of the probabilistic structure of statistical mechanics. Failure, after sufficient effort, to construct such a pre-quantum underpinning for QM would support the view that QM is exact, in need perhaps only of a modified interpretation.

Acknowledgments

This work was supported in part by the Department of Energy under Grant #DE-FG02-90ER40542.

References

1. D. Home: *Conceptual Foundations of Quantum Mechanics*, Chapt. 8 (Plenum, New York 1997)
2. L. E. Ballentine, Rev. Mod. Phys. **42**, 358 (1970)
3. R. Penrose: *The Emperor's New Mind* (Oxford University Press, New York 1989)

4. F. Mandl: *Quantum Mechanics*, 2nd. edn., Chapt. 4, Sec. 16, pp. 69-70 (Buttersworth, London 1957)
5. A. J. Leggett, Suppl. Prog. Theor. Phys. **69**, 80 (1980)
6. H. Everett, III, Rev. Mod. Phys. **29**, 454 (1957)
7. B. S. DeWitt and N. Graham: *The Many-Worlds Interpretation of Quantum Mechanics* (Princeton University Press, Princeton 1973)
8. R. Omnès, Rev. Mod. Phys. **64**, 339 (1992)
9. R. B. Griffiths and R. Omnès, Physics Today **52**, No. 8, Part 1, 26 (1999)
10. D. Dürr, S. Goldstein, and N. Zanghí, J. Stat. Phys. **67**, 843 (1992)
11. S. Goldstein, Physics Today **51**, No. 3, 42 (1998) **51**, No. 4, 38 (1998)
12. J. Ax and S. Kochen, 'Extension of Quantum Mechanics to Individual Systems', quant-ph/9905077
13. C. W. Gardiner: *Handbook of Stochastic Methods*, Chapt. 4 (Springer, Berlin 1990)
14. L. P. Hughston, Proc. Roy. Soc. Lond. A **452**, 953 (1996)
15. S. L. Adler and L. P. Horwitz, 'Structure and Properties of Hughston's Stochastic Extension of the Schrödinger Equation', quant-ph/9909026
16. G 't Hooft, J. Stat. Phys. **53**, 323 (1988) 'Quantum Gravity as a Dissipative Deterministic System', gr-qc/9903084 'Determinism and Dissipation in Quantum Gravity', hep-th/0003005
17. S. L. Adler, Nucl. Phys. B **415**, 195 (1994) S. L. Adler and A. C. Millard, Nucl. Phys. B **473**, 199 (1996) S. L. Adler and A. Kempf, J. Math. Phys. **39**, 5083 (1998)

Bohmian Mechanics

Detlef Dürr

Mathematisches Institut
Universität München
Theresienstr. 39
80333 München, Germany

Abstract. This is a short review of Bohmian Mechanics with special emphasis on the role of probability within this deterministic quantum theory without observers. I discuss the equations of motion and the statistical mechanics of this theory.

1 The Theory

1.1 Equations

Bohmian mechanics is a quantum theory without observers [1,2]. This means that neither the act of observation nor the notion of observer play any role in defining the theory – the theory is not about observers and observation – and it explains all non relativistic quantum phenomena. The theory is about something primitive[1], the basic ontology, and the laws for that are given.

Bohmian mechanics is a deterministic theory of point particles. Like Newtonian mechanics it is invariant under Galilei transformations, but unlike Newtonian mechanics it is a first-order theory – acceleration is not a concept entering the law of motion. Rather this law directly determines the velocities of the particles as follows.

For an N-particle system the positions of the N particles form the configuration space variable $Q = (\boldsymbol{Q}_1, \ldots, \boldsymbol{Q}_N) \in \mathbb{R}^{3N}$, $\boldsymbol{Q}_k \in \mathbb{R}^3$ being the position of k-th particle. The motion of the configuration is determined by a time-dependent wave function on configuration space $\psi(q,t) \in \mathbb{C}$, $q = (\boldsymbol{q}_1, \ldots, \boldsymbol{q}_N) \in \mathbb{R}^{3N}$, that obeys Schrödingers equation (m_k is the mass of the k-th particle and V is the potential)

$$\mathrm{i}\hbar \frac{\partial \psi(q,t)}{\partial t} = -\sum_{k=1}^{N} \frac{\hbar^2}{2m_k} \Delta_k \psi(q,t) + V(q)\psi(q,t). \tag{1}$$

The wave function's role is to induce a velocity-vectorfield on configuration space $v^{\psi}(q)$, the integral curves of which are the trajectories of the particles

$$\frac{\mathrm{d}\boldsymbol{Q}_k}{\mathrm{d}t} = \boldsymbol{v}_k^{\psi}(Q) = \frac{\hbar}{m_k} \frac{\mathrm{Im}(\psi^* \nabla_k \psi)}{\psi^* \psi} (\boldsymbol{Q}_1, \ldots, \boldsymbol{Q}_N). \tag{2}$$

[1] In contrast, observers are complicated objects, made of many parts: for example one may get a haircut and still be functioning as observer.

To solve (1), (2) one needs an initial configuration (i.e. the initial positions of all the particles) and an initial wave function.

In [3] we show how the law of motion may be seen as emerging in a natural manner from the requirement that it be invariant under Galileian transformations.

Non Newtonian Character. Bohmian mechanics is a first-order theory: Given the velocity field, specifying the positions of the particles at one time fixes the trajectories. Newtonian mechanics needs both position and velocity specified for the motion to be determined. Differentiating (2) with respect to time t does not render the theory second order, no more than does a Newtonian theory become a third order theory by differentiating $m\ddot{\boldsymbol{x}} = \boldsymbol{F}$ with respect to time. Nevertheless David Bohm, who worked this theory out in 1952 [4,5], presented this first-order theory also in the Newtonian looking second-order form. It is clearly useful to take the first-order character seriously. For example configuration space trajectories can never *cross* each other. Bell [1] also emphasized the first-order character. He writes the right hand side of (2) as the quotient of the quantum flux $\boldsymbol{j}^{\psi}$ and the density $\rho = |\psi|^2$, the significance of which I shall address later.

Nonlocal Character. The wave function is a function on configuration space. That space is not 'anschaulich' like physical space, it is abstract. But since the wave function is part of the law of the motion it is important for the understanding of Bohmian mechanics that the abstract configuration space be taken seriously. The role played by the configuration space wave function in defining the motion makes Bohmian mechanics nonlocal: Looking at (2), the velocity of each particle depends on the position of all the other particles no matter how far the other particles are away. The amount of action at a distance depends only on the degree of *entanglement* of the wave function on configuration space, that is, the extent to which the wave function differs from a product of single particle wave functions.

Because of the observed violations of Bell's inequalities [1] (see [6] for various examples), nature is nonlocal. A fundamentally correct physical theory must therefore be nonlocal. Bohmian mechanics shows that this can naturally be achieved – by incorporating the *wave function on configuration space* as physical field, which is naturally a nonlocal object.

Names. The theory goes under various names, but it's always Bohmian mechanics: Causal Interpretation [4], de Broglie-Bohm pilot wave theory (Bell [1]), or de Broglie-Bohm theory, or pilot wave theory (Valentini in this volume) or simply Bohm's theory. De Broglie's name appears because de Broglie presented reluctantly the equations of Bohmian mechanics (which even earlier had been written down by Madelung, for a fluid picture of the wave evolution) at the Solvay conference 1927, but in the discussion with Pauli at this conference, he found himself unable to defend his ideas. A stochastic version of Bohmian mechanics, where the trajectories are perturbed by white noise, exists, namely, stochastic mechanics (sometimes called Nelson's stochastic mechanics [7]).

Point Particle Theories. Other deterministic point particle theories are for example Newtonian mechanics and the relativistic Wheeler–Feynman Electromagnetism [8] of point charges – a theory without fields. (For a relativistic theory of extended charges and fields see the contribution by Michael Kiessling in this volume.)

While in these theories physicists agree that the particles are ontological, the particles in Bohmian mechanics are often seen with distrust and given funny attributes (Adler in this volume calls them hidden 'particles'). On the one hand, this is understandable because orthodox quantum mechanics forbids us to talk about the trajectories of particles. On the other hand, the point particles are ontological in Bohmian mechanics – they are, in fact, what the theory is about. Ignoring them, the theory becomes a theory about nothing, pretty much like orthodox quantum mechanics, where one resorts to the dubious notion of observer as fundamental. To understand Bohmian mechanics and how quantum phenomena emerge from it, it is necessary that the particles be ontological – i.e., be taken seriously.

Other Quantum Theories Without Observers. A pure wave function ontology, with no particles as part of the ontology, lies behind the idea of the collapse models of quantum mechanics, and they are discussed in this volume by Alberto Rimini and Gian Carlo Ghirardi. So I say nothing more.

The program of Decoherent Histories or Consistent Histories arises in my understanding also from the wish to have facts in the quantum world without the action of some observer's mind. Roland Omnès reports about that in this volume.

1.2 Frequently Asked Questions

I answer briefly some questions that may arise immediately when the theory is presented as above. Typically the questions are of the nature: *But isn't there a problem with this and that?* and the answer is: *No, there is no problem.*

What about Spin? Can Bohmian mechanics deal with spin?

Instead of the wave function $\psi(q,t) \in \mathbb{C}$ let $\psi(q,t) \in$ (spinspace $=$) $\mathbb{C}^k, k = 1,2,3,\ldots$, i.e. the wave function is a spinor, which defines then the velocity field in an obvious way, namely by using the inner product in spinspace in (2) [9]. The Schrödinger equation (1) is then replaced by the Pauli equation, involving electromagnetic fields and Pauli matrices. Spin is thus a property of the wave function.

What about Fermions or Bosons? Can Bohmian mechanics describe indistinguishable particles?

In (1) and (2) the particles are labeled, as if they were distinguishable (for example by different masses). To formulate Bohmian mechanics for indistinguishable particles one may demand that the law be invariant under permutation of the (artificially introduced) labels, or one may formulate Bohmian mechanics from the beginning on the *right* configuration space of indistinguishable particles. As one might expect both approaches establish that only symmetric or antisymmetric wave functions can be used in the formulation of the law [12].

What about the Newtonian Limit? Does Newtonian mechanics emerge from Bohmian mechanics?

The Newtonian limit or classical limit of Bohmian mechanics is a limit in which the Bohmian trajectories become Newtonian. There are various physical situations in which this is the case. The question itself presents no conceptual difficulties. Concerning the transition of a first-order theory to a second order theory, one may note that the Bohmian trajectories do have at a given time (say the initial time) a position and a velocity, and that the velocity is determined by the wave function. The extra degree of freedom needed for a second order theory (the free choice of an initial velocity) resides in the wave function. (Note also that a one dimensional Bohmian world is special since Bohmian trajectories can never cross, which in one dimension presents a 'topological' barrier for the particle motion, while classical trajectories do not have that. Here classical motion still emerges from the motion of narrow wave packets.) For a recent discussion of the classical limit of quantum mechanics from a Bohmian perspective see [14].

Concerning the appearance of masses (and the potential V) in (1) and (2) we note that actually only naturalized quantities $\mu_k = \frac{m_k}{\hbar}$ (or $\frac{V}{\hbar}$) appear in the law, which only a posteriori, in the Newtonian limit of Bohmian mechanics, are recognized as connected with the inertial mass (or classical potential) [9].

What about Relativity? Can Bohmian mechanics be turned into a relativistic theory?

There are various ideas behind this question. One idea is that relativistic quantum theory is a quantum theory of fields, so, superficially, quantum field theory shows that particles cannot be ontological. The phenomena of pair creation and annihilation support that argument. On the other hand, pair creation and annihilation is a particle phenomenon, and the aim of quantum field theory is to describe particles (for example in scattering situations). Therefore under further scrutiny it may well be the case that particles are still part of the ontology. To come to grips with these aspects of relativistic quantum phenomena may in fact be easy compared to the following: Nature is nonlocal and any relativistic theory of nature must respect that, i.e. it must account in a Lorentz invariant way for faster than light effects to achieve action at a distance. For toy models in which this is achieved see [13], and see also the discussion in [10].

There is also the more metaphysical idea that eventually physics will look quite different from the way it looks now: I am thinking here of future quantum gravity and physics beyond the Planck length. From a Bohmian perspective, the metaphysical answer to this is the following: The moral of Bohmian mechanics is not that the ontology must be a particle ontology, but rather that physics, and I mean now in particular quantum physics, can still be (I would rather like to say must be) about something very clear, about *some clear* ontology. Otherwise one ends in endless debates about interpretations of mathematical symbols. Furthermore Bohmian mechanics shows that one should not be afraid of thinking of the simplest ontology, one directly suggested by the phenomena, as the right ontology, despite of the droppings of authorities. All the same, it may also happen that the primitive ontology for quantum theory at deeper levels is more elusive.

2 The Explanation of Phenomena

2.1 Trajectories

Newtonian mechanics applies with a high degree of accuracy over a very wide range of scales: From planetary motion, to apples falling to earth, to bullets shot from a gun, and further on to the motion of gas molecules in a dilute gas. The difference of description of the gas and of the motion of the bullet (or of other macroscopic bodies) is rather striking. In the case of the bullet one is used to taking aim, that is, one thinks of a very precise initial position and velocity of the bullet for computing its trajectory, or, better, for predicting its trajectory, and one could even consider testing such a prediction in experiment, observing for example where and at which time the bullet arrived. Thus the bullet motion is described in detail. However, for the gas a statistical description is used and accepted. I shall come to that in the next section, because what springs to mind first when a mechanical law is written down, is the study of detailed motions like the bullet motion. I will go along with that here but I warn that the computation of particle trajectories in Bohmian mechanics analogous to what one is used to in Newtonian mechanics is not as illuminating. The reason is that in Bohmian mechanics the positions of the particles are, as I shall explain later, typically randomly distributed, given the wave function, just as one would expect from Born's statistical law: At every moment of time, the configuration of the system is $|\psi|^2$ distributed, and that is that, for most of the cases at least[2].

Motion in Ground States. Ground state wave functions can be taken to be real; thus by (2) the velocity field is zero, so $\boldsymbol{Q}(t) = \boldsymbol{Q}(0)$ – the particles do not move. Many find this at first disturbing, in particular when it comes to the ground state of the hydrogen atom: First one learns the Bohr-model,

[2] Exit time, tunneling times and scattering, as well as the classical regime are situations where trajectories are helpful [11,14].

where electrons move on orbits around the nucleus, then one learns that that is false because the existence of trajectories contradict the Heisenberg uncertainty principle, and now Bohmian mechanics says that the electron is at rest. This seems even worse, because the Coulomb force is supposed to act on the charge and standing still is not much of a reaction. So this is a good example to get familiar with the radically non Newtonian character of Bohmian mechanics, in which the action of forces does not play a fundamental role.

Double Slit Experiment. This is one of the experiments that strongly suggests the particle ontology. Particles are sent one at a time through a double slit. There is always only one particle on the way. One can see it coming at the screen, let's say a photo plate! When it arrives it makes a black spot where it lands. At the end, after many particles have passed, the black spots add up to a pattern, which looks like an *interference* pattern of waves, and Bohmian mechanics explains that. Just look at the trajectory picture of the computer solutions (due to Dewdney et al. see [5]) of the equations (1) and (2), where the initial positions of the trajectories are chosen in a random manner. Now note the following: The trajectories curve wildly. Why? Look how the trajectories spread immediately after passing through the slit, the particles moving on straight lines guided by the spherical wave parts originating at the slits, until the two parts of the wave begin to interfere (the typical interference pattern of spherical waves builds up before the photo plate) and the particles, guided by this wave, move so as to reproduce the $|\psi|^2$-distribution. Furthermore if one considers the symmetry line

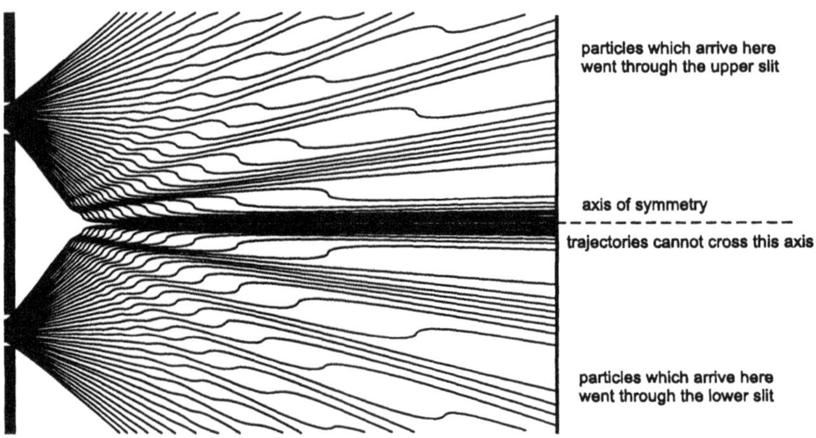

Fig. 1. Possible trajectories in the two slit experiment

in this experiment and recalls that trajectories can't cross, one sees that trajectories starting in the upper half must go through the upper slit and must hit the upper half of the photoplate. (This is one way of observing through which slit the particle has gone without destroying the interference pattern. Just look where it arrives on the photo plate.)

Tunneling. In textbooks tunneling through a barrier is often discussed by stationary wave methods. In one dimension trajectories cannot cross, and one finds immediately that in this stationary picture particle trajectories can only travel towards the barrier. This, one sometimes hears, is bad for Bohmian mechanics, because it does not describe back scattering of particles. But this is wrong. When a particle is sent towards a barrier, it is guided by a wave packet which is not stationary, and that's all. Because trajectories can't cross, if the particle is 'in front' of the packet it goes through the barrier, if it is in the back it turns back. It couldn't be nicer than that.

Fig. 2. Tunneling through a barrier: A potential V

Cats and Cloud Chamber Tracks. A macroscopic Bohmian trajectory, which looks Newtonian, is that of the cat, when it jumps out of the box in which the atom did not decay. Other Bohmian trajectories are made visible to the eye in cloud chamber tracks.

2.2 Statistical Mechanics

Equilibrium. Statistical mechanics and the kinetic theory of gases is by now well established. A famous success is Boltzmann's explanation of the second law of thermodynamics – the meaning of entropy and why it should increase. See Sheldon Goldstein's as well as Herbert Spohn's contributions to this volume. To describe a gas, a statistical description is used, and the regularities derived are not certain but merely overwhelmingly 'probable.' Nevertheless it is clear to most that a statistical description is a matter of convenience, not necessity: it is impractical to specify the initial velocities and positions of all the gas molecules, compute their motion and observe where they are later. We are simply ignorant about these details, about which we don't much care, and that's where the statistical analysis comes to the rescue.

Boltzmann characterized the initial conditions for which the thermodynamic regularities hold as overwhelmingly probable. Nowadays, it is widely recognized that it would be more appropriate to speak in terms of *typicality*. The good initial conditions are typical with respect to a measure of typicality, in the sense that, with respect to this measure, exceptions are rare – they are assigned very small weight by the measure. This measure is determined by the physical law: It must be "equivariant" with the motion of the particles, so that the notion of typicality is independent of time. Typicality is discussed and applied in the contributions by Goldstein, Kiessling and Spohn. I shall return to this in later sections, where I discuss the meaning of typicality statements in more detail.

In the Newtonian case the measure is determined by a strong form of equivariance, namely stationarity, i.e. its density is a stationary solution of the continuity equation for the Hamiltonian flow on phase space, which is called Liouville's equation for Hamiltonian or Newtonian dynamics. Liouville's equation allows for many stationary solutions, and some feel uneasy that the measure of typicality is sometimes not uniquely determined.

Nonequilibrium. But we think that *in principle* we could do without statistics in a classical world, in principle we could know all the velocities and positions of all the gas molecules and we could even displace some or all of the molecules according to our taste, i.e., we could change what looks typical into something that looks atypical. More to the point: We have the impression that we could get rid of randomness altogether if we wished to do so. Why? Because our universe is in a state of global nonequilibrium (see [3])! Typicality with respect to an equivariant measure is in contrast an equilibrium notion.

If one applies this idea of complete control of initial conditions at least for small subsystems to Bohmian mechanics, it is often felt disturbing that one cannot control wave function and position of a Bohmian particle the way one would like. Born's statistical law is a law! If the wave function is ψ, the position is $|\psi|^2$-distributed. A common reaction is then this: *If the position of the particle cannot be experimentally controlled and given a sharp value, no matter what the wave function is, then what are the particles good for? Then they play no physical role!* As superficial as this argument is, we may as well apply it to the universe. *We humans* cannot experimentally control the initial conditions of the stars in the milky way, hence the stars play no physical role. Should we deny then that they *are*?

I think what is felt as disturbing is that now it seems that there is a fact to be explained: Why is it that Born's statistical law holds without fail, while the equidistribution of gas molecules in a container may fail to hold (by prior proper manipulation from outside)[3]?

Boltzmann's answer to this could turn out to be surprisingly simple: Maybe the statistical law of Born arises because it is 'overwhelmingly probable,' or, better, typical. Because if that were the case then it would explain why it does not fail (see also Goldstein's article) – no more than the second law of thermo-

[3] So as long as one does not talk about particles, i.e. as long as Born's statistical law is not about a distribution of particles but about results of 'observations,' it has seemed acceptable. On what grounds? On no grounds.

dynamics, no more than it is possible to build perpetual motion machines. And it would explain why the name *Born's law* is properly chosen.

2.3 Statistical Bohmian Mechanics

In [3] we showed that the empirical import of Bohmian mechanics emerges from typicality. What is typical is the distribution of particle positions. But before I explain more, I recall that in orthodox quantum theory it is often said that randomness is intrinsic. What is meant by that?

Observe that in orthodox quantum theory only the Schrödinger equation (1) appears and that equation has no intrinsic randomness. (The collapse models, as discussed by Rimini in this volume could serve as examples for an intrinsically random theory). The Schrödinger evolution thus cannot account for the random spots on the photo plate in the two slit experiment. (Moreover, it cannot account for any fact, random or nonrandom.) This is related to the measurement problem of orthodox quantum theory, which I describe next.

The Collapse. The evolution of a system coupled to a measurement device in which the wave function of the device (Φ) becomes correlated with particular wave functions ($\psi_k, k = 1, 2, ...$) of the system,

$$\psi_k(x)\Phi_0(y) \longrightarrow \psi_k(x)\Phi_k(y), \tag{3}$$

is called a measurement process, where the arrow indicates the evolution of the wave function of the combined system: the measured system and the apparatus. The Φ_k are wave functions corresponding to distinct pointer positions, and Φ_0 is the ready state. Then by the linearity of the Schrödinger equation

$$\sum_k a_k\psi_k(x)\Phi_0(y) \longrightarrow \sum_k a_k\psi_k(x)\Phi_k(y). \tag{4}$$

That is what comes out of the theory – a macroscopic superposition of pointer wave functions. That is not familiar in nature and that is the measurement problem. It is 'solved' by talk: The *observer* collapses the sum in (4) to only one of the wave functions, let's say to $\psi_{k_0}(x)\Phi_{k_0}(y)$. And he does so with the probability $|a_{k_0}|^2$. But no one knows why he should do that. That is the *intrinsic* part. That is why there is nothing to show. It's intrinsic.

The observer is supposed to do two things. He creates facts and he does so in a random fashion, but at the same time obeying Born's statistical law. So he must be very smart! That is why Bell asked whether the observer must have a degree in physics – who else would know about Born's statistical law? But it is clear that the measurement problem cannot be talked away, it is not a philosophical issue. It is a hardware problem!

It is worthwhile to understand the measurement situation in Bohmian mechanics. The Bohmian trajectories of the system and apparatus particles are in configuration space made of the system-coordinates X and the apparatus-coordinates Y. Then look at the following picture:

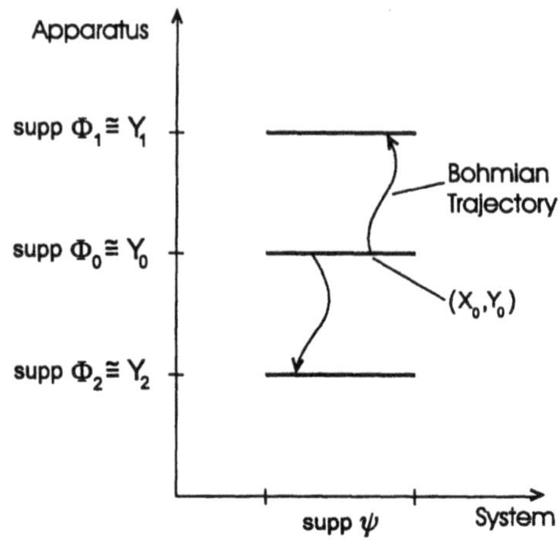

Fig. 3. The apparatus axis represents the configuration space of Y coordinates, the system axis that of the X coordinates. Since the supports of the apparatus wave functions Φ_k are sufficiently disjoint, corresponding, as they do, to macroscopically different pointer positions, only the term in (4) in whose support Y is situated is relevant to the behavior of the system, so that as a practical matter all but one of the terms in (4) may be ignored.

When the apparatus wave functions have sufficiently disjoint supports, then it is very difficult to have them interfere again. This is because of purely thermodynamical considerations concerning the number of degrees of freedom which have to be controlled to achieve interference in the future. Some have difficulties with such arguments, especially if they go further, to include the recordings of the pointer positions on paper. That is, when the wave function is now one which is even more macroscopically entangled, including patterns of ink on paper. Everyone would agree that it would be exceedingly difficult to produce interference which could turn such facts into fakes, but that it is all quantum mechanics, and that it is only because of thermodynamics that these funny things do not happen, that not everyone agrees about. In fact, in the collapse models, discussed by Rimini, these funny things cannot happen. The wave function does collapse by law. But in Bohmian mechanics it is simply a matter of convenience to proceed with the components of the wave function that will guide the particles from now on and forget about the other parts of the wave function. After all, what we really care about is what the particles are doing; this, according to Bohmian mechanics, is what quantum theory is fundamentally about, and it is out of the behavior of the particles that physical phenomena emerge. The collapse is not a physical process, but a matter of pragmatism.[4]

[4] At the same time, it is true that the collapse of the wave function can be regarded as a physical process for Bohmian mechanics, but the wave function that collapses is the effective wave function of a subsystem of a larger system, see Definition 1 below.

This explains the collapse, but the randomness, where does that come from? That is what I address in the last section.

Decoherence. A note on the side: The Schrödinger equation determines the evolution of the wave function of the N-particle system. When this system 'interacts' with another system, of, say, M particles, the wave function for the combined system lives on the configuration space of the $M + N$ particles (as in (3),(4).) This should not be seen as extra requirement supplementing the theoretical description but rather as a consequence of the basic equations (1),(2). I shall return to this point in the next section. In any case, as we just said, the interaction of one system with another brings about the entanglement of their wave functions, and the more complex the entanglement becomes, in particular when the entanglement becomes macroscopic, interference between terms of the resulting superposition becomes practically impossible. In (4) it is certainly the case that interference of the different pointer positions is practically impossible, and this practical impossibility gets stronger with further interactions, since pointers are usually in some room full of radiation, air, flies and all kinds of interacting environments. This aspect of entanglement is now being called 'decoherence'. It is a prerequisite for the collapse to do no harm: For all practical purposes one may forget about macroscopic superpositions. And it is a prerequisite for the formulation of the measurement problem, and in and of itself in no way its solution.

Subsystems. Bohmian mechanics is, like all fundamental physical theories, a theory of the Universe – of a Bohmian particle universe, nonrelativistic and so on. So what we are dealing with in everyday life are subsystems of this universe and subsystems of subsystems, and what we want to have is a good description of subsystems, in fact we would like to have Bohmian subsystems, meaning that there is a Schrödinger wave function which guides the particles of the subsystem, at least for some time. What I imply here is indeed that the wave function which enters (1),(2) as defining equations of Bohmian mechanics is the wave function of the largest system conceivable – the universe. What we need to have is a wave function of a hydrogen atom, of a molecule, of many molecules, of pointers, of cats. How can we get such a thing, a wave function for a system of interest, which is a subsystem, almost always, of a larger system?

Let us denote the mother of all wave functions – the wave function of the universe – by $\Psi(q)$ and let us write x for the system-coordinates and y for the environment-coordinates, describing the rest of the universe, everything which is not the system. So we have a splitting $q = (x, y)$, $Q = (X, Y)$. Recall that the small letters stand for generic variables and the capital letters for the actual configuration of the particles. By inspection of (1), (2) and recalling the last sections, one is quickly lead to the following concept:

Definition. Let

$$\Psi(q) = \varphi(x)\Phi(y) + \Phi^{\perp}(q)$$

with Φ and $\Phi^{\perp}$ having (macroscopically) disjoint y-supports and suppose $Y \in \operatorname{supp} \Phi$. Then φ is the **effective** wave function of the system.

Easy to check: The effective wave function of a system governs the motion of its configuration $X(t)$. Moreover, by virtue of the decoherence effects discussed in the **decoherence** and **collapse** sections, as long as a system is decoupled from its environment, the evolution of its effective wave function will be governed by the Schrödinger equation (1) for that system. The effective wave function is the precise substitute for the collapsed wave function in orthodox quantum theory, so one should feel comfortable with this notion.

Quantum Equilibrium Hypothesis. What comes now is technically very easy but at the same time misunderstood by many. This section is about the foundations of statistical Bohmian mechanics (in fact of the statistical analysis of any physical theory). Where does the randomness come from?

Recall the double slit – or the Stern–Gerlach – experiment. What is striking is that we have regular random outcomes: For the double slit we get relative frequencies of black points on the photo plate which are given by Born's law, and for the Stern Gerlach experiment we can get outcomes as in a coin tossing, i.e., in a long run of sending many particles through the inhomogeneous magnetic field, half go up and half down. How can this lawful random behavior be accounted for? Easy: By the law of large numbers for independent random positions of the particles. Just as one would argue in classical physics: Given some *initial randomness – that is, the statistical ansatz* – the instability of the motion should produce enough independence for a law of large numbers statement to hold (this was made very clear in [15]).

Having said this, a hard question seems to remain: What is responsible for the randomness of the *initial conditions*? Since in a deterministic physical theory there is apparently no room for genuine randomness, it would seem that it must come from 'outside'. Thus the usual answer is: Systems are never closed and from the 'outside' come random disturbances. But after including the 'outside' we are dealing only with a bigger physical system, itself physical and deterministic; thus we have merely shifted the question of the origin of randomness to a bigger system, and so it continues and one hopes that eventually the question will evaporate into empty space. But it does not and only if one continues to ask can one know for sure that the shift to include the 'outside' is a dead end. The biggest system conceivable, the universe, has no outside. It is for the universe that the question of the origin of the randomness must be answered! The answer was given by Boltzmann. It is that there is no need for (and, more to the point, no sense to) randomness in the initial conditions of the universe. Rather, a random pattern of events emerges from the deterministic evolution from a fixed initial condition for the universe – provided that initial condition is typical. I shall explain this a bit more here, but for a deeper understanding the reader should consult [3].

We begin by recalling the well-known fact that the 'quantum equilibrium' distribution $|\psi|^2$ for the configuration of a system has, mathematically, a special status within Bohmian mechanics: It is a quasi-invariant or, more precisely, equivariant distribution, analogous to the microcanonical distribution of statistical mechanics. Consider the continuity equation of the Bohmian flow, analogous to Liouville's equation for a Hamiltonian flow,

$$\partial_t \rho(q,t) + \nabla \cdot \boldsymbol{v}^{\varphi_t} \rho(q,t) = 0. \tag{5}$$

Now $\boldsymbol{v}^{\varphi_t} = \frac{\boldsymbol{j}^{\varphi_t}}{|\varphi_t|^2}$, with $\boldsymbol{j}^{\varphi_t}$ the quantum flux, which satisfies an identity holding for solutions of Schrödinger's equation

$$\partial_t |\varphi_t|^2 + \nabla \cdot \boldsymbol{j}^{\varphi_t} = 0. \tag{6}$$

This is roughly analogous to the fact that a Hamiltonian velocity field is divergence free which – used on the continuity equation for the Hamiltonian flow – yields Liouville's equation, for which the usual stationary solution is obvious.

Here, (6) may be rewritten

$$\partial_t |\varphi_t|^2 + \nabla \cdot \boldsymbol{v}^{\varphi_t} |\varphi_t|^2 = 0 \tag{7}$$

so that $\rho(q,t) = |\varphi(q,t)|^2$ satisfies (5) and thus defines an equivariant distribution: It does not change its form as a functional of φ_t during the evolution ($\equiv$ *equivariance*).

Just as in statistical mechanics, this simple mathematical fact leads naturally to the *Quantum Equilibrium Hypothesis (QEH):*
'If a system has effective wave function φ then its configuration has $|\varphi|^2$.' Much in the spirit of Gibbs, this may be regarded merely as a reasonable statistical ansatz for dealing with subsystems: It is supported by the physical theory, it is simple and it works. As is the common praxis in statistical mechanics, we can think of this distribution as describing an *ensemble of identical subsystems*, obtained by sampling across space and time.

Typicality. We are now at a stage analogous to the present status of statistical mechanics. It is known that the rigorous justification of the statistical hypothesis that systems tend to be microcanonically (or Gibbs) distributed is a very hard and unsolved problem. What we would like to have is a rigorous proof that normal thermodynamic behavior emerges for a sufficiently large class of initial conditions for the universe, or, failing that, for the systems of direct interest to us, at least for some initial conditions for these systems. But even the latter is way beyond our reach (see also Goldstein's contribution to this volume).

For Bohmian mechanics, however, one can provide just such a rigorous justification of the QEH; one can show, in fact, that it emerges for 'most' initial conditions for the universe. I will not go through the proof here but rather comment on misconceptions about this statement and its proof.

The statement that the QEH can be justified is a law of large number statement: What we must look at are the empirical distributions of the configuration for ensembles of small subsystems. Small here means compared with the size of the universe. A laboratory is extremely small.

Consider equations (1) and (2) for the evolution of the *universe*! So Q comprises the coordinates of all the particles of a Bohmian universe and Ψ is the mother of all wave functions: the wave function of the universe. Now let

$$\rho_{\text{emp}}^{N,\varphi}(Q,x,t) = \frac{1}{N}\sum_{i=1}^{N}\delta(X_i(Q,t)-x) \tag{8}$$

be the empirical distribution of the configuration for an ensemble of N similar subsystems (subsystem i having configuration X_i), each subsystem having *effective* wave function φ at a given time t (for simplicity I discuss here spatial ensembles at a given time. In general one has time and space ensembles [3]). Here $X_i(Q,t)$ is determined by (2) and the initial conditions Q. We emphasize that ρ_{emp} is a function of Q. Moreover, the set of Q's relevant to (8) is very much constrained by the requirement that the subsystems have effective wave function φ (see the definition of effective wave function).

Suppose now that we could show the following: There exists one initial condition Q, i.e. one evolution of a Bohmian universe, for which[5]

$$\rho_{\text{emp}}^{N,\varphi}(Q,x,t) \approx |\varphi|^2(x). \tag{9}$$

for all appropriate subsystems, wave functions φ, and times t. In other words, that Born's statistical law holds within that special universe. This would be a fantastic result, clearly demonstrating the compatibility of quantum randomness with absolute determinism. Moreover, the empirical data would encourage us to believe that we are in that universe. Nonetheless, such a result would not explain *why* Born's statistical law holds: we would need to know why we should be in that very special universe. Now take the other extreme: Suppose it turned out that for *all* initial Q' (9) is true. That would mean that Born's statistical law is a theorem following from (1) and (2). Then – provided we believe in the correctness of our theory – we must be in one of these universes and purest satisfaction would result.

But Boltzmann's analysis of the origin of the second law of thermodynamics (see the contribution of Goldstein to this volume) should have taught us that the latter is too much to expect. Moreover, it follows from our analysis in [3] that in fact it is impossible – i.e., that there must be bad initial conditions Q for which the QEH fails (see [9]). What is proven in [3], a law of large numbers theorem for (8), thus cannot be much improved upon.[6]

The usual way of formulating a law of large number result is in terms of typicality: Typically the numbers in the interval $[0,1]$ have a decimal expansion

[5] for precise formulations see [3]

[6] In statistical physics the exploration of mixing, the relaxation of a nonequilibrium distribution on the full phase space towards an equilibrium distribution, has become widespread and has led some to the belief that mixing is central to the justification of

in which every digit comes with relative frequency $\approx \frac{1}{10}$. That is: Most of the numbers in the interval $[0, 1]$ have a decimal expansion in which every digit comes with relative frequency $\approx \frac{1}{10}$. Now, 'most' is here with respect to the Lebesgue measure. That measure is a measure of typicality, and one may ask what its relevance is. In physics, according to Boltzmann, we are in the fortunate situation that the measure of typicality is dictated by physics: *Typicality must be stationary*, which requires the measure to be equivariant. (The stationarity of the notion of typicality makes it equilibrium-like, hence 'Quantum Equilibrium'.) Thus for Bohmian mechanics typicality is defined using the $|\Psi|^2$ distribution. It gives us a notion of 'most' and 'few' in this situation, where counting is not possible.

Warning: *We used $|\varphi|^2$ to formulate the statistical hypothesis, in which it represents an empirical distribution. Now we refer to it (or, rather, to $|\Psi|^2$) to define the measure of typicality, an entirely different concept. Moreover, the wave function Ψ of the universe and the wave function φ of a subsystem are also rather different objects, and it is in a sense an accident that they seem so similar and obey the same equations – and (for good reasons) go by the same name. But while for ensembles of subsystems of one universe empirical statistics are relevant, a statistical ensemble of universes is quite another matter.*

We show that for typical solutions of (2) (the universal wave function is assumed to be given) Born's law (9) is valid: the set of exceptions is assigned very small measure by the $|\Psi|^2$ distribution. The failure to distinguish between $|\varphi|^2$ as an empirical distribution and $|\Psi|^2$ as a measure of typicality leads to the following reformulation of our analysis: Consider a probability distribution of universes with density $|\Psi|^2$. Then the configuration of a subsystem with effective wave function φ is $|\varphi|^2$ distributed.[7] But this is not a very useful result, since sampling across universes is neither possible for nor relevant to humans.

More sensible, but still besides the point, is the following. Consider coin tossing. I make the following model. The outcomes of every toss are independent identically distributed random variables. I take the product measure on the probability space of sequences of outcomes. Then the law of large numbers holds: Typically, the relative frequency of heads is approximately $\frac{1}{2}$. That is 'trivially' true, by construction. It is important to appreciate how this appeal to typicality differs from Boltzmann's and from ours in [3]. Note in particular that there is no simple relationship between the initial configuration Q of the universe and the empirical statistics that emerge from Q via the dynamics (2). Note also that in general we make no claim about empirical statistics at the initial universal time, since it need not be the case that a decomposition into subsystems with the same effective wave function (or any effective wave function at all) is possible at this time.

---

the equilibrium hypothesis. This belief is particularly attractive, since it is connected with notions like chaoticity and independence. It is nonetheless quite wrong. For more on this see the contribution of Goldstein to this volume.

[7] In [16] this is referred to as the nesting property.

This result justifies for me the QEH; it explains why Born's law holds – or, what amounts to more or less the same thing, it explains why it should be expected to hold. To go further we would need to analyze just what is meant by explanation; see the contribution of Goldstein to this volume, where this is touched upon.

I remark that the operator formalism of quantum theory emerges (with surprising ease) from statistical Bohmian mechanics [4,17,18].

3 Simplicius Simplicissimus

What is typicality? It is a notion for defining the smallness of sets of (mathematically inevitable) exceptions and thus permitting the formulation of law of large numbers type statements. Smallness is usually defined in terms of a measure. What determines the measure? In physics, the physical theory. Typicality is defined by a measure on the set of 'initial conditions' (eventually by the initial conditions of the universe), determined, or at least strongly suggested, by the physical law. Are typical events most likely to happen? No, they happen because they are typical. But are there also atypical events? Yes. They do not happen, because they are unlikely? No, because the are atypical. But in principle they could happen? Yes. So why don't they happen then? Because they are not typical.

Using typicality one may define probability in terms of law of large numbers type statements, i.e., in terms of relative frequencies and empirical statistics. What is meant by: In a Stern–Gerlach experiment the probability for spin up is $\frac{1}{2}$? The same as for heads turning up in a coin tossing. So what is meant? The law of large numbers: That in a long run of repetitions of the experiment the relative frequency of the outcome in which the spin is up or the coin shows heads will typically be close to the value $\frac{1}{2}$.

Is there another sense of probability? Probably there is! Does explanation via appeal to typicality require deeper conceptual analysis? Certainly. Can appeal to typicality be entirely eliminated from scientific explanation? Very unlikely!

4 Acknowledgements

It is a pleasure to thank Sheldon Goldstein and Roderich Tumulka for their very critical reading of the manuscript leading to substantial improvements of the presentation.

References

1. J. Bell: *Speakable and Unspeakable in Quantum Mechanics* (Cambridge University Press, Cambridge 1987)
2. S. Goldstein: Physics Today, **51**, 3, pp. 42-47 and 4, pp. 38-42 (1998)
3. D. Dürr, S. Goldstein, N. Zanghì: Journal of Stat. Phys. **67**, pp. 843–907 (1992)

4. D. Bohm: Phys. Rev. **85** pp. 166–193 (1952),
5. D. Bohm and B. Hiley: *The Undivided Universe* (Routledge, London and New York 1993)
6. J. Baggott: *The Meaning of Quantum Theory* (Oxford Science Publication, Oxford 1992)
7. E. Nelson: *Quantum Fluctuations* (Princeton University Press, Princeton 1985)
8. J.A. Wheeler, R.P. Feynman: Rev. Mod. Pys. **17**, 157 (1945), Rev. Mod. Phys. **21**, 425 (1949)
9. D. Dürr, S. Goldstein, N. Zanghì: 'Bohmian Mechanics as the Foundation of Quantum Mechanics' , in *Bohmian Mechanics and Quantum Theory: An Appraisal*, ed. by J. Cushing, A. Fine, S. Goldstein (Kluwer Academic Publishers, Dordrecht 1986)
10. T. Maudlin: In *Bohmian Mechanics and Quantum Theory: An Appraisal*, ed. by J. Cushing, A. Fine, S. Goldstein (Kluwer Academic Publishers, Dordrecht 1986)
11. D. Dürr, S. Goldstein, S. Teufel, N. Zanghì: Physica A 279, pp. 416-431 (2000)
12. D. Dürr, S. Goldstein, N. Zanghì: 'Bohmian Mechanics, Identical Particles, Anyons and Parastatistics', in preparation
13. D. Dürr, S. Goldstein, K. Münch-Berndl, N. Zanghì: Phys. Rev. A 60, pp. 2729-2736 (1999),
14. V. Allori, D. Dürr, S. Goldstein, S. Teufel, N. Zanghì: 'Bohmian Mechanics and the Classical Limit of Quantum Mechanics', in preparation
15. M. Smoluchowski: Die Naturwissenschaften, **17**, pp. 253-263 (1918), see also M. Kac: *Probability and Related Topics in Physical Sciences*, Lectures in Applied Mathematics, American Mathematical Society (1991)
16. J. Cushing: *Quantum Mechanics* (The University of Chicago Press, Chicago 1994)
17. D. Dürr, S. Goldstein, N. Zanghì: 'On the Role of Operators in Bohmian Mechanics', in preparation
18. D. Dürr: *Bohmsche Mechanik als Grundlage der Quantenmechanik*, to be published by Springer

Chance of Reduction as Chance of Spontaneous Localisation

Alberto Rimini

Università di Pavia, Dipartimento di Fisica Nucleare e Teorica, Pavia, Italy

Abstract. This is a short review of the Spontaneuos Localisation Program, which yields quantum theories without observers, without adding extra variables to the wave-function. After a short review of the various interpretations of quantum mechanics, I dicuss the discrete version of the spontaneous localisation process in connection with the measurement problem, and after introducing the fundamental notion of "compoundation invariance" I turn to the description of a continuous version of the localisation process.

1 Introduction

By theory of quantum measurement I mean a theoretical framework in which the system **S**+**A** (or **S**+**A**+**E**) is considered, **S** being a quantum system, **A** an apparatus measuring some physical quantity of **S** (and **E** the environment of **S**+**A**, if its consideration is deemed necessary). The aim is to obtain, from the treatment of such a system, results in agreement with the standard postulates of quantum mechanics concerning measurement.

My task here is to describe quantum mechanics with spontaneous localization. If by reduction theory we mean a theory incorporating the reduction of the wave function in the principles of quantum mechanics without introducing an a priori fuzzy distinction between microscopic (quantum) and macroscopic (classical) systems, then quantum mechanics with spontaneous localization, in its various forms, is the only known consistent reduction theory. Thereby, it constitutes one possible way to found a theory of quantum measurement. To the aim of positioning this approach with respect to others, I shall briefly review some relevant formulations of quantum mechanics, pointing out their main features with reference to the theory of quantum measurement in the sense indicated above.

Copenhagen Interpretation. Its characteristic feature is that it identifies measurements as processes of a special nature because of the presence of the apparatus **A** which is assumed to be non–describable by quantum mechanics. In other words the Copenhagen interpretation rejects the program of the theory of quantum measurement. On the other hand, the special nature of the measurement process allows the assumption of the reduction principle, and, as a consequence, the wave function (alone) describes individual quantum systems.

Ensemble Interpretation [1]. It assumes that the wave function describes an ensemble of systems, not an individual system, and never suffers reductions. It further assumes that, given the wave function at a time, the distribution of the results of a measurement performed at that time on the ensemble of systems is given by the standard quantum probability law. As a consequence of the lack of reductions, if a measurement (an ensemble of measurements) is performed by **A** on **S** at a certain time, the wave function of **S**+**A**+**E** after the measurement contains all entangled terms corresponding to the possible results. But, if any measurement is successively performed on the ensemble of systems **S**+**A**, decoherence due to **E** prevents the various terms from interfering, so that the distribution of the results is the standard one. In this sense the ensemble interpretation allows to set up a theory of quantum measurement. However, no element of description of the system corresponding to the result of an individual measurement is there, and this fact causes problems of consistency, typically if subensembles selected according to the results of a measurement are considered [2].

History Formulations [3]. These are developments of Everett's theory of the universal wave function. Again the wave function always evolves according to the Schrödinger equation, no reduction takes place, and decoherence plays an essential role. However, the formulation is such that an element of description of the system, the branch label, is used together with the wave function and is in correspondence with the result of an individual measurement. As a consequence, no consistency problem of the type encountered in the ensemble interpretation arises. The problem here, in my opinion, is that branch labels are used as elements of description of the system, but, at the level of principles only their existence is assumed, without declaring what they specify and when. Such a declaration takes place in the course of the treatment, as a consequence both of the dynamical properties of the system and, unfortunately, of arbitrary choices [4].

Bohmian Mechanics [5]. In this theory a definite element of description of the system, the configuration, is put beside the wave function and both together describe an individual system. The wave function suffers no reduction, but the configuration is in correspondence with the result of an individual measurement. As a consequence of this, a consistent theory of measurement can be set up [6]. Since there is no reduction decoherence must be taken into account.

Reduction Theories. Here the wave fuction remains the sole element of description of the system, but a universally occurring stochastic reduction process is incorporated into the principles of the theory. A consistent theory of quantum measurement can be set up, in which the wave function after an individual measurement is in correspondence with the result of the measurement. The time evolution given by the Schrödinger equation is altered by the universal reduction process.

It is seen that there are several strong similarities between reduction theories and the Copenhagen interpretation: in both reduction is there, the wave function is the sole element of description, and it describes an individual system. Of course there are also important differences. Among these, the role of chance. In the Copenhagen interpretation chance intervenes only in exceptional, or at least special, circumstances – measurements. In reduction theories chance is a universal, always present feature of time evolution, which becomes more or less important according to the specific situation.

Before tackling the task of expounding quantum mechanics with spontaneous localization as the only known consistent reduction theory, I notice that I have used several times the expression *element of description*. Someone could ask me: what do you mean by that? is it a synonymous of element of reality? or at least does the introduction of an element of description entail the existence of an element of reality? The answers belong to the sphere of my privacy, so that I let the questions drop.

2 Reduction Theories and the Measurement Problem

As I have already said, a reduction theory is characterized by the fact that no element of description of the system is added to the wave function and that, correspondingly, the Schrödinger equation is altered in such a way that reduction is described as a physical process affecting the wave function. It is understood that this alteration must be ruled by definite, universally valid, precise mathematical statements. It must rapidly suppress superpositions originated by the dynamics of measurement. For the rest, it must have practically unobservable consequences, so that predictions of standard quantum mechanics remain valid. It will share with ordinary reduction stochastic and nonlinear features.

The idea of a stochastic modification of the Schrödinger equation aimed at solving the measurement problem is an old one [7]. The following difficulties were encountered: i) the basis problem – which are the states to which the stochastic process leads? ii) the trigger problem – how can the process be effective in measurement situations and ineffective for the rest? In spontaneous localization theories a definite answer is given to the first question and the answer to the second one follows.

Consider the awkward superposition arising in measurement

$$|\psi\rangle|A_0\rangle \longrightarrow |\varphi_1\rangle|A_1\rangle|\cdots\rangle + |\varphi_2\rangle|A_2\rangle|\cdots\rangle + \dots .$$

The spontaneous–localization solution to the measurement problem is based on the assumption that, in any measurement, the apparatus contains a macroscopic part which is located in macroscopically–separated positions in the different final states $|A_1\rangle$, $|A_2\rangle$, ... Then the proposed stochastic localization process, acting on such a macroscopic part of the apparatus, reduces the unwanted superposition to one of its terms. It is seen that the question about the basis selected by the process is answered directly for the apparatus part of the total system, indirectly and dependently on the structure of the superposition, for the measured–system part.

The approach sketched above has nice features. First, the measurement problem is solved in a conceptually simple way. Secondly, the wave function remains the sole tool for describing the system and it describes an individual system. On the whole, the conceptual apparatus is very near to textbook quantum mechanics.

One can also list disturbing features. First of all, the Schrödinger equation is altered (may be one is very reluctant to that). Furthermore, a satisfactory Lorentz invariant version has not been found as yet. Finally, one can feel uneasy because ordinary decoherence, which certainly is there, becomes redundant. I would not include among the disturbing features the fact that chance becomes a universal trait of the quantum–mechanical behaviour.

3 Spontaneous Localization of a System of Distinguishable Particles

I begin describing the first proposed (GRW) spontaneous localization process [8], even though it is presently overcome, because it is much more intuitively intelligible than the more recent and refined theoretical models.

It is assumed that each particle i of a system of N distinguishable particles experiences, at Poisson distributed random times with mean frequency λ_i, a spontaneous sudden process. In the time intervals between two successive sudden processes the system evolves according to the Schrödinger equation. The spontaneous sudden process is a localization described by

$$|\psi\rangle \longrightarrow |\psi^i_{\boldsymbol{x}}\rangle = |\phi^i_{\boldsymbol{x}}\rangle / \||\phi^i_{\boldsymbol{x}}\||, \qquad |\phi^i_{\boldsymbol{x}}\rangle = \hat{L}^i_{\boldsymbol{x}} |\psi\rangle,$$

where $\hat{L}^i_{\boldsymbol{x}}$ is a positive, selfadjoint, linear operator localizing particle i around the point $\boldsymbol{x}$, normalized so that $\int \mathrm{d}^3\boldsymbol{x}\, \left(\hat{L}^i_{\boldsymbol{x}}\right)^2 = 1$. The probability density for the occurrence of $\boldsymbol{x}$ is assumed to be

$$\mathrm{pr}(\boldsymbol{x}) = \||\phi^i_{\boldsymbol{x}}\||^2 .$$

The operator $\hat{L}^i_{\boldsymbol{x}}$ is chosen to have the Gaussian shape

$$\hat{L}^i_{\boldsymbol{x}} = G^{(3)}(\hat{\boldsymbol{q}}^i - \boldsymbol{x}) = \mathcal{N}^3 \exp\left(-\tfrac{1}{2}\left((\hat{\boldsymbol{q}}^i - \boldsymbol{x})/\alpha\right)^2\right), \qquad \mathcal{N} = \left(\frac{1}{\pi\alpha^2}\right)^{\frac{1}{4}} .$$

The width α measures the accuracy of the localization. On the whole, the process depends on the width α and, for each particle, on the frequency λ_i. For sufficiently long times, the effectiveness of the localization process of particle i depends on the ratio λ_i/α^2.

When we started to study the model theory I have just described we had the vague idea of assuming it as a fundamental, universally valid mechanical theory, at least in the nonrelativistic domain. By suitably choosing the frequency parameters, one could obtain weak (for microscopic particles or strong (for macroscopic objects) localization effects. I shall discuss whether this point of view is tenable, disregarding for the moment the fact that the theory cannot be applied to systems with identical particles.

4 Compoundation Covariance

In fundamental mechanical theories, such as classical or quantum mechanics, a dynamical principle (the Newton law or the Schrödinger equation) is applied to *particles*, which are regarded as the constituents of the considered physical system. The constituents being particles means first that their linear dimensions are small enough, with respect to other lengths relevant to the problem, so that they can be treated as pointlike with reference to a chosen degree of accuracy of the description of the system. Furthermore, the possible internal structures of the particles must be sufficiently hard, so that their modifications in the considered dynamical conditions are irrelevant for the motion of the particles as wholes, again with reference to the chosen degree of accuracy of the description.

Several levels of description, i.e. several different identifications of the objects to be considered as the pointlike constituents of the system, are generally possible. Given an acceptable coarse description, a finer description in which the particles of the coarse description are regarded as objects compound of smaller particles, is certainly acceptable too. As an example, consider a Li^+ ion moving in a weak electrostatic field. If the field is weak enough, a coarse description in which a single charged particle is the sole constituent of the system is possible. But a finer description, in which two electrons and a ^7Li nucleus (in a suitable initial condition) are the constituent particles of the system, is also possible. And an even finer description, in which two electrons, three protons, and four neutrons (in a suitable initial condition) are the constituent particles, is possible too.

Both within classical and quantum mechanics, *each coarse level of description can be derived from the finer level* (in suitable dynamical conditions concerning forces and initial conditions) and *the dynamical principle has the same form in the different levels.* I refer to this property of a mechanical theory as *compoundation covariance.*

If compoundation covariance were not there, a problem of consistency would arise. Therefore, I state the following general principle: any mechanical theory not declaring to which elementary constituents it applies must enjoy compoundation covariance. If this principle is assumed, it constitutes a criterion of acceptability when one proposes a new mechanical theory or a modification of an existing theory.

Let me investigate whether the GRW model of spontaneous localization enjoys compoundation covariance.

5 Compoundation Covariance of the GRW Process

Consider a system of N particles forming an object. The wave function has the form

$$\psi(q,s) = \Psi(\boldsymbol{Q})\,\chi(r,s),$$

where $\Psi(\boldsymbol{Q})$ is the centre–of–mass wave function and $\chi(r,s)$ is the structural wave function, r being the internal coordinates and s the spin quantum numbers. The localization of particle i results in the wave function

$$\hat{L}^i_{\boldsymbol{x}}\,\Psi(\boldsymbol{Q})\,\chi(r,s) = G^{(3)}(\boldsymbol{q}^i - \boldsymbol{x})\,\Psi(\boldsymbol{Q})\,\chi(r,s).$$

The structural wave function $\chi(r,s)$ is strongly peaked around a value r_0 of the internal coordinates, the width Δr of $\chi(r,s)$ being typically of the order of 10^{-8} cm or less. The only exception to this situation is that of superconducting states, which must be considered separately. Let α be such that

$$\Delta r \ll \alpha.$$

In such conditions, considering the position $\bar{\boldsymbol{q}}^i$ of particle i with respect to the center of mass,

$$\boldsymbol{q}^i = \bar{\boldsymbol{q}}^i(r) + \boldsymbol{Q},$$

one can put in the Gaussian function appearing in the definition of $\hat{L}^i_{\boldsymbol{x}}$

$$\boldsymbol{q}^i \simeq \bar{\boldsymbol{q}}^i(r_0) + \boldsymbol{Q},$$

so that

$$\begin{aligned}\hat{L}^i_{\boldsymbol{x}}\,\Psi(\boldsymbol{Q})\,\chi(r,s) &\simeq G^{(3)}(\,\boldsymbol{Q} - (\boldsymbol{x} - \bar{\boldsymbol{q}}^i(r_0))\,)\,\Psi(\boldsymbol{Q})\,\chi(r,s)\\ &= G^{(3)}(\boldsymbol{Q} - \boldsymbol{X}^i)\,\Psi(\boldsymbol{Q})\,\chi(r,s) = \big(\hat{L}^{\text{c.m.}}_{\boldsymbol{X}^i}\,\Psi(\boldsymbol{Q})\big)\chi(r,s).\end{aligned}$$

The probability density for the occurrence of $\boldsymbol{X}^i = \boldsymbol{x} - \bar{\boldsymbol{q}}^i(r_0)$ is

$$\text{pr}(\boldsymbol{X}^i) = \text{pr}(\boldsymbol{x}) = \|L^i_{\boldsymbol{x}}\Psi\chi\|^2 \simeq \|(\hat{L}^{\text{c.m.}}_{\boldsymbol{X}^i}\,\Psi)\chi\|^2 = \|\hat{L}^{\text{c.m.}}_{\boldsymbol{X}^i}\,\Psi\|^2.$$

One can see that, in the considered conditions, the localization process does not affect the structural wave function of the system, and the localization of a single constituent particle entails a localization of the same type of the centre of mass. The process enjoys compoundation covariance, the frequency of the localization of the centre of mass being given by

$$\lambda_{\text{c.m.}} = \sum\nolimits_i \lambda_i\,.$$

A possible choice of the width and frequency parameters satisfying the requirements of sect. 2 is

$$\alpha \approx 10^{-5}\,\text{cm}\,,$$

$$1/\lambda_{\text{nucleon}} \approx 10^{16}\,\text{s} \approx 3\cdot 10^{8}\,\text{years} \qquad \text{i.e.} \qquad 1/\lambda_{(1\text{g})} \approx 2\cdot 10^{-8}\,\text{s}\,.$$

6 Physical Consequences of the GRW Process

A single microscopic particle suffers localizations with frequency λ. Since λ is tremendously small, the Schrödinger equation remains fully valid even if the wave function of the particle is extended over a macroscopic region.

In the case of a system of N independent particles, e.g. a beam, the wave function has the form

$$\psi(q,s) = \psi_1(\boldsymbol{q}_1,s_1)\dots\psi_i(\boldsymbol{q}_i,s_i)\dots\psi_N(\boldsymbol{q}_N,s_N).$$

Then

$$\hat{L}^i_{\boldsymbol{x}}\psi(q,s) = \psi_1(\boldsymbol{q}_1,s_1)\dots\left(\hat{L}^i_{\boldsymbol{x}}\psi_i(\boldsymbol{q}_i,s_i)\right)\dots\psi_N(\boldsymbol{q}_N,s_N),$$

so that a localization of particle i affects only that particle. The overall frequency is $\sum_i\lambda_i$, but the relevant quantity is the fraction of particles which suffer a localization in a time unit, i.e. $\sum_i\lambda_i/N$, which again is small. Interference experiments with the beam are undisturbed.

If N particles form an object, the internal wave function is unaffected, as we have seen. If the object is macroscopic, since λ is large, its position as a whole is rapidly made sharply definite. This is just the result desired in connection with the measurement problem.

7 A Familiy of Continuous Stochastic Processes in Hilbert Space

I look now for a continuous stochastic process in Hilbert space having the same physical effects as the previous discontinuous process. Let me start with the (Itô) simple stochastic differential evolution equation

$$\mathrm{d}|\psi\rangle = \left[-\frac{\mathrm{i}}{\hbar}\hat{H}\,\mathrm{d}t + g\,\hat{A}\,\mathrm{d}B\right]|\psi\rangle$$

where $\hat{A}$ is a selfadjoint operator, and $B(t)$ is a Wiener process such that

$$\overline{\mathrm{d}B(t)} = 0, \qquad \overline{(\mathrm{d}B(t))^2} = \mathrm{d}t. \tag{1}$$

The above equation does not conserve the norm of $|\psi\rangle$. To get norm conservation, one can first introduce a counterterm such that the norm is conserved in mean,

$$\mathrm{d}|\psi\rangle = \left[-\frac{\mathrm{i}}{\hbar}\hat{H}\,\mathrm{d}t + g\,\hat{A}\,\mathrm{d}B - \tfrac{1}{2}g^2\hat{A}^2\,\mathrm{d}t\right]|\psi\rangle,$$

then further modify it so that the norm is conserved individually, by replacing $\hat{A}$ with the nonlinear operator

$$\hat{A}_\psi = \hat{A} - \langle\psi|\hat{A}|\psi\rangle,$$

getting finally [9]

$$\mathrm{d}|\psi\rangle = \left[-\frac{\mathrm{i}}{\hbar}\,\hat{H}\,\mathrm{d}t + g\,\hat{A}_\psi\,\mathrm{d}B - \tfrac{1}{2}\,g^2(\hat{A}_\psi)^2\,\mathrm{d}t \right] |\psi\rangle. \tag{2}$$

This is precisely the infinite frequency limit of a discontinuous process of the previously considered type in which the quantity A is made sharp through the repeated action with mean frequency λ of the Gaussian operator

$$\hat{L}_a^A = G(\hat{A} - a),$$

the limit being taken with the prescription

$$\lambda/\alpha^2 \xrightarrow[\lambda\to\infty]{} g^2$$

where α is the accuracy parameter appearing in G.

One can prove directly (without reference to the corresponding discontinuous process) that the evolution equation with the stochastic term alone

$$\mathrm{d}|\psi\rangle = \left[g\,\hat{A}_\psi\,\mathrm{d}B - \tfrac{1}{2}\,g^2(\hat{A}_\psi)^2\,\mathrm{d}t \right] |\psi\rangle$$

has the effect

$$|\psi\rangle \xrightarrow[t\to\infty]{} \hat{P}_e|\psi\rangle/\|\hat{P}_e\psi\|, \qquad \Pr(e) = \|\hat{P}_e\psi\|^2, \tag{3}$$

where $\hat{P}_e$ is the projection operator on the eigenspace of $\hat{A}$ belonging to the eigenvalue e (with obvious modifications in the case of continuous spectrum). If both the Schrödinger term and the stochastic term are kept, the net result will depend on the competition of the two evolution processes.

It will be necessary in the following to induce several compatible quantities to have sharp values. This can be done in different ways.

Let $\hat{A}^m$ be a set of commuting selfadjoint operators. Associate with it a set of independent Wiener processes $B_m(t)$ such that

$$\overline{\mathrm{d}B_m(t)} = 0, \qquad \overline{\mathrm{d}B_m(t)\mathrm{d}B_{m'}(t)} = \delta_{mm'}\mathrm{d}t, \tag{4}$$

and replace the single stochastic term in eq. (2) by a sum of similar terms. The resulting evolution equation is

$$\mathrm{d}|\psi\rangle = \left[-\frac{\mathrm{i}}{\hbar}\,\hat{H}\,\mathrm{d}t + \sum\nolimits_m g_l\,\hat{A}_\psi^m\,\mathrm{d}B_m - \tfrac{1}{2}\sum\nolimits_m g_m^2(\hat{A}_\psi^m)^2\,\mathrm{d}t \right] |\psi\rangle. \tag{5}$$

If the Schrödinger term is dropped the limit for large t of $|\psi\rangle$ is given by eq. (3) where P_e are the projections on the common eigenspaces of the operators $\hat{A}^m$.

A different structure of the evolution equation is obtained if a unique Wiener process $B(t)$ with the property (1) is associated to the set of commuting selfadjoint operators $\hat{A}^l$. The operator $g\hat{A}$ is replaced in eq. (2) by the selfadjoint sum $\sum_l g_l \hat{A}^l$. The resulting evolution equation is

$$\mathrm{d}|\psi\rangle = \left[-\frac{\mathrm{i}}{\hbar}\,\hat{H}\,\mathrm{d}t + \sum\nolimits_l g_l\,\hat{A}_\psi^l\,\mathrm{d}B - \tfrac{1}{2}\left(\sum\nolimits_l g_l\hat{A}_\psi^l\right)^2\mathrm{d}t \right] |\psi\rangle. \tag{6}$$

If the Schrödinger term is dropped the limit for large t of $|\psi\rangle$ is given by eq. (3) where P_e are the projections on the eigenspaces of the operator $\sum_l g_l \hat{A}^l$.

In the case of a set of commuting selfadjoint operators depending on two index variables $\hat{A}^{lm}$, one can proceed in the second way for the index l and in the first way for index m, i.e. associating a set of independent Wiener processes $B_m(t)$ with the property (4) to the set of operators $\sum_l g_{lm} \hat{A}^{lm}$. The resulting evolution equation is

$$\mathrm{d}|\psi\rangle = \left[-\frac{\mathrm{i}}{\hbar} \hat{H}\,\mathrm{d}t + \sum_m \sum_l g_{lm}\, \hat{A}^{lm}_\psi\, \mathrm{d}B_m - \tfrac{1}{2} \sum_m \left(\sum_l g_{lm} \hat{A}^{lm}_\psi \right)^2 \mathrm{d}t \right] |\psi\rangle. \tag{7}$$

If the Schrödinger term is dropped the limit for large t of $|\psi\rangle$ is given by eq. (3) where P_e are the projections on the common eigenspaces of the operators $\sum_l g_{lm} \hat{A}^{lm}$.

As an example of the formalism described above, let me get the continuous version of the GRW process. In the discontinuous version there is an independent localization process for each particle and for each space axis. The corresponding continuous process is obtained from the evolution equation (5) through the identifications

m	$\longrightarrow$	i	particle index,
$\hat{A}^m$	$\longrightarrow$	$\hat{\boldsymbol{q}}^i$	position operator of particle i,
$B_m(t)$	$\longrightarrow$	$\boldsymbol{B}_i(t)$	a vector Wiener process for each i, $\overline{\mathrm{d}\boldsymbol{B}_i(t)} = 0$, $\overline{\mathrm{d}\boldsymbol{B}_i(t)\mathrm{d}\boldsymbol{B}_{i'}(t)} = \delta_{ii'} \mathbf{I}\, \mathrm{d}t$,
g_m	$\longrightarrow$	g_i	$g_i^2 = \lambda_i / \alpha^2$,

where $\mathbf{I}$ is the unit tensor. The resulting evolution equation is

$$\mathrm{d}|\psi\rangle = \left[-\frac{\mathrm{i}}{\hbar} \hat{H}\,\mathrm{d}t + \sum_i g_i\, \hat{\boldsymbol{q}}^i_\psi \cdot \mathrm{d}\boldsymbol{B}_i - \tfrac{1}{2} \sum_i g_i^2 \left(\hat{\boldsymbol{q}}^i_\psi \right)^2 \mathrm{d}t \right] |\psi\rangle\,. \tag{8}$$

Its physical consequences, for sufficiently long times, are the same as those of the discontinuous process.

8 Spontaneous Localization of a System of Identical Particles

In the case of identical particles, the positions $\hat{\boldsymbol{q}}_i$ of individual particles are no more dynamical variables of the system. Correspondingly, the GRW process and its continuous version (which both act making sharp the $\hat{\boldsymbol{q}}_i$'s) do not respect the correct symmetry of the wave function.

The simplest way to introduce a localization process in the case of identical particles (a single kind, for the moment) is replacing the operators "positions of the various particles" by the operators "number densities around the various space points" as the operators which are made sharp by the process. This can be obtained from eq. (5) through the identifications

m	$\longrightarrow$	$\boldsymbol{x}$	space point,
$\hat{A}^m$	$\longrightarrow$	$\hat{N}^{\boldsymbol{x}}$	number–density operator around $\boldsymbol{x}$,
$B_m(t)$	$\longrightarrow$	$B_{\boldsymbol{x}}(t)$	stochastic field such that $\overline{\mathrm{d}B_{\boldsymbol{x}}(t)} = 0$, $\overline{\mathrm{d}B_{\boldsymbol{x}}(t)\mathrm{d}B_{\boldsymbol{x}'}(t)} = \delta^{(3)}(\boldsymbol{x}-\boldsymbol{x}')\,\mathrm{d}t$,
g_m	$\longrightarrow$	g	independent of $\boldsymbol{x}$ for the sake of translational invariance.

The resulting evolution equation is

$$\mathrm{d}|\psi\rangle = \left[-\frac{\mathrm{i}}{\hbar}\hat{H}\,\mathrm{d}t + g\int \mathrm{d}^3\boldsymbol{x}\,\hat{N}^{\boldsymbol{x}}_{\psi}\mathrm{d}B_{\boldsymbol{x}} - \tfrac{1}{2}g^2\int \mathrm{d}^3\boldsymbol{x}\left(\hat{N}^{\boldsymbol{x}}_{\psi}\right)^2\mathrm{d}t\right]|\psi\rangle\,. \tag{9}$$

The number–density operator $\hat{N}^{\boldsymbol{x}}$ can be defined, in second–quantized language, as

$$\hat{N}^{\boldsymbol{x}} = \sum\nolimits_s \int \mathrm{d}^3\boldsymbol{x}'\mathrm{G}^{(3)}(\boldsymbol{x}'-\boldsymbol{x})\,a^{+}(\boldsymbol{x}',s)a(\boldsymbol{x}',s), \tag{10}$$

where

$$\mathrm{G}^{(3)}(\boldsymbol{x}) = \mathcal{N}^3\exp\left(-\tfrac{1}{2}(\boldsymbol{x}/a)^2\right), \qquad \mathcal{N} = \left(\frac{1}{2\pi a^2}\right)^{\frac{1}{2}}. \tag{11}$$

The stochastic terms in eq. (9) with $\hat{N}^{\boldsymbol{x}}$ given by (10) and (11) define the spontaneous localization process for identical particles (GPR) [10]. It is strictly equivalent to a previous proposal [11] introduced as a modification of eq. (8), without consideration of the number–density operator.

The width a of the Gaussian (11) must be large enough to avoid fluctuations on a microscopic scale, and small enough to allow significant variations on a macroscopic scale of the resulting density. In spite of the different meaning, its order of magnitude must be nearly the same as that of the width α appearing in the localization operator of the GRW process. A sensible assumption is

$$a \approx 10^{-5}\,\mathrm{cm}\,.$$

The parameter g measures the intensity of the coupling to the stochastic field. We shall see that a reasonable value is

$$g^2 \approx 10^{-30}\,\mathrm{cm}^3\,\mathrm{s}^{-1} \qquad\qquad \text{(for nucleons)}\,.$$

I shall discuss compoundation covariance of theories of this type in the case of several kinds of identical particles.

9 Spontaneous Localization of a System of Several Kinds of Identical Particles

There exist two natural generalizations of the GPR process to the case of several kinds of identical particles.

In the first case one introduces independent stochastic fields for the different kinds of particles. Each particle field is coupled to its own stochastic field via its own coupling constant. The evolution equation is obtained from eq. (5) through the identifications

$$
\begin{array}{lcll}
m & \longrightarrow & \begin{cases} k \\ \boldsymbol{x} \end{cases} & \begin{array}{l}\text{kind–of–particle index,}\\ \text{space point,}\end{array} \\
\hat{A}^m & \longrightarrow & \hat{N}^{k\boldsymbol{x}} & \text{number–density operator of particles } k \text{ around } \boldsymbol{x}, \\
B_m(t) & \longrightarrow & B_{k\boldsymbol{x}}(t) & \text{set of stochastic fields such that} \\
 & & & \overline{\mathrm{d}B_{k\boldsymbol{x}}(t)} = 0,\ \overline{\mathrm{d}B_{k\boldsymbol{x}}(t)\mathrm{d}B_{k'\boldsymbol{x}'}(t)} = \delta_{kk'}\delta^{(3)}(\boldsymbol{x}-\boldsymbol{x}')\,\mathrm{d}t, \\
g_m & \longrightarrow & g_k & \text{independent of } \boldsymbol{x} \\
 & & & \text{for the sake of translational invariance.}
\end{array}
$$

The resulting equation is

$$
\mathrm{d}|\psi\rangle = \left[-\frac{\mathrm{i}}{\hbar}\hat{H}\,\mathrm{d}t + \sum\nolimits_k g_k \int \mathrm{d}^3\boldsymbol{x}\,\hat{N}_\psi^{k\boldsymbol{x}}\mathrm{d}B_{k\boldsymbol{x}} - \tfrac{1}{2}\sum\nolimits_k g_k^2 \int \mathrm{d}^3\boldsymbol{x}\left(\hat{N}_\psi^{k\boldsymbol{x}}\right)^2 \mathrm{d}t\right]|\psi\rangle\,. \tag{12}
$$

Equation (12) defines the independent–fields spontaneous localization (IFSL) process.

Alternatively, one introduces a universal stochastic field to which each particle field is coupled via its own coupling constant. The evolution equation is obtained from eq. (7) through the identifications

$$
\begin{array}{lcll}
l & \longrightarrow & k & \text{kind–of–particle index,} \\
m & \longrightarrow & \boldsymbol{x} & \text{space point,} \\
\hat{A}^{lm} & \longrightarrow & \hat{N}^{k\boldsymbol{x}} & \text{number–density operator of particles } k \text{ around } \boldsymbol{x}, \\
B_m(t) & \longrightarrow & B_{\boldsymbol{x}}(t) & \text{universal stochastic field such that} \\
 & & & \overline{\mathrm{d}B_{\boldsymbol{x}}(t)} = 0,\ \overline{\mathrm{d}B_{\boldsymbol{x}}(t)\mathrm{d}B_{\boldsymbol{x}'}(t)} = \delta^{(3)}(\boldsymbol{x}-\boldsymbol{x}')\,\mathrm{d}t, \\
g_{lm} & \longrightarrow & g_k & \text{independent of } \boldsymbol{x} \\
 & & & \text{for the sake of translational invariance.}
\end{array}
$$

The resulting equation is

$$
\mathrm{d}|\psi\rangle = \left[-\frac{\mathrm{i}}{\hbar}\hat{H}\,\mathrm{d}t + \int \mathrm{d}^3\boldsymbol{x} \sum\nolimits_k g_k\,\hat{N}_\psi^{k\boldsymbol{x}}\mathrm{d}B_{\boldsymbol{x}} - \tfrac{1}{2}\int \mathrm{d}^3\boldsymbol{x}\left(\sum\nolimits_k g_k\,\hat{N}_\psi^{k\boldsymbol{x}}\right)^2 \mathrm{d}t\right]|\psi\rangle\,. \tag{13}
$$

Equation (13) defines the universal–field spontaneous localization (UFSL) process.

10 Compoundation Covariance of IFSL and UFSL Processes – Mass-Density Process

Let the system consist of objects, of various kinds labelled by the Greek index κ, each object consisting in turn of constituent particles, of various kinds labelled by the Latin index k. We assume that objects of all kinds are microscopic, in the sense that their linear dimensions are smaller than the width a appearing in the definition of densities.

The IFSL and UFSL theories, when applied to constituent particles, are specified by the couplings g_k of constituent particles of kind k to their own stochastic fields $B_{k\boldsymbol{x}}(t)$ in the case of the IFSL theory, to the universal stochastic field $B_{\boldsymbol{x}}(t)$ in the case of the UFSL theory. Similarly, when objects are considered as constituent particles, the same theories are specified by the couplings g_κ of object particles of kind κ to their own stochastic fields $B_{\kappa\boldsymbol{x}}(t)$ in the case of the IFSL theory, to the universal stochastic field $B_{\boldsymbol{x}}(t)$ in the case of the UFSL theory.

Compoundation covariance imposes relations between the couplings g_k and g_κ in terms of the numbers $n_{\kappa k}$ of constituent particles of kind k in objects of kind κ [12].

In the IFSL case the condition of compoundation covariance is

$$\sum_k n_{\kappa k}\, n_{\kappa' k}\, g_k^2 = g_\kappa^2\, \delta_{\kappa\kappa'}.$$

For $\kappa \neq \kappa'$ the right hand side is zero, so that, except for the case in which there is only a single kind of objects, the only solution is that in which all g's are zero. Therefore, the IFSL process does not enjoy compoundation covariance.

In the UFSL case the condition of compoundation covariance is

$$\sum_k n_{\kappa k}\, g_k = g_\kappa. \tag{14}$$

Therefore the UFSL process enjoys compoundation covariance provided the g's are additive. Since in nonrelativistic theories masses are additive,

$$m_\kappa = \sum_k n_{\kappa k}\, m_k,$$

a natural solution of condition (14) is

$$g_k = g_0\, \frac{m_k}{m_0}, \qquad g_\kappa = g_0\, \frac{m_\kappa}{m_0},$$

m_0 being a reference mass and g_0 a universal constant, i.e.

$$g_K = g_0\, \frac{m_K}{m_0}$$

for any kind K of microscopic particle, compound or not.

If that choice is made, a process is obtained in which the mass densities around each space point are made sharp. The evolution equation is

$$\mathrm{d}|\psi\rangle = \left[-\frac{\mathrm{i}}{\hbar}\,\hat{H}\,\mathrm{d}t + \frac{g_0}{m_0}\int \mathrm{d}^3\boldsymbol{x}\, \hat{M}_\psi^{\boldsymbol{x}} \mathrm{d}B_{\boldsymbol{x}} - \frac{1}{2}\left(\frac{g_0}{m_0}\right)^2 \int \mathrm{d}^3\boldsymbol{x}\left(\hat{M}_\psi^{\boldsymbol{x}}\right)^2 \mathrm{d}t\right]|\psi\rangle, \tag{15}$$

where

$$\hat{M}^{\boldsymbol{x}} = \sum_k m_k \hat{N}^{k\boldsymbol{x}}.$$

A reasonable assumption for the width a appearing in the definition of densities and for the coupling g_0 to the stochastic field is

$$a \approx 10^{-5}\,\text{cm}, \qquad g_0^2 \approx 10^{-30}\,\text{cm}^3\,\text{s}^{-1},$$

having taken the proton mass as the reference mass m_0.

11 Physical Effects of the Mass-Density Process

Consider first a single (elementary or compound) microscopic particle of the kind K. It is understood that its linear dimensions are smaller than a. It can be shown [10] that, for the single particle, the mass–density process is equivalent to its localization within a volume of linear dimensions a with an effective mean frequency equal to

$$\lambda_K = g_0^2 \left(\frac{m_K}{m_0}\right)^2 \left(\frac{1}{4\pi a^2}\right)^{\frac{3}{2}}.$$

Using the values of a and g_0 proposed in sect. 10, one finds for example

$$\lambda_{\text{electron}} \approx 10^{-23}\,\text{s}^{-1}, \qquad \lambda_{\text{proton}} \approx 10^{-17}\,\text{s}^{-1}, \qquad \lambda_{A=250} \approx 10^{-12}\,\text{s}^{-1}.$$

The localization process is negligible for all microscopic particles.

For any macroscopic object whose structure is such that the component particles are well localized in comparison with a, it is easily shown [10] that the structural wave function is practically unaffected by the process. On the other hand a localization process acts on the centre–of–mass wave function. Two kinds of effects must be estimated in connection with the localization of the centre of mass.

First we are concerned with the reduction effects. It can be shown [10] that the off–diagonal elements of the centre–of–mass statistical operator in the position representation are damped exponentially at a time rate Γ given by

$$\Gamma = g_0^2 \left(\frac{\varrho}{m_0}\right)^2 V_{\text{out}},$$

where ϱ is the mass density of the macroscopic object and V_{out} is the volume of the object which is uncovered in the two positions to which the off–diagonal element refers. Reduction is in all cases very rapid. For example, for $\varrho = 1\,\text{g}\,\text{cm}^{-3}$, $V_{\text{out}} = 10^{-5}\,\text{cm}^3$, one gets

$$\Gamma \approx 10^{17}\,\text{s}^{-1}.$$

A second effect of the localization process is a certain amount of stochasticity that it necessarily introduces in the motion of macroscopic objects. This kind of effect can be intuitively understood by noting that squeezing position implies widening momentum, so that successively propagation enlarges the spread of final positions. In order that the classical behaviour of macroscopic objects be not contradicted, it has to checked that the effect is negligible. It can be measured by a quantity having the meaning of momentum diffusion coefficient, whose value in the direction of $\boldsymbol{n}$ is [10]

$$D_{\mathrm{p}} = g_0^2 \hbar^2 \left(\frac{1}{4\pi a^2}\right)^{\frac{1}{2}} \left(\frac{\varrho}{m_0}\right)^2 S_\perp$$

where $S_\perp$ is the section of the object transverse to $\boldsymbol{n}$. This formula was derived for a parallelepiped and for $\boldsymbol{n}$ parallel to its edges, but it is anyway meaningful as an indication of order of magnitude. For $\varrho = 10\,\mathrm{g\,cm^{-3}}$, $S_\perp = 1\,\mathrm{cm^2}$ one gets

$$D_{\mathrm{p}} \approx 10^{-30} \left(\mathrm{g\ cm\ s^{-1}}\right)^2 \mathrm{s^{-1}}.$$

It is seen that the effect is very small.

The effect of spontaneous localization on superconducting internal states, where electron wave functions are extended, must be studied separately. It was shown [12] that suppression of supercurrents is negligible and that reduction of superpositions of different supercurrents due to spontaneous localization is slow compared to reduction due to other causes.

12 Conclusion

I have shown that the program outlined in sects. 1 and 2 of working out a reduction theory could be realized. Spontaneous localization played an essential role in that, because of its characteristic property of cumulating, in the case of macroscopic objects, effects which are negligible at the level of microscopic constituents. In conjunction with the quite realistic assumption that apparatus always contain macroscopic parts located in macroscopically–separated positions in the different final states, it provides a satisfactory theory of quantum measurement.

I want to emphasize the role, in working out the theory, of the principle of compoundation covariance. In the case of the GRW process, the exigence of checking its validity led to the discovery of the cumulative property. In the case of the mass–density process, it provided a convincing argument to favour such a process rather than other possibilities.

People, including myself, can believe in the whole edifice, or not, or be agnostic. In any case, I think it was worth while to show that it could be done.

References

1. K. Gottfried, *Quantum Mechanics* (Benjamin, New York, 1966).
2. G. Peruzzi and A. Rimini, Found. Phys. Lett. **9**, 505 (1996).
3. R. B. Griffiths, J. Stat. Phys. **36**, 219 (1984); R. Omnès, J. Stat. Phys. **53**, 893 (1988); M. Gell-Mann and J. B. Hartle, in *Foundations of Quantum Mechanics*, ed. by S. Kobayashi, H. Ezawa, Y. Murayama, and S. Nomura (The Physical Society of Japan, Tokyo, 1990), p. 321.
4. I. Giardina and A. Rimini, Found. Phys. **26**, 973 (1996).

5. D. Bohm, Phys. Rev. **85**, 166, 180 (1952).
6. J. S. Bell, in *Quantum Gravity 2*, ed. by C. Isham, R. Penrose, and D. Sciama (Clarendon Press, Oxford, 1981), p. 611; D. Dürr, S. Goldstein, and N. Zanghì, J. Stat. Phys. **67**, 843 (1992).
7. P. Pearle, Phys. Rev. D **13**, 857 (1976); **33**, 2240 (1986).
8. G. C. Ghirardi, A. Rimini, and T. Weber, Phys. Rev. D **34**, 470 (1986); **36** 3287 (1987).
9. N. Gisin, Phys. Rev. Lett. **52**, 1657 (1984); P. Pearle, *ibid.* **53**, 1775 (1984).
10. G. C. Ghirardi, P. Pearle, and A. Rimini, Phys. Rew. A **42**, 78 (1990).
11. P. Pearle, Phys. Rev. A **39**, 2277 (1989).
12. M. Buffa, O. Nicrosini, and A. Rimini, Found. Phys. Lett. **8**, 105 (1995).

Probabilities, Decohering Histories, and the Interpretation of Quantum Mechanics

Roland Omnès

Laboratoire de Physique Théorique
(Unité Mixte de Recherche UMR - CNRS N° 8627)
Bâtiment 210, Université de Paris-Sud, 91405 Orsay (France)

Abstract. This is a short review of the consistent histories approach to quantum mechanics. After a short description of Bohr's and von Neumann's approach to quantum mechanics the notion of history will be defined and the determination of probabilities of histories will be discussed.

1 Introduction

One will be concerned here with three questions about probabilities and the interpretation of quantum mechanics. The first one goes back to Thomas Bayes, when a logical inference "a implies b" is expressed by means of a conditional probability

$$p(b|a) = 1 , \tag{1}$$

one may then ask the question:

1. Is the logic of quantum mechanics Bayesian?

The second question has to do with the opposition of probabilism versus determinism. When Max Born discovered the randomness of quantum mechanics, the question "why probabilities ?" was immediately raised. Emphasis has changed by now. Quantum mechanics has been vindicated so universally and precisely that one would rather look at the macroscopic world as obeying quantum laws and ask:

2. Why is there determinism?

A third question is concerned with the interpretation of probability calculus and it was apparently first asked by Émile Borel:

3. Do very small probabilities make sense?

This neglected question came back recently to the forefront of physics with the interest for decoherence, because the discussions about the meaning of this effect always amounted to the meaning of very small probabilities.

These three questions are central in the interpretation of quantum mechanics by means of consistent (or decohering) histories, and I will therefore also summarize this interpretation [1].

Though originally an independent approach, consistent histories were soon found to yield answers for the problems of interpretation very close to the orthodox ones. A convenient way for introducing histories is therefore to pin down a significant difference between two main versions of the "orthodox" interpretation. One of them, the Copenhagen interpretation by Bohr, Heisenberg and Pauli, deals directly with hard problems such as the meaning of measurements or the relation between macroscopic and microscopic physics. A second version of interpretation, by von Neumann, is more directly concerned with logic. The theory of histories[2–4] can be seen as a missing link between the two approaches: it relies on von Neumann's well-defined assumptions and, after having solved a few non-trivial problems, it can turn most of the Copenhagen rules into so many theorems, and even correct a few of them.

2 The Bohr Approach

The basic statements of the Bohr interpretation are well known:

1. *The state* of a physical system is described by a wave function (or a state vector).
2. *Classical physics is taken for granted.* One cannot apply the quantum formalism to a macroscopic system, whose main features are described on the contrary by classical physics.
3. *Correspondence.* The opposition between classical and quantum physics is slightly alleviated by a (rather fuzzy) assumption of continuity between them: the correspondence principle.
4. *Complementarity.* This is again a principle, stating that two different representations of a quantum system by means of classical concepts cannot hold together (at least generally). One must choose to speak either of a wave or a particle and, when speaking of a particle, one must either consider its position or its momentum. The relevant choice is dictated by the experimental setup. Complementarity has drastic logical consequences since it disagrees with standard realism.
5. *Measurements* are essential for an understanding of quantum physics (This is a consequence of Point 2: measurements provide the sole link between microscopic quantum physics and macroscopic classical physics).
6. *Randomness.* The possible results of a measurement are occurring at random. The relevant probabilities are given by Born's rule. Notice however that according to Rule 2, classical determinism is valid at a macroscopic level.
7. *Reduction.* Bohr's construction is crowned by the reduction rule (or "wave function collapse"), which was often believed to express a genuine physical effect.

3 The Von Neumann Approach

Von Neumann's approach is substantially different from Bohr's, particularly through the following *epistemic principle* standing behind it:

1. There is only one kind of physical laws and they are given by the quantum principles.

These principles involve essentially *(i)* the Hilbert space mathematical formalism ; *(ii)* relativistic – or galilean – invariance under a change of reference frame and *(iii)* Heisenberg-Schrödinger dynamics for the observables or the wave function.

Von Neumann tried to derive a measurement theory from these principles, because his epistemic approach implied that they should generate their own interpretation. He also made an important remark concerning the language of physics and its relation with the mathematical formalism:

2. Every property of a quantum system asserts that "the value of some observable $A(t)$ is in some range Δ of real numbers". This property (or "elementary predicate" as he called it) is associated with a projection operator in Hilbert space (namely the operator

$$P = \sum_{a \in \Delta} |a, n\rangle\langle a, n| \,, \tag{2}$$

an eigenvector of $A(t)$ with eigenvalue a being denoted by $|a, n\rangle$).

Von Neumann encountered however three major problems:

(i) He found that macroscopic entanglements are generated by a quantum measurement, a remark to become famous with the example of Schrödinger's cat.
(ii) He could not extend the quantum predicates to the case of classical properties, as they hold for a macroscopic system.
(iii) The elementary predicates did not obey the standard rules of logic.

In contrast with the philosophically minded Bohr approach, von Neumann's one is more promising because it ends up with three well-defined problems. One has "only" to solve them. This task has now been achieved: Problem *(i)* is solved by the decoherence effect [5–9]. Problem *(ii)* is solved by using coarse graining [4,10] or, more conveniently and more directly (in my opinion) by the powerful mathematical techniques of microlocal analysis [11]. Finally, Problem *(iii)* is the task of "consistent" histories.

In my own view, the first two ideas are essential for a real understanding of interpretation. The case of histories is different. I would agree that one can understand quantum mechanics without them, just as much as one can dig a big hole with a shovel rather than an excavating machine. Histories make interpretation simpler, because they provide a language for it. This language is sensible, because it agrees with conventional logic, and universal, because it covers both quantum (microscopic) and classical (macroscopic) physics.

4 The Notion of History

Almost every textbook or experimental paper is full of histories. It describes an experiment by explaining which kind of experimental devices have been used and how they worked; it also mentions which kind of physical events occurred or were supposed to occur among atoms and particles.

The idea of histories [2] was to describe an experiment as a series of microscopic events, using von Neumann's framework. Every event is expressed by a projection operator $P_k(t_k)$ where t_k denotes the time at which a relevant property is supposed to occur. The whole experiment, at least as far as microscopic events are concerned, is therefore described by a sequence of projections $\{P_1(t_1), P_2(t_2) \cdots P_n(t_n)\}$. This is called a history.

But the idea is not restricted to microscopic events. Histories have been extended to include the essential features and the detailed working of experimental devices, after solving von Neumann's Problem *(ii)*. It goes as follows: A classical property differs from a quantum property by asserting simultaneously the values of a position observable in some interval with a width Δq and of the corresponding momentum in an interval Δp (with $\Delta q \cdot \Delta p \ll h$). This property can still be expressed by a projection operator, or rather a set of equivalent projections.

There are three ways for obtaining this result:

- The most general method relies on microlocal analysis, which is the most powerful mathematical technique for the study of linear operators [11].
- One can also follow Klaus Hepp [12] and use coherent states $|q,p\rangle$, i.e. Gaussian wave functions where q and p denote the average values of position and momentum. The construction of the operator P is particularly simple since it is given by
$$P = \int_{\Delta q \times \Delta p} |q,p\rangle\langle q,p| \, \mathrm{d}q \, \mathrm{d}p/h \; , \tag{3}$$
- Finally, one can use a coarse-grained description of matter in the macroscopic device, decoherence implying in that case a classical behavior [4,10].

Whichever approach is used, it shows that the classical description of macroscopic physics is a special case of the quantum description so that, from there on, I will consider histories as including every relevant feature (microscopic or macroscopic) in an experiment.

5 Probability of Histories

One can define a probability for a history such as $\{P_1(t_1), P_2(t_2) \ldots P_n(t_n)\}$ (to be denoted by a). To do so, one must introduce a *complete* family of mutually *exclusive* histories. (In some of these histories, for instance, a scattering of two particles occurs whereas, in another history, their collision ends up in a reaction). One can then define the probability of history a: Let the product $C_a = P_1(t_1) \cdot P_2(t_2) \cdots P_n(t_n)$ denote a "history operator" and ϱ denote the density operator for the initial state of the system (including the state of the experimental devices) at a time earlier than t_1. Then one has

$$p_a = \mathrm{tr}\left\{C_a^\dagger \, \varrho \, C_a\right\} \; . \tag{4}$$

In order to satisfy the basic properties of probabilities, and particularly their additivity, the family of histories should not be arbitrary. It must satisfy the following *consistency conditions* [2]:

$$\text{Real part of tr}\left\{C_a^\dagger\,\varrho\,C_b\right\} = 0 \tag{5}$$

for two different histories, $a \neq b$.

One may wonder whether expression (4) is unique. Goldstein and Page [13] conjectured it is not but, recently, G. Nistico [14] proved uniqueness under simple assumptions. The idea consists in noticing that the beginning of a history, say from time t_1 to t_k, is also a history. The probability for the complete history a going from t_1 to t_n can then be written as

$$p_a = p_k \cdot p(n|k)\ , \tag{6}$$

where p_k is the probability for the initial part of history a (from t_1 to t_k) and $p(n|k)$ a conditional probability for the later part of history a (from t_{k+1} to t_n), assuming the initial part as given. If the probabilities of histories (including so-called "compound histories") and the conditional probabilities satisfy the axioms of probability calculus, one finds that Eq. (4) is necessarily valid and the consistency conditions (5) must hold. (These conditions, for instance, eliminate as inconsistent a description of an interference experiment trying to assert through which path a particle has gone).

6 Histories and Logic

There is a close connection between our Question 1 (Is logic founded on probabilities?), von Neumann's problem *(iii)* (propositions do not generally satisfy the standard rules of logic) and the meaning of consistent histories.

One can extract a lot of propositions from a consistent family of histories (e.g. the statement of an elementary event $P_k(t_k)$ or a whole history). The logical operations "and, or, not" are easily defined on the set of propositions. Logical inference $a \Rightarrow b$ (a implies b where a and b are two propositions) is then everything needed for using elementary logic. Introducing the conditional probability

$$p(b|a) = p(a \text{ and } b)/p(a)\ , \tag{7}$$

one can show easily that Condition (1) is a correct definition of inference and the standard rules of logic are satisfied.

One then stands on a firm logical ground for interpretation and the answer to Question 1 is that the statements about an experiment are related by Bayesian inference. It should be noticed that most propositions are not expressed by a unique projection, but by a set of projections (as obvious in the case of a history). This is how one avoids the logical difficulties with von Neumann's third problem.

7 Complementarity

Complementarity is no more a fuzzy philosophical principle in the history formalism. It means only that different consistent families of histories can sometimes account for the same microscopic events (if for instance one of them states a position property and the other one a momentum property, for the same object at the same time). It means that histories cannot pretend to restore realism in the quantum world.

I will seize this opportunity for putting an end to a controversy. Bernard d'Espagnat asked whether because of complementarity there is a notion of truth (or reality) in the history description of quantum events (outside the truth of classical data) [15]. I proposed such a definition [16] and it was disproved by Fay Dowker and Adrian Kent [17]. They were right though, fortunately, the truth of data is enough for interpretation [1]. I should mention however that some of Dowker and Kent's conclusions are overstretched because of their reliance on purely algebraic arguments, taking no account of decoherence.

8 The Question of Determinism

There is an important case when complementarity does not enter. It occurs with a purely classical device when no quantum measurement (or anything like it) is acting. Histories can still be used and they involve only classical properties (though still relying on quantum principles). The resulting classical logic is unique and in complete agreement with the observed uniqueness of empirical reality.

What about determinism in that case? Determinism is essentially a logical equivalence between two classical properties holding at different times: one of the properties can state for instance that an apple starts from up a tree with zero velocity whereas the other property states that the apple has fallen below the branch it started from.

Determinism holds in that case (i.e. the logical equivalence it stands for is correct) with a probability p_{det}. This probability is most often very close to 1. The main point is that determinism is then cast into a probabilistic framework (taking into account a very small probability for its failure).

The mathematics of semiclassical physics standing behind this result are rather involved and they rely on a famous theorem in microlocal analysis by Egorov [18]. The proof of this theorem shows exceptions where p_{det} is not close to 1 and determinism does not hold. Examples are provided by a chaotic motion extending down to the Planck scale h in phase space or the existence of very narrow potential barriers. These exceptions, which can be directly read out from the proof of Egorov's theorem, confirm obviously the soundness of the approach.

9 Decoherence

One can learn much by looking at the trace in (4) for the probabilities or (5) for the consistency conditions. It is convenient to consider the trace as a summation over three different kinds of degrees of freedom: *(i)* those of atoms or particles undergoing a measurement *(ii)* the degrees of freedom for the main features of experimental devices (i.e. those of classical physics), *(iii)* the very large number of "ignored" degrees of freedom describing the atomic substance of the apparatus and the external environment (atmospheric molecules, external light and so on). Equivalently, one might formally consider the whole physical system as made of three interacting subsystems: the relevant particles, a collective (classical) system and a so-called "environment" involving the innumerable degrees of freedom on which there is no information.

The trace can then be thought of as made in three steps:

- First on the environment. The result is most often decoherence.
- Then a trace is made over the collective degrees of freedom, resulting in external conditions on the particles.
- Finally, the trace over the particles degrees of freedom is performed, as it was first introduced by Griffiths.

Decoherence is essentially a destruction of macroscopic interferences, an effect which has been much investigated theoretically and recently observed experimentally [9]. It removes macroscopic entanglements and its consequences are essentially the same as a reduction process.

10 Small Probabilities

Two main aspects of the history interpretation are the inclusion of classical physics (with determinism) in the quantum framework and the replacement of reduction by decoherence. These assertions are so important that there is no surprise in hearing controversies about them. As far as I know, all the controversies turn around the meaning of very small probabilities or they can be brought back to it. Here are some examples:

- Determinism and classical statements are approximate: there is always a small probability for their violation.
- Measurement data are never absolutely disentangled: there is always a tiny probability for incomplete decoherence.

These small probabilities are unavoidable when quantum mechanics is supposed universal. It is clear for instance that in a toy model of the universe containing, say, 33 particles, there could be some determinism or decoherence but not much. The probabilities for violating determinism or maintaining macroscopic entanglements are on the contrary very small in the practice of physics (they are given typically by an exponential whose exponent contains some power of $\hbar^{-1}$). They have no consequence on observation and they can only influence, in some sense, our vision of the world, i.e. the philosophy of science.

Let us define for convenience a very small probability as so small that no man-made apparatus can measure it (or even an apparatus containing less matter than the observable universe). Some investigation on this question can be made, involving of course a precise calculation of the relevant probabilities and the possibility of measuring them. What is then the meaning of very small probabilities? This is the main question for deciding whether the present new version of interpretation is satisfactory.

I only know as an answer the one by Émile Borel [19]: Very small probabilities are empirically and philosophically meaningless and they should be considered as strictly zero. This is the Zero-th axiom by Borel in his interpretation of probability calculus. Too small probabilities not only correspond to nearly impossible events but, were one of these events to happen, it would be impossible to reproduce it. The interpretation of probabilities seem therefore to stand prior to the interpretation of quantum mechanics, though one is then obviously dealing with philosophical options with no practical consequence.

The question: "do very small probabilities make sense?" is certainly of some interest in a meeting such as the present one. Much of the philosophical aspects of the interpretation of quantum mechanics may hinge on it. Does one need for instance microscopic realism as Bohmian mechanics assumes? Does one need some physical reduction mechanism? Certainly not if very small probabilities are physically meaningless so that one may be completely satisfied with a world obeying standard quantum mechanics.

References

1. For reviews of the history interpretation in the present spirit, see R. Omnès: Rev. Mod. Phys. **64**, 339 (1992), *The Interpretation of Quantum Mechanics* (Princeton University Press, 1994) *Understanding Quantum Mechanics* (Princeton University Press, 1999)
2. R. G. Griffiths: J. Stat. Phys. **36**, 219 (1984)
3. R. Omnès: J. Stat. Phys. **53**, 893 (1988)
4. M. Gell-Mann, J. B. Hartle: In *Complexity, Entropy, and the Physics of Information* ed. by W. H. Zurek (Addison-Wesley, Redwood City, CA 1991)
5. N. G. van Kampen: Physica **20**, 603 (1954)
6. H. D. Zeh: Found. Phys. **1**, 69 (1970)
7. W. H. Zurek: Phys. Rev. **D26**, 1862 (1982) Physics Today **44** (10), 36 (1991)
8. R. Omnès: Phys. Rev. **A 56**, 3383 (1997)
9. M. Brune, E. Hagley, J. Dreyer, X. Maître, A. Maali, C. Wunderlich, J. M. Raimond, S. Haroche: Phys. Rev. Letters **77**, 4887 (1996)
10. M. Gell-Mann, J. B. Hartle: Phys. Rev. **D 47**, 3345 (1993)
11. R. Omnès: J. Stat. Phys. **57**, 357 (1989) J. Math. Phys. **38**, 697 (1997)
12. K. Hepp: Helv. Phys. Acta **45**, 237 (1992)
13. S. Godstein, D. Page: Phys. Rev. Lett. **74**, 3715 (1995)
14. G. Nistico: Assigning probabilities to quantum histories. Preprint 10/98, Departimento di Mathematica, Universita de Calabria, Arca Vacata di Rende, Italia (1998)
15. B. d'Espagnat: J. Stat. Phys. **56**, 747 (1989)
16. R. Omnès: J. Stat. Phys. **62**, 841 (1991)
17. F. Dowker, A. Kent: J. Stat. Phys. **82**, 1575 (1996)
18. Yu. V. Egorov: Uspehi Mat. Nauk. **24**, 5, 235 (1969) see L. Hörmander: *The analysis of partial differential operators* (Springer, Berlin 1985)
19. E. Borel: *Valeur pratique et philosophie des probabilités* (Gauthier-Villars, Paris 1937)

Space Time and Probability

Simon W. Saunders

Sub-Faculty of Philosophy, University of Oxford, 10 Merton St., Oxford OX1 4JJ, UK

Abstract. Within the consistent history formalism we show how non-epistemic probability can enter in Everett's Interpretation within a relativistic context.

1 Introduction

Special relativity is most naturally formulated as a theory of spacetime geometry, but within the spacetime framework probability appears to be a purely epistemic notion. It is possible that progress can be made with rather different approaches – covariant stochastic equations, in particular – but the results to date are not encouraging. However, it seems a non-epistemic notion of probability can be made out in Minkowski space on Everett's terms.

I shall work throughout with the consistent histories formalism. I shall start with a conservative interpretation, and then go on to Everett's.

2 Probability in Consistent Histories

In the consistent histories approach histories are represented by ordered products of Heisenberg-picture projection operators, of the form

$$C(\underline{\alpha}(n,-m)) = P_{\alpha_n}(t_n)...P_{\alpha_1}(t_1)P_{\alpha_0}(t_0)P_{\alpha_{-1}}(t_{-1})...P_{\alpha_{-m}}(t_{-m}). \qquad (1)$$

Here $\underline{\alpha}(n,-m)$ is the ordered sequence of variables $\langle\alpha_n, ...\alpha_{-m}\rangle$; its values are *histories.* (As is usual, variables will on occasion stand for values as well. I shall also write $\underline{\alpha}, \underline{\alpha}'$, etc. where beginning and end points of histories do not need to be made explicit.) Each variable α_k, where $k \in \{n, ... -m\}$, ranges over subspaces of Hilbert space as defined by a resolution of the identity, i.e. a pairwise disjoint set of projections $\{P_{\alpha_k}(t_k)\}$ which sum to the identity. Histories, therefore, are ordered sequences of sub-spaces of Hilbert space. The question of which set of projections is to be selected at each time is a version of *the preferred basis problem* of quantum mechanics. I shall come back to this presently; for the time being, for the sake of concreteness, suppose the spectrum of each projection is a subset of the configuration space of the system (call it an *outcome*). The probability of the history $\underline{\alpha}(n,-m)$ given state ρ (which may be pure or mixed) is:

$$\Pr(\underline{\alpha}(n,-m)) = \mathrm{tr}(C(\underline{\alpha}(n,-m))\rho C(\underline{\alpha}(n,-m))^*). \qquad (2)$$

This quantity can be obtained by repeated application of Luders' rule, supposing the projections of (1) to be measured sequentially, conditionalizing the state on the outcome in each case.

The conditional probability of history $\underline{\alpha}(n,1)$, given history $\underline{\alpha}(0,-m)$ ("future outcomes conditional on the present and the past") – assuming the latter has non-zero probability – is:

$$\Pr(\underline{\alpha}(n,1)|\underline{\alpha}(0,-m)) = \frac{\operatorname{tr}(C(\underline{\alpha}(n,1))C(\underline{\alpha}(0,-m))\rho C(\underline{\alpha}(0,-m))^* C(\underline{\alpha}(n,1))^*)}{\operatorname{tr}(C(\underline{\alpha}(0,-m))\rho C(\underline{\alpha}(0,-m))^*)} \tag{3}$$

These quantities are real numbers in the interval $[0,1]$. The sum of the probabilities for all such histories (holding the conditions fixed) is equal to one. Neither property holds for the retrospective conditionals, of the form:

$$\Pr(\underline{\alpha}(-1,-m)|\underline{\alpha}(0)) = \frac{\operatorname{tr}(C(\underline{\alpha}(0))C(\underline{\alpha}(-1,-m))\rho C(\underline{\alpha}(-1,-m))^* C(\underline{\alpha}(0))^*)}{\operatorname{tr}(C(\underline{\alpha}(0))\rho C(\underline{\alpha}(0))^*)} \tag{4}$$

("past outcomes conditional on the present"). A condition that ensures that these too are correctly normalized and sum to one is the so called *consistency condition* (also called the *weak decoherence condition*) [5,10]:

$$\underline{\alpha} \neq \underline{\alpha}' \Rightarrow \operatorname{tr}(C(\underline{\alpha})\rho C(\underline{\alpha}')^*) + \operatorname{tr}(C(\underline{\alpha}')\rho C(\underline{\alpha})^*) = 0. \tag{5}$$

Consistency is a restriction on the preferred basis. As it stands it is a weak constraint [2]; it is more stringent if it is to hold on variation of the state. Certain variables, and associated spectral decompositions – integrals of densities that obey local conservation laws, for example – habitually decohere [6].

The consistency condition is not needed to ensure normalization and additivity of probability for histories in the predictive case. We can dispense with it in the retrodictive case as well; we need only replace the denominator in (4) by:

$$\sum_{\langle \alpha_{-1} \dots \alpha_{-m} \rangle} \operatorname{tr}(C(\underline{\alpha}(0,-m))\rho C(\underline{\alpha}(0,-m))^*). \tag{6}$$

But this strategy is seriously deficient when we consider the algebraic structure which histories inherit from their definition. For example – the one we have at the back of our minds – consider outcomes as subsets of configuration space. As such they inherit the structure of a Boolean algebra. One outcome can be contained in another by set inclusion, and this generalizes naturally to histories:

$$\alpha_n \subseteq \alpha'_n, \dots, \alpha_1 \subseteq \alpha'_1 \Longleftrightarrow \underline{\alpha}(n,-m)) \subseteq \underline{\alpha}'(n,-m). \tag{7}$$

The Boolean operations of intersection and union extend to histories as well. If a probability measure on this space of histories is to respect this Boolean algebra, then:

$$\Pr(\underline{\alpha} \cup \underline{\alpha}') = \Pr(\underline{\alpha}) + \Pr(\underline{\alpha}') - \Pr(\underline{\alpha} \cap \underline{\alpha}'). \tag{8}$$

This condition is not in general satisfied when probabilities are defined by (1, 2); nor does it follow on introducing normalization factors such as (6). Using (1) and (2), from (8) one obtains (5); equation (8) is in fact a form of the consistency condition. It ensures that the probability measure over histories respects the natural set-theoretic relations that follow from coarse-graining.

In physical terms, an outcome consists in values of certain variables (in our case, configuration space variables) having values in designated intervals of reals. On partitioning these intervals one obtains new, finer-grained outcomes, and new, finer-grained histories. The sums of probabilities for non-intersecting histories should equal the probability of their union; this is the consistency condition. The analogous condition is automatically satisfied by histories in the pilot-wave theory, where – because deterministic – it is equivalent to an additivity requirement for probabilities of disjoint subsets of configuration space at a single time.

Consider now the interpretation of probability. Although the notation for histories, including the time t_0, is suggestive, it does not in fact imply that "the present" is in any way preferred; it only reflects, let us say, our particular location in a given history, the one which is actual. We suppose it is a consistent history. We suppose, moreover, that although the available data is necessarily approximate, telling us that variables take values in certain intervals, there is all the same a unique, maximally fine-grained, history of events – the one which actually occurs. We proceed to assign probabilities for all such fine-grained histories, conditional on the coarse-grained history fixed by the available data. The conditional probabilities of (3) and (4) are special cases of them. But if the actual history is maximally fine-grained – no matter that we do not know what it is – then the probability of every possible outcome at every time, conditional on the actual history, is either zero or one. The only non-trivial notion of probability available can only concern incompletely specified histories, histories which can be further fine-grained (whilst still satisfying the consistency condition). Probability is epistemic.

This framework as it stands cannot be cast in relativistic form, but there is a near neighbour to it, where the preferred basis concerns subsets of the spectra of self-adjoint quantities built out of algebras of local fields. The various measures and the consistency condition can in turn be expressed in terms of path integrals [7]. There seems to be no fundamental obstacle to extending this notion of spacetime probability to relativistic quantum theory. But the probabilities arrived at in this way are all epistemic.

3 Probability in the Everett Interpretation

There is no non-epistemic notion of probability consistent with relativity theory because relativity requires the tenseless spacetime perspective, and that in turn forces an epistemic notion of probability.

There has been plenty of debate in the philosophy literature about this argument [12,9,16]; here I wish only to show that there is certainly a loophole in it – if one is prepared to follow the logic of the treatment of tense, in moving to the spacetime perspective, to include the treatment of the determinate and indeterminate as well.

Suppose as before that the present is not in any way privileged – that the question of what is "now", like the question of what is "here", is simply a matter of where one happens to be located (in space-time). Now extend this analysis to "determinateness". We are to view these terms as what philosophers call *indexicals*, terms whose reference depends on the context of utterance. But in conformity with relativity, suppose that such contexts are ineluctably *local*; what is determinate is in the first instance is what is here and now – and, derivative on this, what has probability one relative to the here and now. If we are to calculate these probabilities using (3), (4) (and supposing our history space is in fact a quasiclassical domain [4]), we will obtain pretty much the same results as using quantum mechanics as standardly interpreted: events in the past which would have left a record in the present will be determinate; events of a similar kind, but in the future, will in general be indeterminate; events remote from the here and now will likewise, in general, be indeterminate This, typically, is our determinate vicinity. So one would expect on general physical grounds.

We arrive, given the preferred basis, not at a space of histories, but at a space of vicinities. They are all of them possibilities; the question is which of them are realized. But it would be quite impossible to suppose that only one of them is; that would be akin to solipsism, an egocentric or at best an anthropocentric view of reality. No more can there be only a single "here". One might suppose that there is only a single "now"; there are those who believe that the past and future are not real, or not in the sense that "the now" is real. But it had better be a global, universal now, or else there is some local *spatial* vicinity which has this special status as well – and we are back to the conflict with special relativity (according to which there *is* no privileged global notion of "now"). In non-relativistic physics we have a half-way house, in which moments but not places can be singled out as real; it is not available in the relativistic case.

We might link up these possible local vicinities, in spacelike and timelike directions, and thereby arrive at the set of all possible histories. We would be led back, that is to say, to the framework previously considered, the consistent histories formalism – and to the epistemic notion of probability already considered. The only remaining alternative is to suppose that *all* possible local states of affairs exist. This I take to be Everett's approach. All that there is is the set of vicinities, with all the conditional probabilities defined between them. In physics we do not only take up an *atemporal* perspective, situated at no particular moment in time, we take up an *acontingent* one, independent of any contingent event too. But it is a *local* perspective. It says nothing about which history a particular vicinity is part of; indeed, one and the same vicinity will be part of many distinct histories – depending on the probabilities. There is no hint of the epistemic notion of probability that we had before.

We should be clear on the difference between this and a *many histories* interpretation. Reifying all the histories of a given history space does not materially change the situation regarding the interpretation of probability. Probability remains as before, epistemic. The actual history is the one in which one happens to be; probability, as before, concerns the measure of histories consistent with a given coarse-grained description. It makes no difference if they are fictitious or real.

What is so different about the Everett approach? It is that there is no univocal criterion for identity over time. There is no fact of the matter as to what a given actuality will later turn into. Transition probabilities can be defined as before ((3, 4), where the outcomes now concern local events), but there is no further criterion of identity over time. This criterion is determinate – one has a unique future event – only insofar as the probability for that event is one (in which case it is already part of the vicinity, according to our earlier definition).

It is has been objected by many that the notion of probability is incoherent in the Everett interpretation [8], but one can hardly insist that for the notion of probability to make any sense the future must already be settled. The approach is moreover much less extravagant – and stays within the bounds of what can be locally defined – than the many histories approach, according to which there are vast numbers of qualitatively identical local vicinities, a different one for every distinct history that can accommodate it (a point well-known in the philosophy literature [1]).

But here I am concerned with the intelligibility of the approach, not its plausibility. Does probability make any sense on its terms? In every other solution to the problem of measurement, including many-histories, the future is univocal. There is a unique fact of the matter as what will actually be. It is by no means clear that this statement can be coherently denied, or that any meaningful notion of probability can be made out in its absence.

It is not a conceptual necessity, however. It is obviously unnecessary in the case of ordinary objects. Whether the Ship of Theseus, parts of which are steadily replaced over time, remains really the same, or whether it is the ship built from its parts that is really the original, has been a worry for philosophers for millennia. No one else has been much concerned by it. If we have any very pressing *a priori* convictions on this, it seems they come into play at the level of *personal* identify; that there must be a unequivocal notion of identity when we come to *human* beings. But this too has been the subject of a long and inconclusive debate in philosophy.

The problem can be starkly posed [14]. The human brain has remarkable bilateral symmetry. It can indeed be divided – an operation known as a commissurotomy – without obvious deleterious effects (although not as far as the brain stem and the rectilinear formation). But as shown by a series of classic experiments [15] the two hemispheres can be made to operate as distinct and non-communicating centres of consciousness. It is a stretch to take the further step – and to consider the two hemispheres as physically separated altogether, so that they may function as separate persons – but it is hard to see why objections on the grounds of medical feasibility should be germane to the conceptual difficulty. So let us take this further step. In the process, suppose too that there is rather greater functional symmetry between the two hemispheres than is in fact typical. It follows that following the process of fision there are two perfectly normal persons each with equal right – and claiming that right – to count themselves the same as the person prior to the division. Prior to division, it seems there can be no fact of the matter as to which of the two one is going to be.

How are we to understand this? One can hardly expect to have some sort of shared identity, that the two persons that result will somehow think in tandem. It is unreasonable to expect death – each of the two that result will certainly *say* that they have survived; one would be inclined to call it survival if only a single hemisphere were to remain. So what does one expect? On any behavioral criterion, the situation is exactly the same as a probabilistic event with two possible outcomes. We can speak of probability, even though there is no fact of the matter as to what one will be.

This is a distinctively philosophical thought experiment, but I do not think it differs fundamentally from physical thought experiments. Unlike many of the latter, it may even turn out to be practically possible. And I do not see why the concept of unequivocal identity over time should be retained as sacrosanct, when so many other basic concepts of space and time have been so radically revised in physical science.

There remain of course other difficulties with Everett's ideas – the preferred basis problem among them. This problem is particularly severe for a one-history interpretation of history space, where the ontology – the question of what exists – depends on the answer to it. On reflection, for the same reason, it is just as bad in a many-histories interpretation. It is ameliorated but not removed in the Everett approach, where the preferred basis does not determine what ultimately exists, but only the constitution of particular observers – of which vicinities in the universal state are habitable [17,13]; but this is a much larger question than the one that I have been concerned with.

References

1. Albert, D. and B. Loewer [1988] 'Interpreting the Many-Worlds Interpretation', *Synthese* **77**, 195-213.
2. Dowker, F., and A. Kent [1996] 'On the Consistent Histories Approach to Quantum Mechanics', *Journal of Statistical Physics* **82**, 1575-1646
3. Everett III, H. [1957] 'Relative State Formulation of Quantum Mechanics', *Reviews of Modern Physics* **29**, 454-62.
4. Gell-Mann, M. and J.B. Hartle [1993] 'Classical Equations for Quantum Systems', *Physical Review* **D47**, 3345-3382
5. Griffiths, R. [1984] 'Consistent Histories and the Interpretation of Quantum Mechanics', *Journal of Statistical Physics*, **36**, 219-72.
6. Halliwell, J. [1998] 'Decoherent Histories and Hydrodynamical Equations', *Physical Review D***58,** 105015
7. Hartle, J. [1989] 'The Quantum Mechanics of Cosmology', in *Quantum Cosmology and Baby Universes: Proceedings of the 1989 Jerusalem Winter School for Theoretical Physics*, ed. S. Coleman, J. Hartle, T. Piran, and S. Weinberg, World Scientific, Singapore, 1991, pp.65-157.

8. Loewer, B.: [1996] 'Comment on Lockwood', *British Journal for the Philosophy of Science*, **47**, 229-32.
9. Maxwell, N. [1985] 'Are Probabilism and Special Relativity Incompatible?', *Philosophy of Science*, **52**, 23-43.
10. Omnes, R.: [1988] *Journal of Statistical Physics*, **53,** 933.
11. Pearle, P. [1990] 'Towards a Relativistic Theory of Statevector Reduction', in A. Miller, ed., *Sixty-Two Years of Uncertainty*, Plenum Press, New York, p.193-214.
12. Putnam, H. [1967] 'Time and Physical Geometry', *Journal of Philosophy*, **64**, 240-47, reprinted in Philosophical Papers, Vol. 1, Cambridge University Press, Cambridge, 1975, 198-205.
13. Saunders, S. [1993], 'Decoherence and Evolutionary Adaptation', *Physics Letters* **A184**, 1-5.
14. Saunders, S. [1998] 'Time, Quantum Mechanics, and Probability', *Synthese*, **114**, p.405-44, 1998.
15. Sperry, R. [1982] 'Some Effects of Disconnecting the Cerebral Hemispheres', *Science*, **217**, 1223-26.
16. Stein, H. [1991] 'On Relativity Theory and the Openness of the Future', *Philosophy of Science*, **58**, 147-67.
17. Zurek, W. [1994] 'Preferred States, Predictability, Classicality, and the Environment-Induced Decoherence', in *The Physical Origins of Time Asymmetry*, J.J. Halliwell, J. Perez-Mercader and W.H. Zurek, eds., Cambridge University Press, Cambridge.

Hidden Variables, Statistical Mechanics and the Early Universe

Antony Valentini

Present address: Theoretical Physics, Blackett Laboratory, Imperial College, London SW7 2BZ, England, and Center for Gravitational Physics and Geometry, Department of Physics, The Pennsylvania State University, University Park, PA 16802, USA.
Permanent address: Augustus College, 14 Augustus Road, London SW19 6LN, England.

Abstract. One of the central mysteries of quantum theory is that it seems to be fundamentally nonlocal – and yet the nonlocality cannot be used for practical signalling at a distance. The consistency of modern physics seems to depend on a 'conspiracy', in which nonlocality is hidden by quantum equilibrium noise . It is as if there is an underlying nonlocality which we are unable to control because of the statistical character of quantum events. I explore the possibility of quantum nonequilibrium for Hidden Variables Theories like the pilot-wave theory of de Broglie and Bohm in the context of nonlocality.

1 Introduction

One of the central mysteries of quantum theory is that it seems to be fundamentally nonlocal – and yet the nonlocality cannot be used for practical signalling at a distance. As argued elsewhere [11–14], the consistency of modern physics seems to depend on a 'conspiracy', in which nonlocality is hidden by quantum noise. It is as if there is an underlying nonlocality which we are unable to control because of the statistical character of quantum events.

A natural explanation for this peculiar state of affairs is the hypothesis that quantum probability distributions are not fundamental, but merely represent a special state of equilibrium in which nonlocality happens to be masked by statistical noise [11–14]. In quantum theory a system with wavefunction ψ has an associated probability distribution given by the Born rule $\rho = |\psi|^2$. This is usually regarded as a fundamental law. If instead we regard $\rho = |\psi|^2$ as analogous to, say, the Maxwell distribution of molecular speeds for a classical gas in thermal equilibrium, the above 'conspiracy' takes on a different light: it is seen to be an accidental, contingent feature of the 'quantum equilibrium' distribution $\rho = |\psi|^2$. In a universe that is in (global) thermal equilibrium, it is impossible to convert heat into work; this is the classical thermodynamic heat death. In a universe that is everywhere in quantum equilibrium, it is impossible to use nonlocality for signalling; this is the 'subquantum heat death' which, we claim, has actually occurred in our universe.

The pilot-wave theory of de Broglie and Bohm [4,2,3] offers a concrete model of this scenario.[1] A system with wavefunction ψ is assumed to have a definite configuration $x(t)$ at all times whose velocity is determined by the de Broglie guidance equation $\pi = \partial S/\partial x$, where π is the canonical momentum and S is the phase of ψ (given locally by $S = \hbar \operatorname{Im} \ln \psi$). The wavefunction ψ is interpreted as an objective 'guiding field' in configuration space, and satisfies the usual Schrödinger equation.[2]

At the fundamental hidden-variable level, pilot-wave theory is nonlocal. For example, for two entangled particles A and B the wavefunction $\psi(x_A, x_B, t)$ has a non-separable phase $S(x_A, x_B, t)$ and the velocity $\mathrm{d}x_A/\mathrm{d}t = \nabla_A S(x_A, x_B, t)/m$ of particle A depends instantaneously on the position x_B of particle B.[3] However at the *quantum* level, where one considers an ensemble with distribution $\rho(x_A, x_B, t) = |\psi(x_A, x_B, t)|^2$, operations at B have no statistical effect at A: as is well known, quantum entanglement cannot be used for signalling at a distance.

It is worth emphasising that this 'washing out' of nonlocality by statistical noise is peculiar to the distribution $\rho = |\psi|^2$. If one considers an ensemble of entangled particles at $t = 0$ with distribution $\rho_0(x_A, x_B) \neq |\psi_0(x_A, x_B)|^2$, it may be shown by explicit calculation that changing the Hamiltonian at B induces an instantaneous change in the marginal distribution $\rho_A(x_A, t) \equiv \int \mathrm{d}x_B\, \rho(x_A, x_B, t)$ at A. For a specific example it was found that a sudden change $\hat{H}_B \to \hat{H}'_B$ at B – say a change in potential – leads after a short time ε to a change $\Delta\rho_A \equiv \rho_A(x_A, \varepsilon) - \rho_A(x_A, 0)$ at A given by [12]

$$\Delta\rho_A = -\frac{\varepsilon^2}{4m}\frac{\partial}{\partial x_A}\left(a(x_A)\int \mathrm{d}x_B\ b(x_B)\frac{\rho_0(x_A, x_B) - |\psi_0(x_A, x_B)|^2}{|\psi_0(x_A, x_B)|^2}\right) \tag{1}$$

(Here $a(x_A)$ depends on ψ_0; $b(x_B)$ also depends on $\hat{H}'_B$ and is zero if $\hat{H}'_B = \hat{H}_B$.)

For nonequilibrium ensembles $\rho_0 \neq |\psi_0|^2$, there are genuine *instantaneous signals* at the statistical level. This has recently been shown to be a general feature, independent of pilot-wave theory. Any deterministic hidden-variables theory that reproduces quantum theory leads, for hypothetical nonequilibrium ensembles, to instantaneous signals at the statistical level [17].

Note that our inability to see the trajectories – or hidden variables – is also a contingent feature of equilibrium: the uncertainty principle holds if and only if

[1] Note that in 1927 at the Fifth Solvay Congress de Broglie proposed the full pilot-wave dynamics in configuration space for a many-body system, not just the one-body theory; and unlike Bohm's reformulation of 1952, de Broglie's new approach to dynamics in the 1920s was always based on velocities [16]. The proceedings of the Fifth Solvay Congress are being translated into English [1].

[2] Pilot-wave dynamics may be applied to fields as well as particles – even to fermion fields [13,14,16] and non-Abelian gauge theories [16]. Note that the natural spacetime structure associated with this velocity-based dynamics is Aristotelian spacetime $E \times E^3$, which has an inbuilt natural state of rest [15].

[3] It is being assumed here that the position of particle B could in some sense be changed while keeping the wavefunction fixed.

$\rho = |\psi|^2$ [12].[4] Heuristically, it is natural to compare our limitations with those of a Maxwell demon in thermal equilibrium with a gas, whose attempts to sort fast and slow molecules fail. The common objection to hidden variables – that their detailed behaviour can never be observed, making their existence doubtful – is seen to be misguided: for the theory cannot be blamed if we happen to live in a state of statistical equilibrium that masks the underlying details. There is no reason why nonequilibrium could not exist in the remote past or in distant regions of the universe [13], in which case the details of the 'hidden-variable level' would not be hidden at all.

Our central theme, then, is the subquantum 'heat death of the universe' at the hidden-variable level. How does one account for the quantum noise – encapsulated by the Born rule $\rho = |\psi|^2$ – that pervades our observed universe at the present time?

2 'Empirical' Approach to Statistical Mechanics

It will be shown how one may set up an analogue of classical statistical mechanics, based on pilot-wave dynamics in configuration space [11,13,16]. We shall focus in particular on an analogue of the classical coarse-graining approach. But first, we shall address some foundational issues that were not considered in the author's original papers.

In *what sense* might one 'explain' the probability distribution $\rho = |\psi|^2$, in the context of a deterministic (and time-reversal invariant) pilot-wave theory?

Say the Moon is now at position P (in phase space) at time t. This might be 'explained' – using Newton's laws – by the fact that the Moon was (earlier) at position Q at time t_0; and if t_0 is in the remote past, one would in practice *deduce* that the Moon must have been at Q at time t_0, from the observed position P today. That one has had to deduce the past (Q) from the present (P) would not change our physical intuition that the Moon may reasonably be said to be at P now 'because' it was at Q at t_0.

This example may seem remote from statistical mechanics. However, many workers would object – in the author's opinion wrongly – that Q at t_0 is a mere deduction, and *cannot* be regarded as a satisfactory explanation for P today. The argument would be that one would have a satisfactory explanation only if it could be shown that *all* – or at least 'most' – possible positions of the Moon at t_0 evolve into the Moon being at P today. Only then would one have an 'explanation' for the present, as opposed to a mere deduction about the past.

If this seems an unfair portrayal of what many workers in the foundations of statistical mechanics would claim, then consider what is attempted in that field. The 'Holy Grail' has always been to explain relaxation to thermal equilibrium by showing that all or 'most' initial states do so. And the great stumbling block has

[4] Note, however, that the approximately classical functioning of the experimenter also plays a role here. A subquantum demon – an automaton that functions at the hidden-variable level – would see subquantum trajectories even for an equilibrium ensemble [16].

always been that, by time-reversal invariance, for every initial state that evolves towards equilibrium there is another that evolves away from it. One must then place restrictions on initial states, such as an absence of correlations, or a lack of fine-grained microstructure, and so on, restrictions that are violated by the unwanted 'time-reversed', entropy-decreasing states. One then tries to argue that the restricted set of initial states is 'natural', or that 'most' initial states – with respect to some suitable measure – satisfy the required conditions. As is well known, this program is as controversial today as it was a century ago.

To avoid such controversy one might adopt a more modest – and more realistic – 'empirical' approach [14,16]. Going back to the example of the Moon, one would try to deduce – or perhaps guess – where the Moon must have been before to explain its position today. In the same sense, for a box of gas evolving to thermal equilibrium, one should try to deduce or guess what the initial (micro-)state must have been like to yield the observed behaviour – without trying (in vain) to show that all or 'most' initial states would do so. Of course, in the case of a gas there will be a whole class of microstates yielding the required behaviour; one will not attempt to deduce the exact initial state uniquely (unlike in the case of the Moon). Further, because there are so many variables one resorts to statistical methods. Thus, one tries to construct a class of initial conditions that yields the observed behaviour, and one tries to understand the evolution of that class towards a unique equilibrium state on the basis of a general mechanism (without having to solve the exact equations of motion).

In the case at hand, what is to be explained is the observation of equilibrium today to within a certain experimental accuracy, $\rho = |\psi|^2 \pm \varepsilon$. For example, for a large number of independent Hydrogen atoms each with ground-state wavefunction ψ_{100}, one might measure the distribution of electron positions (with respect to the nuclei, in a nonrelativistic model) as accurately as current technology allows, and obtain a result $\rho = |\psi_{100}|^2 \pm \varepsilon$. The initial or earlier conditions of the universe must have been such as to reproduce this result today.

One possible initial condition is just equilibrium itself (which is preserved by the equations of motion). That is, the universe could have started in equilibrium, leading to $\rho = |\psi|^2$ exactly today. But initial equilibrium is only one possibility among (uncountably) many: even without any proofs or assumptions about relaxation $\rho \to |\psi|^2$, it is clear on grounds of continuity alone that if initial equilibrium evolves to equilibrium today then an infinite class of initial states sufficiently close to equilibrium must evolve to $\rho = |\psi|^2 \pm \varepsilon$ today; and given the violence of the early universe, one expects there to be a large class of initial states *far* from equilibrium that evolve to $\rho = |\psi|^2 \pm \varepsilon$ today as well.

What sort of past conditions could have evolved to $\rho = |\psi|^2 \pm \varepsilon$ today? Below, we shall hypothesise a class of possible initial (nonequilibrium) states and give a general mechanism explaining their approach to equilibrium, as is done in classical statistical mechanics.

It is not true, of course, that all initial states evolve towards equilibrium; time-reversal invariance forbids. Nor does it need to be true: 'bad' initial conditions are ruled out on *empirical*, observational grounds.

3 Subquantum H-Theorem

There are many approaches to classical statistical mechanics. Here we focus on a pilot-wave analogue of the classical coarse-graining H-theorem [11]. This is not because we believe coarse-graining to be the best approach; it has advantages and disadvantages, like the others.

For a classical isolated system, both the probability density p and the volume element $\mathrm{d}\Omega$ (on *phase* space) are preserved along trajectories. The classical H-function $H_{\text{class}} = \int \mathrm{d}\Omega \; p \ln p$ is constant in time. If we replace the fine-grained p by the coarse-grained $\overline{p}$ and assume that $\overline{p}_0 = p_0$ at $t = 0$, then $\overline{H}_{\text{class}}(t) \leq \overline{H}_{\text{class}}(0)$ – the classical coarse-graining H-theorem of the Ehrenfests. The decrease of $\overline{H}_{\text{class}}$ corresponds to the formation of structure in p and the consequent approach of $\overline{p}$ to uniformity. (See for example [10].)

Consider now an ensemble of complicated many-body systems, each with wavefunction Ψ, the configurations X distributed with a probability density P. The gradient ∇S of the phase of Ψ determines a velocity field $\dot{X}$ in configuration space. (For a low-energy system of N particles of mass m, we have 3-vector velocities $\dot{X}_i = \nabla_i S/m$ where $i = 1, 2, ..., N$.) The continuity equations

$$\frac{\partial P}{\partial t} + \nabla \cdot (\dot{X} P) = 0 \tag{2}$$

(which holds by definition of P) and

$$\frac{\partial |\Psi|^2}{\partial t} + \nabla \cdot (\dot{X} |\Psi|^2) = 0 \tag{3}$$

(which follows from the Schrödinger equation) imply that the ratio $f \equiv P/|\Psi|^2$ is preserved along trajectories: $\mathrm{d}f/\mathrm{d}t = 0$, where $\mathrm{d}/\mathrm{d}t = \partial/\partial t + \dot{X} \cdot \nabla$. Further, like $\mathrm{d}\Omega$ classically, the element $|\Psi|^2 dX$ is preserved along trajectories too (where $\mathrm{d}X$ is the volume element in configuration space). This suggests replacing $p \to f$ and $\mathrm{d}\Omega \to |\Psi|^2 \mathrm{d}X$ in H_{class}, yielding the *subquantum H-function*

$$H = \int \mathrm{d}X \; P \ln(P/|\Psi|^2) \tag{4}$$

This is in fact the relative negentropy of P with respect to $|\Psi|^2$.

The above continuity equations now imply that $H = H(t)$ is constant: $\mathrm{d}H/\mathrm{d}t = 0$, as in the classical case. But if one divides configuration space into cells of volume δV and averages P and $|\Psi|^2$ over the cells, yielding coarse-grained values $\overline{P}$ and $\overline{|\Psi|^2}$, one may define a coarse-grained H-function

$$\overline{H} = \int \mathrm{d}X \; \overline{P} \ln(\overline{P}/\overline{|\Psi|^2}) \tag{5}$$

If we assume the initial state has 'no fine-grained microstructure' at $t = 0$,

$$\overline{P}_0 = P_0 \; , \quad \overline{|\Psi_0|^2} = |\Psi_0|^2 \tag{6}$$

then it may be shown that

$$\overline{H}_0 - \overline{H}(t) = \int \mathrm{d}X \; |\Psi|^2 \left(f \ln(f/\tilde{f}) + \tilde{f} - f \right) \geq 0 \tag{7}$$

(where $\tilde{f} \equiv \overline{P}/\overline{|\Psi|^2}$), so that

$$\overline{H}(t) \leq \overline{H}_0 \tag{8}$$

for all t [11].

It is instructive to examine the time derivatives of $\overline{H}(t)$ at $t = 0$. Use of the continuity equations shows that $\left(\mathrm{d}\overline{H}/\mathrm{d}t\right)_0 = 0$ and [13]

$$\left(\frac{\mathrm{d}^2\overline{H}}{\mathrm{d}t^2}\right)_0 = -\int \mathrm{d}X \frac{\overline{|\Psi_0|^2}}{\overline{f_0}} \left(\overline{(\dot{X}_0 \cdot \nabla f_0)^2} - \overline{(\dot{X}_0 \cdot \nabla f_0)}^2 \right) \leq 0 \tag{9}$$

The quantity in brackets is just the (non-negative) variance $\mathrm{var}_{\delta V}(\dot{X}_0 \cdot \nabla f_0)$ of $(\dot{X}_0 \cdot \nabla f_0)$ over a coarse-graining cell δV. If we suppose that $\mathrm{var}_{\delta V}(\dot{X}_0 \cdot \nabla f_0) \neq 0$ then we have the strict inequality $\overline{H}(t) < \overline{H}_0$ and $\overline{H}(t)$ *must decrease* over at least a finite time interval $(0, T)$ immediately after $t = 0$ [16].

Because $x \ln(x/y) + y - x \geq 0$ for all real x, y, the coarse-grained H-function is bounded below by zero, $\overline{H} \geq 0$; further, $\overline{H} = 0$ if and only if $\overline{P} = \overline{|\Psi|^2}$ everywhere [11]. A decrease of $\overline{H}$ towards its minimum value then corresponds to an approach of $\overline{P}$ towards $\overline{|\Psi|^2}$.

The decrease of $\overline{H}$ corresponds to a 'stirring' of the two 'fluids' P and $|\Psi|^2$ by the same velocity field $\dot{X}$, making P and $|\Psi|^2$ less distinguishable on a coarse-grained level. (This is similar to the classical Gibbs stirring of two liquids.) As in the corresponding classical case, the H-theorem gives us an insight into the mechanism whereby equilibrium is approached (for the particular class of initial conditions specified by (6)). Whether or not equilibrium is actually reached will depend on the system. It must be assumed that the initial steps towards equilibrium described by the H-theorem are actually completed in Nature, for appropriately complex systems, so that $\overline{H}(t) \to 0$ and $\overline{P} \to \overline{|\Psi|^2}$.

Given (coarse-grained) equilibrium $\overline{P} = \overline{|\Psi|^2}$ for our ensemble of many-body systems, it is straightforward to show that if a single particle is extracted from each system and prepared with wavefunction ψ, the resulting ensemble of particle positions has a (coarse-grained) distribution $\overline{\rho} = \overline{|\psi|^2}$ [11]. It is therefore to be envisaged that the equilibrium distribution seen today for ensembles of single particles arose via relaxation processes in the complex systems of which the particles were once part.[5]

4 An Estimate for the Relaxation Timescale

It is important first of all to have a rough, order-of-magnitude estimate for the timescale over which relaxation to quantum equilibrium takes place. What we

[5] For further details see [11–13,16].

have in mind here is not the time taken to reach equilibrium but the timescale over which there is a significant *approach* to equilibrium.[6]

We may define a relaxation timescale τ in terms of the rate of decrease of $\overline{H}(t)$ near $t = 0$. Because $\left(\mathrm{d}\overline{H}/\mathrm{d}t\right)_0$ vanishes one has to consider $\left(\mathrm{d}^2\overline{H}/\mathrm{d}t^2\right)_0$. Thus we define τ by $1/\tau^2 \equiv -\left(\mathrm{d}^2\overline{H}/\mathrm{d}t^2\right)_0/\overline{H}_0$ [13], where (9) gives $\left(\mathrm{d}^2\overline{H}/\mathrm{d}t^2\right)_0$ in terms of the initial state. Expanding $\dot{X}_0 \cdot \nabla f_0$ in a Taylor series within each coarse-graining cell of volume $\delta V = (\delta x)^{3N}$, it is found that $\left(\mathrm{d}^2\overline{H}/\mathrm{d}t^2\right)_0 = -I(\delta x)^2/12 + O\left((\delta x)^4\right)$ where $I \equiv \int \mathrm{d}X \; (|\Psi_0|^2/f_0)|\nabla(\dot{X}_0 \cdot \nabla f_0)|^2$ [16]. Thus

$$\tau = \frac{1}{\delta x}\sqrt{\frac{12\overline{H}_0}{I}} + O(\delta x) \tag{10}$$

For δx small compared to the lengthscale over which $\dot{X}_0 \cdot \nabla f_0$ varies, $\tau \propto 1/\delta x$. Taking $\overline{H}_0 \sim 1$ (a mild disequilibrium) and crudely estimating I one obtains

$$\tau \sim \frac{1}{\delta x}\frac{m\hbar^2}{(\Delta P_0)^3} \tag{11}$$

where ΔP_0 is the (quantum) momentum spread of Ψ_0 [16]. Given $\tau \propto 1/\delta x$ this formula may also be obtained on dimensional grounds. Note that this result is merely a crude estimate.

5 Numerical Simulations

A numerical simulation of relaxation has been performed for the simplest possible case of an ensemble of independent particles in a one-dimensional box [13]. The results were surprisingly good, given that the particles cannot move past each other. This simple model is of course an unrealistic setting for relaxation: one should really consider an ensemble of complicated systems with many degrees of freedom. Nevertheless the model is instructive.

Consider, then, a one-dimensional box with infinite barriers at $x = 0, L$ and energy eigenfunctions $\phi_n(x) = \sqrt{2/L}\sin(n\pi x/L)$ $(n = 1, 2, 3, ...)$ with eigenvalues $E_n = \frac{1}{2}(\pi n/L)^2$ (units $m = \hbar = 1$). The box contains an ensemble of independent particles, each guided by the same wavefunction $\psi(x,t)$. At $t = 0$ the wavefunction was taken to be a superposition of the first M eigenfunctions, with amplitudes of equal modulus but randomly-chosen phases θ_n:

$$\psi_0(x) = \sum_{n=1}^{M} \frac{1}{\sqrt{M}} \phi_n(x) \exp(i\theta_n)$$

The particles were taken to be uniformly distributed at $t = 0$: thus, $\rho_0(x) = 1/L$ and of course $\rho_0(x) \neq |\psi_0(x)|^2$. These initial conditions are sketched in Fig. 1, for the case $M = 10$ and $L = 100$.

[6] Cf. the scattering time of classical kinetic theory.

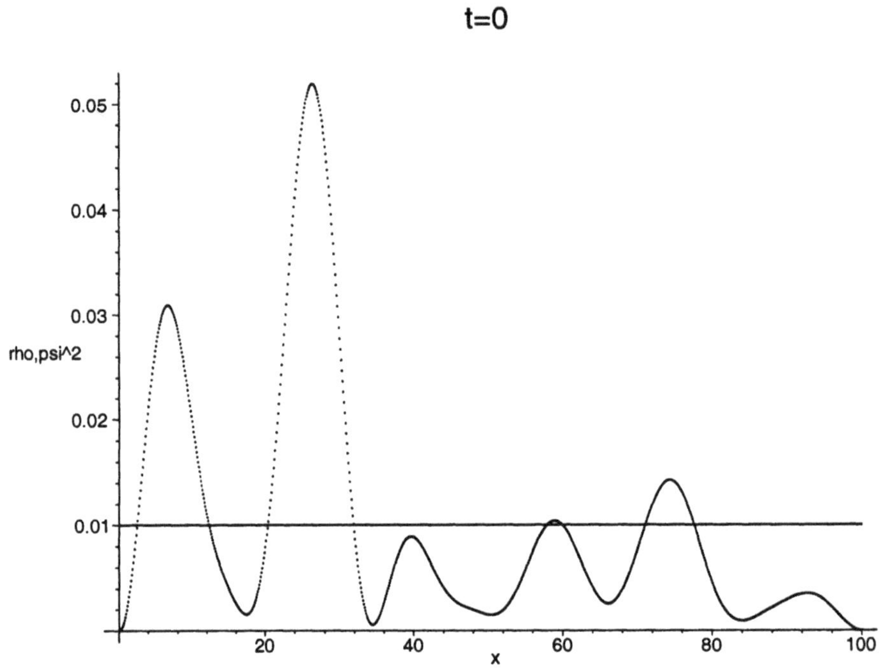

Fig. 1. Initial conditions at $t = 0$. We have plotted $|\psi|^2$ for a superposition of ten modes with amplitudes of equal modulus and random phase. The initial density ρ is taken to be uniform

By numerical calculation of the trajectories (given by $\dot{x} = (\partial/\partial x)\,\mathrm{Im}\ln\psi$), the distribution $\rho(x,t)$ at later times may be determined, while the Schrödinger evolution of $\psi(x,t)$ is just

$$\psi(x,t) = \sum_{n=1}^{M} \frac{1}{\sqrt{M}} \phi_n(x) \exp i(\theta_n - E_n t)$$

For $L = 100$, $\psi(x,t)$ is periodic in time with period $T_P = 2\pi/E_1 \approx 12{,}700$ in our units. Because the particles cannot move past each other, each trajectory must recur with period T_P (to ensure that the equilibrium distribution recurs) and so any $\rho_0(x) \neq |\psi_0(x)|^2$ recurs as well. While this simple model exhibits a strong form of recurrence, nevertheless for times smaller than the recurrence time the approach to equilibrium is significant.

An example is shown in Fig. 2, again for ten modes ($M = 10$). At $t = 120$, ρ has developed sharp peaks which coincide with the (smooth) maxima of $|\psi|^2$. An experimenter with a 'blunt' measuring device, say with resolution $\delta x = 10$,

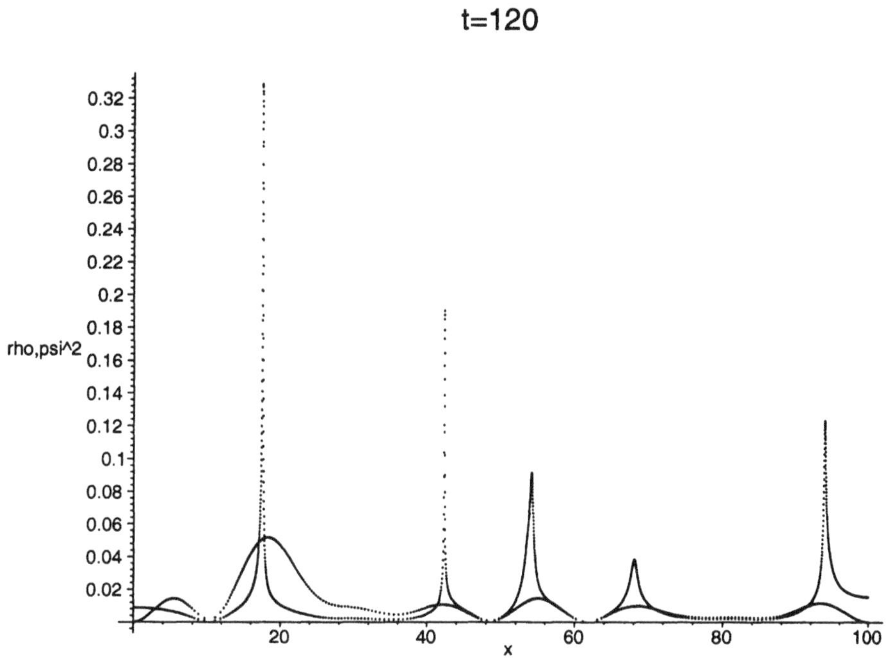

Fig. 2. At $t = 120$. There is a strong coincidence between the sharp peaks of ρ and the smooth peaks of $|\psi|^2$, indicating an approximate approach to equilibrium on a coarse-grained level

would conclude that $\overline{\rho}$ and $\overline{|\psi|^2}$ are roughly equal. The approach to equilibrium (at $t \ll T_P$) is surprisingly good for such a simple, highly constrained model.[7]

The evolution of $\overline{H}(t)$ has also been calculated[8]. An example is shown in Fig. 3, once again for $M = 10$, where $\overline{H}$ is defined with a coarse-graining length $\delta x = 1$. The curve $\overline{H} = \overline{H}(t)$ shows a strict decrease soon after $t = 0$ (as it must); thereafter it decreases steadily but not monotonically, for $t \ll T_P$. (Of course, as the recurrence time is approached, $\overline{H}$ necessarily increases as the initial conditions are restored.)

It is clear from Fig. 3 that $\overline{H}$ initially decreases on a timescale of order ~ 100. This result compares well with our crude estimate (11). For here $\Delta P_0 \approx (\pi/\sqrt{3})M/L$ and with $L = 100$ and $\delta x = 1$ our prediction is $\tau \sim 2 \times 10^5/M^3$. For $M = 10$ we have $\tau \sim 200$ – in fair agreement with the numerical calculation.

The timescale over which $\overline{H}$ decreases – which may be quantified in terms of the time $t_{5\%}$ taken for $\overline{H}$ to decrease by 5% – has been calculated (with

[7] For further details see [13,16]. Note that here the quantum timescale for the evolution of $|\psi|^2$ is $\Delta t \equiv \hbar/\Delta E \sim 70$, where ΔE is the quantum energy spread.

[8] with C. Dewdney and B. Schwentker

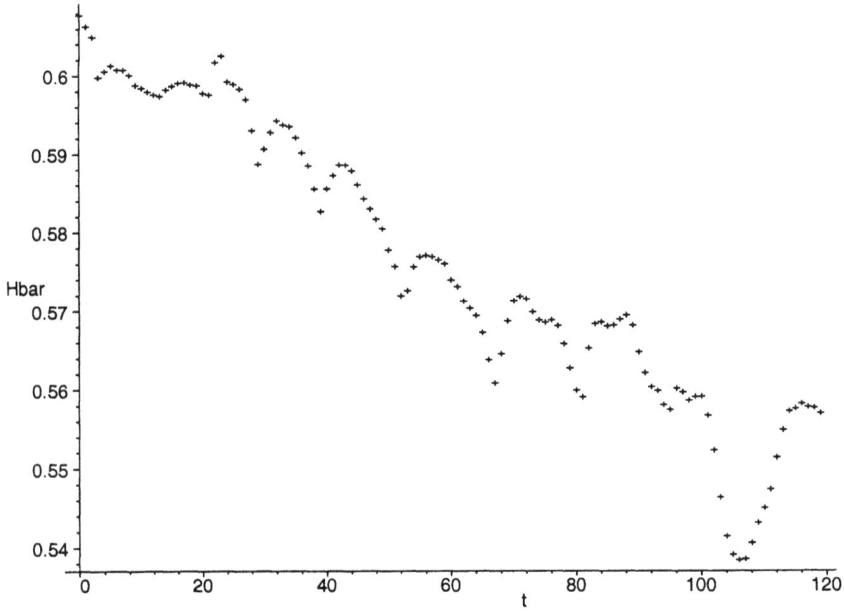

Fig. 3. Plot of the coarse-grained $\overline{H}$-function against time

C. Dewdney and B. Schwentker) for different numbers of modes, $M = 10,\ 15,$ ....., 40 (with fixed $\delta x = 1$), and for different coarse-graining lengths ranging from $\delta x = 0.2$ up to $\delta x = 2$ (with fixed $M = 20$). The results agree quite well with our predictions that $t_{5\%} \propto 1/M^3$ and $t_{5\%} \propto 1/\delta x$ [5,16].[9]

The above model serves an illustrative purpose only. The particles cannot move past each other, and the approach to equilibrium is limited. Higher-dimensional models should show a much better approach to equilibrium.

Calculations analogous to the above are currently being performed by Dewdney, Schwentker and Valentini for a two-dimensional box. Bearing in mind the generally chaotic nature of the trajectories for this system [8], an efficient relaxation to equilibrium is to be expected.

6 Comments on 'Typicality'

A completely different approach to explaining quantum equilibrium is based on a notion of 'typicality' for the initial configuration X_0^{univ} of the whole universe [6,7]. Here, if Ψ_0^{univ} is the initial wavefunction of the universe, $|\Psi_0^{\mathrm{univ}}|^2$ is taken to

[9] In the first case, upon averaging over many trials with different random initial phases, a plot of $\ln t_{5\%}$ versus $\ln M$ was found to have a slope of -2.6, compared with the predicted value of -3. In the second case, there are departures from linearity for $\delta x \gtrsim 0.5$ due to the large variation in the velocity field over some coarse-graining cells, the result $\tau \propto 1/\delta x$ being valid only for small δx.

be the 'natural measure' on the space of initial configurations. It is shown that quantum distributions for measurements on subsystems are obtained for 'almost all' $X_0^{\rm univ}$, with respect to the measure $|\Psi_0^{\rm univ}|^2$.

While Dürr *et al.* give a general proof, their approach may be illustrated by the case of a universe consisting of an ensemble of n independent subsystems (which could be complicated many-body systems, or perhaps just single particles), each with wavefunction $\psi_0(x)$. Writing $\Psi_0^{\rm univ} = \psi_0(x_1)\psi_0(x_2).....\psi_0(x_n)$ and $X_0^{\rm univ} = (x_1, x_2, x_3, ..., x_n)$, a choice of $X_0^{\rm univ}$ determines – for large n – a distribution $\rho_0(x)$ which may or may not equal $|\psi_0(x)|^2$.

Now it is true that, with respect to the measure $|\Psi_0^{\rm univ}|^2$, as $n \to \infty$ almost all configurations $X_0^{\rm univ}$ yield equilibrium $\rho_0 = |\psi_0|^2$ for the subsystems. It might then be argued that, as $n \to \infty$, disequilibrium configurations occupy a vanishingly small volume of configuration space and are therefore intrinsically unlikely. However, for the above case, with respect to the measure $|\Psi_0^{\rm univ}|^4$ almost all configurations $X_0^{\rm univ}$ correspond to the *dis*equilibrium distribution $\rho_0 = |\psi_0|^4$. This has led to charges of circularity: that an equilibrium probability density $|\Psi_0^{\rm univ}|^2$ is in effect being assumed for $X_0^{\rm univ}$; that the approach amounts to inserting quantum noise into the initial conditions themselves [14].[10]

However, there is a certain rationale to Dürr *et al.*'s approach. It has been argued (S. Goldstein, private communication) that in statistical mechanics one always rules out unwanted initial conditions in some way, usually by deeming them exceptional (or 'untypical') with respect to a specified measure. This is perfectly true. But in the author's opinion, a class of initial conditions should be selected on *empirical* grounds. Even today we know only that, for all subsystems probed so far, $\rho = |\psi|^2 \pm \varepsilon$ where ε is some experimental accuracy. There is certainly *no* reliable empirical evidence that $\rho = |\psi|^2$ exactly – or even approximately – in the very early universe, near the big bang.[11]

Note also that pilot-wave theory in exact equilibrium may never be susceptible to an experimental test of the basic law of motion (the de Broglie guidance equation); details of the trajectories may well be forever hidden from us, unless the human brain has an unexpected sensitivity to hidden variables [16]. Restricting pilot-wave theory to quantum equilibrium is as artificial as it would be to restrict classical mechanics to thermal equilibrium; and a huge amount of potentially new physics is lost thereby.

[10] Note that if the word 'typicality' is replaced by 'probability', the result of Dürr *et al.* [6,7] becomes equivalent to the 'nesting' property proved in [11], which states that an equilibrium probability for a many-body system implies equilibrium probabilities for extracted subsystems, as briefly discussed above.

[11] In fact, the horizon problem in cosmology (the uniformity of the microwave background over regions larger than the classical causal horizons of a Friedmann universe) suggests that nonequilibrium $\rho \neq |\psi|^2$ may have existed at early times, the resulting nonlocality playing a role in homogenising the early universe [12,13,16]. Inflation attempts to remove the horizon problem by changing the early expansion; but as yet there is no satisfactory inflationary theory. Early quantum disequilibrium offers an alternative approach.

Finally, Dürr *et al.* focus on initial configurations for the whole universe while in Sect. 2 we emphasised initial distributions for subsystems. One might think that the former description is more correct because in the early universe there are no independent subsystems (that is, subsystems with their own wavefunction). But even leaving aside the point that the asymptotic freedom of interactions at high energies allows one to treat particles at very early times as essentially free, the two modes of description are really interchangeable: even if there are no independent subsystems, one can imagine (theoretically) what *would* happen if (say) single particles were extracted from a gas of interacting particles, prepared with a certain wavefunction ψ, and had their positions measured. Whether the measured distribution ρ is equal to $|\psi|^2$ or not will depend on the configuration of the whole system prior to the extraction. By this (purely theoretical) device, one could characterise the total configuration as 'equilibrium' or 'nonequilibrium' [16]. Our Sect. 2 could then be rephrased in terms of configurations rather than distributions: our 'empirical' approach would amount to trying to find out what sort of initial configuration could explain the statistics observed today – and again, while an initial 'equilibrium' configuration is a possibility, there are clearly many 'nonequilibrium' configurations at $t = 0$ that explain the observations just as well.

7 The Early Universe. Suppression of Relaxation at Early Times

At first sight one might think that, even if the universe did begin in quantum nonequilibrium, relaxation $\rho \to |\psi|^2$ would take place so quickly (in the extreme violence of early times) as to erase all record of nonequilibrium ever having existed. However, closer analysis reveals a plausible scenario whereby quantum nonequilibrium could survive to the present day, for some types of particle left over from the big bang, leading to observable violations of quantum theory.

The scenario has three stages: (i) depending on how the relaxation timescale τ scales with temperature T, it is possible that relaxation will be *suppressed* at very early times by the very rapid expansion of space, in which case (ii) those particles that decouple at very early times will still be out of quantum equilibrium at the time of decoupling, after which (iii) the free spreading of particle wavefunctions *stretches* any residual nonequilibrium up to larger lengthscales that are within reach of experimental test. This scenario will now be briefly sketched, in this section and the next.[12]

In the standard hot big bang model, the early universe contains what is essentially an ideal gas of effectively massless (relativistic) particles. It is instructive to recall the standard reasoning about how thermal equilibrium between distinct particle species is achieved: the interaction timescale t_{int} – that is, the mean free time between collisions – must be smaller than the expansion timescale $t_{\text{exp}} \equiv a/\dot{a} \sim (1 \text{ sec})(1 \text{ MeV}/kT)^2$, where $a(t) \propto t^{1/2}$ is the scale factor and t is

[12] For further details see [16,18].

the time since the beginning.[13] As is well known, this condition need not always be satisfied: for instance, if the cross section $\sigma(T)$ falls off faster than $1/T$ at high temperatures then $t_{\rm int} \sim 1/nc\sigma$ – with a number density $n \sim T^3$ (in natural units where $\hbar = c = 1$) – will fall off slower than $1/T^2$ so that, at sufficiently high temperatures, $t_{\rm int} \gtrsim t_{\rm exp}$ and the particles do not thermalise. The functions $\sigma = \sigma(T)$ are determined by particle physics; once they are known, it is possible to determine the thermal history of the universe, and in particular to deduce when particle species fall out of thermal contact with each other.

Similar reasoning applies to the relaxation $\rho \to |\psi|^2$. There will be a temperature-dependent timescale $\tau = \tau(T)$ over which relaxation takes place. There will also be a *competing* effect due to the expansion of space: as space expands wavefunctions are stretched, and it is easy to see in pilot-wave theory that this results in a proportionate expansion of the disequilibrium lengthscale.

To see this, consider an initial distribution $\rho_0(x)$ with disequilibrium on a small lengthscale δ_0. We may write $\rho_0(x) = |\psi_0(x)|^2 f_0(x)$, where $\psi_0(x)$ has some width Δ_0 and $f_0(x) \neq 1$ varies over distances $\delta_0 \ll \Delta_0$ (so that coarse-graining over distances much larger than δ_0 yields an equilibrium distribution). If the wavefunction expands up to a width $\Delta(t)$, then because f is conserved along trajectories, and because particles initially separated by a distance δx_0 are later separated roughly by $\delta x(t) \sim (\Delta(t)/\Delta_0)\,\delta x_0$, it follows that deviations $f \neq 1$ occur on an expanded lengthscale $\delta(t) \sim (\Delta(t)/\Delta_0)\,\delta_0$. This is true irrespective of whether the wavefunction expands because of the expansion of space or simply because of its own natural free spreading over time.[14]

Thus relaxation $\rho \to |\psi|^2$ occurs if $\tau \lesssim t_{\rm exp}$, while it is *suppressed* if $\tau \gtrsim t_{\rm exp}$. The issue now depends on how τ scales with T. Unfortunately, there are as yet no reliable calculations that can tell us this. But we can make a crude estimate.

Given $\tau \propto 1/\delta x$, for massless particles τ may be estimated on purely dimensional grounds to be $\tau \sim (1/\delta x)\hbar^2/c(\Delta p)^2$ where Δp is the particle momentum spread. Taking $\Delta p \sim kT/c$ we have $\tau \sim (1/\delta x)\hbar^2 c/(kT)^2$ which is larger than $t_{\rm exp} \sim (1\ {\rm sec})(1\ {\rm MeV}/kT)^2$ for all T if $\delta x \lesssim 10 l_P$ (where l_P is the Planck length). However, it would be more realistic to apply our estimate for τ to a coarse-graining length δx that is relevant to the energetic processes at temperature T, the natural choice being the thermal de Broglie wavelength $\delta x \sim \hbar c/kT$ (the typical width of particle wavepackets). We then have $\tau \sim \hbar/kT$ – a plausible result, according to which relaxation is suppressed ($\tau \gtrsim t_{\rm exp}$) if $kT \gtrsim 10^{18}$ GeV$\approx 0.1 kT_P$ or $t \lesssim 10 t_P$ (where T_P and t_P are the Planck temperature and time). Our simple estimate suggests that relaxation to quantum equilibrium began one order of magnitude below the Planck temperature.

Note that in the standard Friedman expansion considered here, the rate $1/t_{\rm exp}$ at which space is expanding becomes infinite as $t \to 0$. Of course, our relaxation rate $1/\tau \propto T$ also tends to infinity as $t \to 0$ (and $T \to \infty$), but this

[13] See any standard cosmology text.

[14] This is consistent with the H-theorem, for H measures negentropy relative to $|\psi|^2$: if $|\psi|^2$ itself expands, the absolute disequilibrium lengthscale can increase without increasing H.

is offset by the expansion rate $1/t_{\text{exp}} \propto T^2$ which tends to infinity even faster. Note also that while our estimate is very crude and should not be taken too seriously, it does illustrate the key point that, if τ decreases with temperature more slowly than $1/T^2$, then relaxation will be suppressed at very early times.

It is therefore plausible that any particles that decouple soon after the Planck era will *not* have had time to reach quantum equilibrium, the extreme violence of that era being offset by the even more extreme expansion of space. And at the time of decoupling, such particles are expected to show deviations $\rho \neq |\psi|^2$ on a lengthscale $\delta x \sim \hbar c/kT_P = l_P \approx 10^{-33}$ cm.

8 Residual Disequilibrium Today. Experimental Tests

After decoupling, particle wavefunctions undergo a huge expansion – partly due to the expansion of space itself and partly due to the free spreading of the wavepacket through space. This results in a proportionate stretching of disequilibrium to much larger lengthscales, as shown above.[15]

For example, relic gravitons are believed to decouple at $kT_{\text{dec}} \sim 10^{19}$ GeV or at redshift $z_{\text{dec}} \sim T_{\text{dec}}/T_{\text{now}} \sim 10^{32}$. The subsequent expansion of space alone, by a linear factor of $\sim 10^{32}$, will stretch the disequilibrium lengthscale up to ~ 1 mm. Of course, there seems to be little hope of detecting the $\sim 1^0$ K graviton background directly in the near future – still less of testing it for violations of quantum theory. However, it is expected that there are other, more exotic particles that decoupled soon after T_P. (Supersymmetry and string theory predict a plethora of new particles at high energies.) These may be no easier to detect directly. However, some of them might decay at later times into more easily detectable particles such as photons, perhaps by annihilation $X + \overline{X} \to 2\gamma$ or by a supersymmetric decay $X \to \gamma + \tilde{\gamma}$ into a photon and a photino. Nonequilibrium for the parent particles should yield nonequilibrium for the decay products as well. Thus, we would suggest testing the Born rule $\rho = |\psi|^2$ for photons produced by the decay of exotic relic particles from the Planck era. For example, in a two-slit interference experiment, such photons might produce an interference pattern that deviates from the quantum prediction.

Experiments are under way searching for exotic relic particles supposed to make up the 'dark matter' pervading the universe. Some of these experiments involve searching for decay photons. However, the usual dark matter candidates (such as neutrinos, neutralinos, axions or gravitinos) are expected to decouple much later than t_P. Nevertheless, our knowledge of particle physics beyond the

[15] Decoupling is of course never exact. Because of small residual interactions at all times, the wavefunctions of relic particles contain tiny scattering terms that perturb the trajectories – possibly 're-mixing' the expanding disequilibrium. Calculation of the effect of scattering by a tenuous medium shows that re-mixing does *not* happen: the trajectory perturbations grow only as $t^{1/2}$ and cannot overcome the linear ($\propto t$) free expansion of the disequilibrium lengthscale [16,18]. We may therefore safely ignore such residual interactions.

standard model is so uncertain that, for all we know, there might exist an appropriate relic particle X – that decouples soon after t_P and partially accounts for dark matter. If such particles were detected, they or their decay products would be our prime candidates for particles violating the Born rule.

In the meantime one might also consider the relic photons that make up the microwave background. But why consider particles that decoupled at $t \sim 10^5$ yr – *much* later than t_P – when our estimates suggest that any disequilibrium will have been erased? One answer is that the hugely expanded wavepackets of relic particles provide an interesting test of the Born rule in extreme conditions. On general grounds (irrespective of the suggestions made here about hidden variables and early nonequilibrium), it would be worth testing quantum theory in such unusual circumstances, where quantum probabilities have spread over intergalactic distances. Relic photons decouple at $kT_{\text{dec}} \sim 1$ eV ($z_{\text{dec}} \sim 10^3$) and at decoupling their wavepackets have widths of order $\sim \hbar c/(1\ \text{eV}) \sim 10^{-5}$ cm. If the packets have (as is widely assumed) spread essentially freely at the speed of light for $\sim 10^{10}$ years then today they will have a width $\sim 10^{28}$ cm and so the packets will have expanded by $\sim 10^{33}$ (ignoring the relatively small effect of the expansion of space by a factor $\sim 10^3$). It can be argued – on general grounds, independent of pilot-wave theory – that for cosmological microwave photons the quantum probability today on lengthscales ~ 1 cm could contain traces of corrections to the Born rule which may have existed (for whatever reason) on the Planck scale at the time of decoupling [16,18].[16]

As suggested elsewhere [14], it would then be worthwhile to test quantum theory for photons from the microwave background. In a two-slit interference experiment with single relic photons, the usual quantum interference pattern – given by $\rho = |\psi|^2$ – might show an *anomalous blurring.*[17]

Generally, corrections to the Born rule for any photons from deep space would produce a number of observable anomalies. In particular, the functioning of some astronomical instruments might be affected: diffraction-grating spectrometers could produce unreliable readings, and the diffraction-limited performance of some telescopes might be impaired.[18]

9 Outlook

It is clear that much remains to be done to develop the above ideas fully. Other approaches to subquantum statistical mechanics (based for example on external perturbations) remain to be developed. Properties of the trajectories such as ergodicity and mixing (in a rigorous sense) should be investigated.

[16] Our point here is that the line of argument given in this paper has led us to propose experimental tests that are actually worthwhile in their own right, because they would probe quantum theory in new and extreme conditions.

[17] Incoming photons will have very nearly plane wavepackets, so there is no ambiguity as to what quantum theory would predict in such an experiment. Note also that it has recently become possible to detect single photons in the far-infrared region [9].

[18] For details see [16].

The process of relaxation to quantum equilibrium should also be studied further, in particular in the early universe. It remains to be seen if the relaxation timescale τ really does decrease with temperature more slowly than $1/T^2$, so that relaxation is suppressed at very early times. An experimental test of the Born rule $\rho = |\psi|^2$ for relic cosmological particles seems feasible, in particular for photons from the microwave background [16,18].

As for the issue of chance in physics, the central conclusion of this work is that, like the cosmic microwave background, *quantum noise is a remnant of the big bang.* And just as the microwave background has been found to have small nonuniformities in temperature, so the 'quantum background' – the quantum noise that pervades our universe – may have small deviations from the Born rule $\rho = |\psi|^2$, and should be probed experimentally.

Acknowledgements

I am grateful to audiences at the Universities of Portsmouth and Utrecht for their comments, to Janneke van Lith for a critical reading of section 2, and to Chris Dewdney for preparing the figures.

References

1. G. Bacciagaluppi and A. Valentini (2001), in preparation.
2. D. Bohm (1952a), A suggested interpretation of the quantum theory in terms of 'hidden' variables. I, *Physical Review* 85, 166–179.
3. D. Bohm (1952b), A suggested interpretation of the quantum theory in terms of 'hidden' variables. II, *Physical Review* 85, 180–193.
4. L. de Broglie (1928), La nouvelle dynamique des quanta, in: *Electrons et Photons*, Gauthier-Villars, Paris.
5. C. Dewdney, B. Schwentker and A. Valentini (2001), in preparation.
6. D. Dürr, S. Goldstein and N. Zanghì (1992a), Quantum equilibrium and the origin of absolute uncertainty, *Journal of Statistical Physics* 67, 843–907.
7. D. Dürr, S. Goldstein and N. Zanghì (1992b), Quantum mechanics, randomness, and deterministic reality, *Physics Letters* A 172, 6–12.
8. H. Frisk (1997), Properties of the trajectories in Bohmian mechanics, *Physics Letters* A 227, 139–142.
9. S. Komiyama, O. Astafiev, V. Antonov, T. Kutsuwa and H. Hirai (2000), A single-photon detector in the far-infrared range, *Nature* 403, 405–407.
10. R. Tolman (1938), *The Principles of Statistical Mechanics*, Oxford.
11. A. Valentini (1991a), Signal-locality, uncertainty, and the subquantum H-theorem. I, *Physics Letters* A 156, 5–11.
12. A. Valentini (1991b), Signal-locality, uncertainty, and the subquantum H-theorem. II, *Physics Letters* A 158, 1–8.
13. A. Valentini (1992), On the pilot-wave theory of classical, quantum and subquantum physics, PhD thesis, SISSA/ISAS, Trieste, Italy.
14. A. Valentini (1996), Pilot-wave theory of fields, gravitation and cosmology, in: *Bohmian Mechanics and Quantum Theory: an Appraisal*, 45–66 (eds. J.T. Cushing *et al.*), Kluwer.
15. A. Valentini (1997), On Galilean and Lorentz invariance in pilot-wave dynamics, *Physics Letters* A 228, 215–222.

16. A. Valentini (2001a), *Pilot-Wave Theory: an Alternative Approach to Modern Physics*, Springer, to be published.
17. A. Valentini (2001b), to be submitted.
18. A. Valentini (2001c), to be submitted.

Perspectives of the Dynamical Reduction Program

Gian Carlo Ghirardi

Department of Theoretical Physics, University of Trieste
International Centre for Theoretical Physics, Trieste, Italy

Abstract. I will concentrate my attention on the so called Dynamical Reduction Program (DRP) some aspects of which deserve to be stressed. In particular I will mention some relatively recent results on the subject which are probably not familiar to the audience and I will focus on some crucial points which remain open and have to be faced before the program itself can be considered as a satisfactory and *exact* (in J.S. Bell's sense [1]) alternative to the so-called orthodox interpretation.

1 Introduction

It is a great pleasure for me to take part to this round table on the foundations of quantum mechanics which, I hope, will allow the audience to get a more general picture of the foundational problems of *our best theory*. I will deal with four major points:

- Relativistic aspects of the DRP,
- Experimental tests of its predictions,
- Issues of interpretation (what is the theory about?),
- Relations with other approaches.

2 Relativistic Aspects and Related Difficulties

It seems appropriate to start the discussion of this point by recalling that J.S. Bell has concluded one of his last papers [2] with the following sentence concerning Bohmiam mechanics and the Dynamical Reduction Program:

> The big question, in my opinion, is which, if either, of these two precise pictures can be redeveloped in a Lorentz invariant way.

Let me state that I fully agree with this sentence and let me stress from the very beginning that, in spite of the many efforts which have been devoted to this matter, no fully satisfactory solution has yet been found. For clarity sake, however, I consider it essential to call attention on the fact that the attempts to elaborate relativistic generalizations of the DRP meet two quite different kinds of difficulties which I will denote as *formal* and *conceptual*, respectively. The formal ones have a precise mathematical nature and they are related to the appearence of untractable divergences in the formalism. The conceptual ones are strictly related to the problem of attributing objectively possessed properties to individual physical systems. In particular various authors [3–5]) have stated that the very approach at the basis of the DRP implies that such properties depend in a fundamental way on the reference frame one is taking into account. Here, I will briefly recall some aspects of the relativistic DRP to stress that while the first type of problems (the occurrence of untractable divergences) characterize all attempts which have been presented so far[1], the second kind of difficulties do not affect at all the formalism [7–13]. The statements of the above mentioned authors concerning the problem under discussion derive simply from having failed to grasp some essential points of the relativistic DRP.

2.1 An Example of a Relativistic Dynamical Reduction Model

After the DRP in its nonrelativistic version had been explicitly implemented [14] by the identification of the nonlinear and stochastic modifications to be added to the standard evolution equation to overcome the difficulties of quantum mechanics, a refinement of it, the Continuous Spontaneous Localization (CSL) model has been worked out [15,16]. Its main advantage derives from the fact that, contrary to the GRW-theory, it allows to deal with systems of identical particles. Moreover it replaces the discontinuous jumps characterizing the original theory by the unfolding of a continuous stochastic process (a sort of Brownian motion in Hilbert space), yielding a theory which, even though physically equivalent [16] to appropriate discontinuos generalizations of the GRW-theory, is more elegant.

Soon after, relativistic generalizations of CSL have been considered. Here we will briefly sketch one such model, originally introduced by P. Pearle [7] and subsequently analyzed in [9,10], in which the dynamical reduction mechanism is governed by a skew-hermitian coupling between appropriate field operators and c-number white noise processes. One works in the Interaction Picture and assumes that the fields are solutions of the Heisenberg equations of motion obtained from a Lagrangian density $L_0(x)$. We remark that $L_0(x)$ is not assumed to describe only free fields. The statevector evolves according to the Tomonaga-Schwinger equation:

$$\frac{\delta|\Psi_V(\sigma)\rangle}{\delta\sigma(x)} = \left[L_I(x)V(x) - \lambda L_I^2(x)\right]|\Psi_V(\sigma)\rangle, \tag{1}$$

where $\sigma(x)$ is a space-like surface, $L_I(x)$ is the Interaction Lagrangian density in the Interaction Picture and $V(x)$ are c-number stochastic "potentials" satisfying:

$$\langle\langle V(x)\rangle\rangle = 0, \qquad \langle\langle V(x)V(x')\rangle\rangle = \lambda\delta(t-t')\delta(\boldsymbol{x}-\boldsymbol{x}'). \tag{2}$$

The evolution equation does not preserve the norm of the statevector but it preserves the average square norm. It has to be supplemented by the following prescriptions:

[1] For a general review of the matter see [6].

Given the initial statevector on the space-like surface σ_0, the statevector on the arbitrary space-like surface σ lying entirely in the future of σ_0 is obtained, for a particular occurrence of the stochastic potential V, by solving the above equation (1) with the considered initial conditions and normalizing the resulting statevector. The actual probability $P_C[V]$ of occurrence of the stochastic potential V is determined by resorting to a "cooking procedure" of the "natural" probability $P[V]$ (i.e. the one associated to the white noise processes defined by (2)). The "cooking" involves the norm of the solution of (1) according to:

$$P_C[V] = P[V] \, \|\Psi_V(\sigma)\|^2 . \tag{3}$$

It is important to stress that (1) is integrable [9]. The most simple case which has been discussed is that of a fermion field coupled to a real scalar meson field by a standard trilinear coupling, so that the Lagrangian density $L_0(x)$ contains, besides the terms describing the free fields, an additional term

$$g\overline{\Psi}(x)\Psi(x)\Phi(x), \tag{4}$$

g being a coupling constant. Moreover the choice

$$L_I(x) = \Phi(x) \tag{5}$$

is made for the interaction term appearing in (1).

As discussed in full detail in [9,10], this model assigns to each space-like hypersurface a unique, well defined state vector and reduction occurs as soon as the hypersurface crosses, towards the future, the space-time region where a macroscopic measurement process takes place.

2.2 The Formal Difficulties

With the above premises we can identify the origin of the mathematical difficulties of the formalism. To guarantee stochastic relativistic invariance, the variance of the stochastic potential (i.e., the second term in (2)) must be a Lorentz invariant function of its arguments. The choice (2) of making it white in time, introduces, due to the appearence of all frequencies, divergences which turn out to be unrenormalizable. On the other hand, if the variance would have a more general form, it would not exhibit a delta-like behaviour in space, giving rise to a nonlocal quantum field theory with the well known difficulties accompanying such a situation.

This fact remains the crucial point for the attempts of relativistic generalizations of the CSL theory. In spite of the remarkable efforts devoted to overcome this problem, in particular by P. Pearle, no breakthrough has been found.

2.3 The Alleged Conceptual Difficulties

We come now to discuss the conceptual problems of the theory. At a naive analysis, they seem to arise from the fact, due to the skew-hermitian nature of the coupling in the Tomonaga-Schwinger equation, that the expectation value of a local observable with compact support may depend (at the individual level – unlike at the ensemble level), on which space-like hypersurface, among all those containing its support, is considered. In particular, denoting as A_I such an observable and as σ_1 and σ_2 two space like surfaces containing its support, one can have:

$$\frac{\langle\Psi(\sigma_1)|A_I|\Psi(\sigma_1)\rangle}{\langle\Psi(\sigma_1)|\Psi(\sigma_1)\rangle} \neq \frac{\langle\Psi(\sigma_2)|A_I|\Psi(\sigma_2)\rangle}{\langle\Psi(\sigma_2)|\Psi(\sigma_2)\rangle}, \tag{6}$$

in spite of the fact that, since the support of A_I is space-like with respect to the space-time region (lying between σ_1 and σ_2) where the measurement takes place, A_I itself commutes with the operator describing the evolution from σ_1 to σ_2.

With reference to this situation, B. d'Espagnat [3] has pointed out that

> the "stochastic relativistic invariance" of the (GRW) theory does not immunize it from an inconsistency connected with relativity. This inconsistency consists in the fact that, in a theory bearing (as it is claimed this one does) on "reality" as opposed to "intersubjective appearances" no statement concerning a "physical event" – in the relativistic, i.e., local sense of the word "event" – can be both true in one reference frame and false in another one, so that the ambiguity described is unacceptable.

Maudlin [4] takes a quite similar attitude by stressing the necessity of choosing the "reference frame" in which the collapse occurs:

> The collapse can be instantaneous in at most one reference frame leading to two possibilities: either some feature of the situation picks out a preferred reference frame, with respect to which the collapse is instaneous, or the collapse is not instantaneous at all. The analogy with gravitational theory might suggest the latter course as the most obvious: why not build a delay into the account of wave collapse? Why not, as in general relativity, have collapse propagate along the future light cone of the measurement event, thereby allowing a Lorentz invariant description? The obvious answer is that such a delayed collapse would come too late. Since polarization measurements can be made at space-like separation, and since the results on one side must be influenced by the collapse initiated at the other, delayed collapses won't work.

Finally Dickson [5] raises similar objections also with reference to different macroscopic objects, to conclude (by repeated considerations of the light cones originating from any space-time point in which an ambiguity is present) that the theory implies ambiguities concerning the "status" of practically all objects of the universe.

As we have discussed in great detail in various papers [9–13], the first two authors miss completely the fact that the relativistic dynamical reduction models are formulated in a completely reference free language, so that any remark about objective differences related to different reference frames is simply mistaken. Moreover, all the above criticisms ignore the fact that the "alleged" ambiguities (which are the relativistic analogue of the possibility, characterizing quantum theory, that a microsystem may very well have only "potentialities" and no "actuality" concerning a property) affect exclusively microscopic systems: all macroscopic objects are in definite macroscopic situations at all times along their world lines. Finally, all these authors do not pay sufficient attention to the fundamentally nonlocal character of the theory: when this is taken into account the ensuing picture is perfectly consistent [13].

It is rather instructive to remark that B. d'Espagnat himself, who had vaguely anticipated in [17] criticisms to relativistic CSL quite similar to those referred to above, has stressed with particular emphasis that any attempt to a relativistic dynamical reducution theory must take into account the fundamental analysis of Y. Aharonov and D. Albert [18], something that we had not done. Curiously enough, in [13] we have been able to prove that the relativistic reduction models proposed so far represent the first precise and explicit examples of theories having all features that the analysis of [18] has identified.

Concluding: lot of work is still necessary to get rid of the formal difficulties affecting the relativistic dynamical reduction program, but its very structure seem to appropriately embody all natural features of a relativistic theory describing processes of this type. We believe that a careful reading of [13] might be extremely useful to grasp the subtle points which such a program involves.

3 Experimental Tests

Even though the dynamical reduction models (GRW and CSL) represent [19] *rival* theories of Standard Quantum Mechanics and not simply a reinterpretation of it, it seems extremely difficult, if not practically impossible, to devise *experimenta crucis* allowing to discriminate between them. This, in a sense, is rather natural: to test the predictions of the theory is more or less so difficult as to put into evidence macroscopic quantum coherence effects.

3.1 The Investigations

It is useful to remind that there has been a lot of theoretical work aimed to identify physical systems and/or situations for which the modified dynamics makes predictions which are testably different from those of the standard theory. We refer the reader to the interesting investigations by A. Rae [20], A. Rimini [21], M. Tegmark [22] and by the review paper [23]. The conclusion of these investigations can be summarized by saying that the *genuine* decoherence of the theory is masqued by the *apparent* decoherence due to the coupling with the environment. Probably Prof. Grigolini will have something to say about particular processes which seem to suggest mechanisms quite similar to those characterizing the GRW theory.

Some authors have considered this practical impossibility of testing the theory against the standard one has a drawback. I must say that I do not think this is the case. After all, almost all proposals which have been put forward to overcome the macro-objectification problem and the conceptual difficulties of the standard theory are built in such a way to guarantee a perfect agreement with standard quantum predictions. The interest of the dynamical reduction program resides in its allowing to take a macrorealistic position about natural processes.

It is also useful to remark that the basic feature of the dynamical reduction theories involving an irreducible irreversibility of natural processes might have some interesting implications concerning deep problems such as the one of the emergence of the arrow of time in nature. I think this point will be mentioned by D. Albert in his lecture.

3.2 Some Recent Results

Before concluding this part I would like to mention that, as it has been proved recently by resorting to explicit examples [24], dynamical reduction models can give rise to deviations from the prediction of standard quantum mechanics when the times which are necessary for the basic decoherence mechanisms of the theory to become effective become comparable with the times which are necessary in order that the standard quantum evolution induces appreciable changes in the statevector. For macro-objects this is certainly not the case: the reduction frequency is of the order of millionths of a second, a characteristic time during which the Hamiltonian evolution cannot change in an appreciable way, e.g., the extension of the wave packet describing the centre-of-mass motion of a macroscopic pointer. It is however not difficult to imagine situations in which the two just mentioned characteristic times might become comparable. With reference to this fact we recall that when an analysis has been made [25] of how a definite perception can be triggered by a superposition of states leading, according to the standard formalism, to different perceptual processes, it has been made plausible that the number of particles which are displaced in the process of transmission of the nervous signal from the retina to the higher visual cortex during the characteristic perceptual time is just sufficient to induce the reduction to one of the two nervous signals. The analysis has been rather qualitative, but it is suggestive of possible situations in which the two above mentioned mechanisms might become competitive. As already anticipated, in [24] it has been shown that if one triggers a process by a superposition of states leading (each by itself) to physically different final situations and a dynamics of the GRW or CSl type is active but is such that the reduction takes an appreciable (on the appropriate hamiltonian scale) time to become effective, then there exist the possibility of putting into evidence the effect of the stochastic and nonlinear modifications of the standard dynamics. The very recognition of this fact suggests that it might be appropriate to reconsider, on a new basis, the search of appropriate systems which could, in principle, lead to a confirmation or to the falsification of the ideas underlying the dynamical reduction program.

4 Issues of Interpretation

I would like to recall that, after its formulation, there has been a lively debate about the real meaning of the dynamical reduction theories, i.e., about what precisely the theory is about. Let me mention three different positions which have been taken about such a question.

The first two are due to J.S. Bell himself, the third one has been presented for the first time in [23]. Let us briefly comment about them. At Schrödinger's Centenary Conference at the Imperial College, Bell has presented [26] a very nice sketch of the GRW theory and has also proposed an interpretation for it which makes explicit reference to the space and time variables characterizing the fundamental processes of the theory, i.e., the hittings suffered by all elementary constituents of any physical system:

> There is nothing in the theory but the wavefunction. It is in the wavefunction that we must find an image of the physical world, and in particular of the arrangement of things in ordinary three-dimensional space. But the wavefunction as a whole lives in a much bigger space, of $3N$ dimensions. It makes no sense to ask for the amplitude or phase or whatever of the wavefunction at a point in ordinary space. It has neither amplitude nor phase nor anything else until a multitude of points in ordinary three-dimensional space are specified. However, the GRW jumps (which are part of the wavefunction, not something else) are well localized in ordinary space. Indeed each is centred on a particular space-time point $(\boldsymbol{x}, t)$. So, we can propose these events as the basis of the "local beables" of the theory. These are the mathematical counterparts in the theory to real events at definite places and times in the real world (as distinct from the many purely mathematical constructions that occur in the working out of physical theories, as distinct from things which may be real but not localized, and as distinct from "observables" of other formulations of quantum mechanics, for which there is no use here). A piece of matter then is a galaxy of such events. As a schematic psychophysical parallelism we can suppose that our experience is more or less directly of events in particular pieces of matter, our brains, which events are in turn correlated with events in our bodies as a whole, and they in turn with events in the outer world.

It is interesting to stress that at that time, Bell, even though he had clearly in mind that the wavefunction of a N-particle system lives in a $3N$-dimensional abstract space, proposed an interpretation making explicit reference to events in the real space-time of nonrelativistic theories. Almost one year later, during a meeting at Erice, he changed his mind choosing to make explicit reference, in the interpretation, to the just mentioned abstract $3N$-dimensional space [2]:

> The GRW-type theories have nothing in their kinematics but the wavefunction. It gives the density (in a multidimensional configuration

> space!) of stuff. To account for the narrowness of that stuff in macroscopic dimensions, the linear Schrödinger equation has to be modified, in the GRW picture by a mathematically prescribed spontaneous collapse mechanism.

Even though it is rather intuitive what he had in mind, I had some difficulties in grasping exactly what he meant with the above statements, and we had various exchanges of ideas about its proposal. In one letter addressed to him, I expressed my worries even though plainly declaring that I saw in it the possibility of adopting a macrorealistic position, the main aim who had guided us in elaborating the scheme. I would like to take this opportunity to quote explicitly a sentence from the answer, dated october 3, 1989, to my letter, who proves that he still considered this as the appropriate interpretation for the theory:

> As regards $|\Psi\rangle$ and the density of stuff, I think it is important that this density is in the $3N$-dimensional configuration space. So I have not thought of relating it to ordinary matter or charge density in 3-space. Even for one particle I think one would have problems with the latter. So, I am inclined to the view that you mention after "... as it is sufficient for an objective interpretation...". And it has to be stressed that the stuff is in $3N$-dimensional space – or whatever corresponds in field theory!

Stimulated by his remarks and by various discussions we had on the subject, I continued worrying about the most appropriate ontology for the dynamical reduction models. In the meantime the continuous version of them (CSL) had been worked out and the appropriateness of relating reduction to masses had emerged. As a consequence I have proposed [23], with my collegues R. Grassi and F. Benatti, an interpretation which makes reference to precise elements of the real 3-dimensional space. Since, as we have already remarked, CSL tends to make definite the average mass density, such a quantity is identified with the appropriate one to account for physical reality at the macroscopic level.

The idea is quite simple. Let us consider the averaged (over the characteristic localization volume $10^{-15}\,\mathrm{cm}^3$ of the GRW theory) mass density operator $M(\boldsymbol{r})$:

$$M(\boldsymbol{r}) = \sum_k m_k N^{(k)}(\boldsymbol{r}) \tag{7}$$

where the index k runs on the various kinds of particles (electrons, protons, neutrons, etc) and the operator $N^{(k)}(\boldsymbol{r})$ is given by

$$N^{(k)}(\boldsymbol{r}) = \left(\frac{\alpha}{2\pi}\right)^{\frac{3}{2}} \sum_s \int \mathrm{d}\boldsymbol{q}\, \mathrm{e}^{-(\frac{\alpha}{2})(\boldsymbol{q}-\boldsymbol{r})^2} a_k^\dagger(\boldsymbol{q}, s) a_k(\boldsymbol{q}, s), \tag{8}$$

with $a_k^\dagger(\boldsymbol{q}, s)$ and $a_k(\boldsymbol{q}, s)$ the creation and annihilation operators for a particle of type k at point $\boldsymbol{q}$ with spin component s. Suppose now that $|\Psi(t)\rangle$ is the normalized statevector describing a physical system at time t. In terms of it we define an average mass density c-number function $\mathcal{M}(\boldsymbol{r}, t)$ in ordinary space as:

$$\mathcal{M}(\boldsymbol{r}, t) = \langle\Psi(t)|M(\boldsymbol{r})|\Psi(t)\rangle. \tag{9}$$

Within the standard formalism, such a quantity, when the statevector $|\Psi(t)\rangle$ correspond to the superposition of two states of the considered system in far away regions, is devoid of any physical meaning, in the sense that no physical effect, e.g. the gravitational ones on other systems, can be considered as due to $\mathcal{M}(\boldsymbol{r},t)$. Within the GRW and CSL theories, however, the situation is radically different: in the case of a macroscopic system the reducing dynamics does not tolerate superpositions of states corresponding to appreciably different values of $\mathcal{M}(\boldsymbol{r},t)$. Such a function becomes "objective" in the sense that the ratio of the variance of $M(\boldsymbol{r})$ to its average value is much smaller than 1, a condition guaranteeing [23] that everything goes as if the "actual" mass density distribution in the universe would be the one defined by $\mathcal{M}(\boldsymbol{r},t)$.

The proposed interpretation can then be simply summarized by stating that what the theory is about is precisely the (average) mass density distribution. As discussed in great detail in [23] this move allows to take consistently a macro-realistic position about natural processes. We stress that the quantity $\mathcal{M}(\boldsymbol{r},t)$ lives in the ordinary physical space-time comtinuum of classical physics.

To conclude this part I cannot avoid mentioning that there has been a lively debate about this interpretation. Various authors have considered it unacceptable due to what has been referred to as "the tail problem". It derives from (and, in our opinion, it reduces to) the fact that the configuration representation of, e.g., the wavefunction of the centre of mass of a macroscopic object cannot have compact support. Thus statements about "locations" of macro-objects are, to some extent, vague. We have no time to discuss this matter here. We refer the interested reader to the recent papers [27–32]. However, we consider it appropriate to stress that all the considered criticisms make reference, in a way or another, to the probabilistic interpretation of quantum mechanics. For those who are familiar with the GRW and CSL theories it should be clear that such an interpretation has to be abandoned (as J.S. Bell has repeatedly remarked). The appropriate attitude is to accept that "what is true of the world out there" is precisely the mass density distribution and that one has to analyze all physical situations by taking precisely such a perspective, period.

5 Relations with Other Approaches

The relations existing between the DPR and other attempts to overcome the difficulties affecting the standard theory have been discussed in various papers. Just to mention an example one could stress the privileged role that the theory assigns, analogously to Bohmian Mechanics, to the position variables. Here we want to outline how, incorporating some of the ideas of the DRP within the context of Decoherent Histories, one could make them a more satisfactory theoretical scheme. Recently, some specific and serious drawbacks of the decoherent histories approach have been discussed in great details [33–35]. In brief, the content of the just quoted papers can be summarized as follows. If one considers the following assumptions (which seems to us unavoidable in order that the decoherent histories program could be taken seriously) one easily derives a Kochen-and-Specker-like contradiction:

a. Every *family* of decoherent histories can be (naturally) endowed with a Boolean structure allowing to recover classical reasoning.
b. Within every decoherent *family* it is possible to assign truth values which preserve the Boolean structure (i.e. they form an homomorphism) to its histories.
c. Every decoherent *history* has a unique truth value, independently from the decoherent family to which it may be considered to belong.
d. Any decoherent *family* can be taken into account.

This implies that at least one of them must be rejected in order to avoid inconsistencies. Griffiths [36] rejects assumption c), an unacceptable move in our opinion. On the contrary, we believe that a reasonable, physically and conceptually clean way out can be found by giving up assumption d). We think that, if one accepts only families whose histories deal exclusively with macroscopic systems and "speak" only of, e.g., the mass density distribution (with an appropriate graining) of the whole universe, one can have a decoherent histories–like approach in which (obviously) decoherence is not perfect but it holds to an extremely high degree of accuracy, no contradiction of the Kochen-and-Specker type occurs and "objective" and "classical" statements can be made about what the theory is about, i.e., precisely the mass distribution in the whole universe. Useless to say that if one chooses such a line one has a scheme which resembles closely the GRW and CSL thories.

References

1. J.S. Bell, in: *Quantum Gravity 2*, C. Isham, R. Penrose and D. Sciama (eds.), Clarendon Press, Oxford, 1981.
2. J.S. Bell, in: *Sixty-Two Years of Uncertainty*, Arthur Miller (ed.), New York, Plenum Press, 1990.
3. B. d'Espagnat, *Veiled Reality*, Addison-Wesley, Reading, Mass., 1995.
4. T. Maudlin, *Quantum Non-Locality and Relativity*, Blackwell, Cambridge, Mass., 1994.
5. W.M. Dickson, *Quantum chance and nonlocality*, Cambridge University press, Cambridge, 1998.
6. P. Pearle in: *Open Systems and Measurement in Relativistic Quantum Theory*, H-P Breuer and F. Petruccione (eds.), *Lecture Notes in Physics*, Springer-Verlag, 1999.
7. P. Pearle, in: *Sixty-Two Years of Uncertainty*, Arthur Miller (Ed.), New York, Plenum Press, 1990.
8. P. Pearle in: *Quantum Chaos – Quantum Measurement*, P. Cvitanovic, I. Percival and A. Wirzba (Eds.), Kluwer Academic Publishers, Dordrecht, 1992.
9. G.C. Ghirardi, R. Grassi and P. Pearle, *Found. Phys.*, **20**, 1271 (1990).
10. G.C. Ghirardi, R. Grassi and P. Pearle, in: *Symposium on the Foundations of Modern Physics 1990*, P. Lahti and P. Mittelstaedt (eds.), World Scientific, Singapore, 1991.
11. G.C. Ghirardi, *Found. Phys. Lett.*, **9**, 313 (1996).

12. G.C. Ghirardi in: *Open Systems and Me4asurement in Relativistic Quantum Theory*, H-P Breuer and F. Petruccione (eds.), *Lecture Notes in Physics*, Springer-Verlag, 1999.
13. G.C. Ghirardi, quant-ph/0003149, to appear on *Found. Phys.*
14. G.C. Ghirardi, A. Rimini and T. Weber, *Physical Review* D **34**, 470 (1986).
15. P. Pearle, *Phys. Rev.*, A **39**, 2277 (1989).
16. G.C. Ghirardi, P. Pearle and A. Rimini, *Phys. Rev.*, A **42**, 78 (1990).
17. B. d'Espagnat, *Conceptual Foundations of Quantum Mechanics* 2nd edn., Addison-Wesley, 1989.
18. Y. Aharonov and D.Z. Albert, *Phys. Rev.*, D **21**, 3316 (1980), *ibidem, Phys. Rev.*, D **24**, 359 (1981).
19. B. van Fraassen, in: *Symposium on the Foundations of Modern Physics 1990*, P. Lahti and P. Mittelstaedt (eds.), World Scientific, 1991.
20. A.I.M. Rae, *J. Phys.*, A **23**: L57 (1990).
21. A. Rimini, in: *Advances in Quantum Phenomena*, E. Beltrametti and J.M. Levy-Leblond (eds.), Plenum Press, 1995.
22. M. Tegmark, *Found. Phys. Lett.*, **6**, 571 (1993).
23. F. Benatti, G.C. Ghirardi and R. Grassi, in: *Advances in Quantum Phenomena*, E. Beltrametti and J.M. Levy-Leblond (eds.), Plenum Press, 1995.
24. G.C. Ghirardi, *Phys. Lett.*, A **262**, 1 (1999).
25. F. Aicardi, A. Borsellino, G.C. Ghirardi and R. Grassi, *Found. Phys. Lett.*, **4**, 109 (1991).
26. J. S. Bell, in: *Schrödinger. Centenary celebration of a polymath*, C.W. Kilmister (Ed.), Cambridge, Cambridge University Press, 1987.
27. P. Lewis, *Brit. J. Phil. Sci.*, **48**, 313 (1997).
28. G.C. Ghirardi and A. Bassi, *Brit. J. Phil. Sci.*, **50**, 705 (1999).
29. R. Clifton and B. Monton, *Brit. J. Phil. Sci.*, **50**, 697 (1999).
30. A. Bassi and G.C. Ghirardi, *Brit. J. Phil. Sci.*, **50**, 719 (1999).
31. R. Clifton and B. Monton, *Brit. J. Phil. Sci.*, **51**, 155 (2000).
32. A. Bassi and G.C. Ghirardi, *Counting Marbles:Reply to Clifton and Monton* to appear in: *Brit. J. Phil. Sci.*.
33. A. Bassi and G.C. Ghirardi, *Phys. Lett.*, A **257**, 247 (1999).
34. A. Bassi and G.C. Ghirardi, *J. Stat. Phys.*, **98**, 457 (2000).
35. A. Bassi and G.C. Ghirardi, *Phys. Lett.*, A **265**, 153 (2000).
36. R.B. Griffiths, *Phys. Lett.*, A **265**, 12 (2000).

Relativistic Theory of Continuous Measurements

Heinz–Peter Breuer[1] and Francesco Petruccione[1,2]

[1] Fakultät für Physik, Albert-Ludwigs-Universität Freiburg, Hermann–Herder–Str. 3, D–79104 Freiburg i. Br., Germany
[2] Istituto Italiano per gli Studi Filosofici, Palazzo Serra di Cassano, Via Monte di Dio 14, 80132 Napoli, Italy

Abstract. It is shown that the stochastic wave function representation of continuous, selective measurements of open quantum systems can be given a relativistically covariant form. As an example, a piecewise deterministic process for the state vector of the source of an optical cavity is constructed which is covariant under Lorentz transformations and which describes the stochastic dynamics induced by a continuous monitoring of the radiated photons through a moving detector. The most general case of a quantum dynamical semigroup generated by an arbitrary number of Lindblad operators is also treated. The resulting equation of motion clearly demonstrates that the stochastic formulation of open quantum systems can be generalized to meet the basic postulates of both quantum measurement theory and special relativity.

1 Introduction

Most measurements performed in experiments on open quantum systems are, in fact, *indirect* measurements. According to the general theory [1], an indirect measurement consists of a quantum object to be measured, a quantum probe, which is a quantum system prepared in a well–defined state and which interacts with the quantum object, and, finally, a measuring apparatus that measures an observable of the quantum probe. In the theoretical analysis, quantum object and quantum probe as well as their interaction are treated quantum mechanically, whereas the meter is treated classically. The interaction between quantum probe and meter constitutes the true measuring process leading to the reduction of the state vector according to the specific outcome of the measurement (selective measurement).

A prototypical example of the foregoing is the measurement of the fluorescence radiation of a laser driven two–level atom by a photocounter [2]. Here, the atom is the quantum object, the quantum probe is given by the quantized radiation field interacting with the atom, and the observable that is measured by the photodetector is the photon number operator. If we assume for simplicity second order perturbation theory, which means that only zero- and one-photon states of the radiation field are excited with appreciable amplitude, then each time a photon is detected the application of the reduction postulate projects the state of the radiation field onto the 1-photon subspace, whereas, at the same time, the atom is put back to the ground state. Within the Markov approximation the atom is probed by a well–defined state of the quantum probe which, for the present example, is the Lorentz invariant vacuum state of the electromagnetic field.

Conventionally, a situation of this type is described by an equation of motion, the vacuum optical master equation, for the reduced density matrix that refers to the atomic variables. In the case that the photodetection represents a continuous (in time) measurement of the electromagnetic field the density matrix equation describes the continuous, *non–selective* measurement of an observable (photon number) of the environment (the electromagnetic field) of the open system (the atom).

It has been shown by Wiseman and Milburn [3,4] that such a physical situation can be interpreted in a rather different and physically very fruitful way. If we regard the continuous monitoring of the environment of the quantum object as a *selective* one, that is, if we really take into account the information gained during the whole measurement, the dynamics of the atomic state vector is to be represented by a stochastic process in the atomic Hilbert space. In mathematical terms the resultant stochastic process is a piecewise deterministic process, that is, a stochastic process whose realizations consist of smooth, deterministic pieces, broken by sudden, instantaneous changes of the state vector. In the above example, these so–called quantum jumps occur whenever the measurement of a photon puts the atom into the ground state as a result of the application of the reduction postulate to the direct photodetection.

This stochastic representation of the dynamics of open quantum systems, the stochastic wave function method, has been designed originally in order to analyze and to simulate open systems that arise in quantum optics [5–8]. The stochastic method has been applied with great success to various physical phenomena as, for example, to laser cooling [9], to lasing without inversion [10], to photochemistry [11], and to the description of electron–phonon interactions in semiconductors [12]. Employing a description of the reduced system's state in a doubled Hilbert space, one can also design stochastic algorithms for the determination of arbitrary multitime correlation functions [13]. In addition, a non-Markovian generalization of the method has been developed in [14] and applied to the treatment of the damped Jaynes-Cummings model.

The above applications are restricted to phenomena that can be treated within non–relativistic quantum mechanics. From a more general point of view it is tempting to ask the question whether it is possible to give a formulation of the stochastic wave function method which is relativistically covariant. Our aim is thus to develop a stochastic formulation of continuous measurements which includes a covariant prescription for the reduction of the state vector.

The starting point will be the following postulate. The state vector reduction resulting from a specific outcome of some measurement takes place across a certain spacelike hypersurface σ of the space-time continuum. This hypersurface σ is defined as an equal-time hypersurface in the rest frame of the macroscopic measuring apparatus, that is the state vector reduction is performed instantaneously in the rest frame of the macroscopic measuring device. Since the hypersurface across which the reduction occurs is, in general, not an equal–time hypersurface in other Lorentz frames but rather a flat, spacelike hypersurface, we have to deal, right from the beginning, with a formalism that treats the state vector as a function on the set Σ of such hypersurfaces. When formulating the stochastic dynamics as a piecewise deterministic process the quantum jumps describing the state reduction will take place along these hypersurfaces σ. It will be shown in this paper that this ansatz leads to a fully consistent, relativistically covariant stochastic state vector dynamics. In physical terms this implies that the stochastic representation of open quantum systems along with its physical interpretation in terms of a selective, continuous monitoring of an environmental observable is in full agreement with the relativity principle.

2 The Quantum Mechanical State Space as Hilbert Bundle

In this section we shall develop an appropriate mathematical framework which allows the formulation of a covariant stochastic state vector evolution. To this end, we introduce an appropriate parametrization for the hypersurfaces σ and attach to each hypersurface σ a Hilbert space of state vectors. To be specific we shall consider here the Dirac equation as an example for a covariant single-particle equation. In mathematical terms our state space thus forms a so-called fiber bundle whoose fibers are Hilbert spaces. We therefore may refer to our state space as a bundle of Hilbert spaces or simply as *Hilbert bundle.*

2.1 The Set of Flat, Spacelike Hypersurfaces in Minkowski Space

In the following a point in Minkowski space $\mathbb{R}^4$ will be denoted by

$$x = (x^\mu) = (x^0, x^i) = (x^0, \boldsymbol{x}), \tag{1}$$

where greek indices run from 0 to 3, and latin indices from 1 to 3. The Lorentz scalar (inner) product is then given by

$$x \cdot y \equiv x^\mu y_\mu = x^0 y^0 - \boldsymbol{x} \cdot \boldsymbol{y}. \tag{2}$$

Throughout this paper we chose units such that $c = \hbar = 1$. Lorentz transformations are written as

$$x'^\mu = \Lambda^\mu{}_\nu x^\nu + b^\mu. \tag{3}$$

For the sake of clarity we shall restrict ourselves to the subgroup of inhomogeneous, proper, orthochronous Lorentz transformations characterized by $\det\Lambda = +1$ and $\Lambda^0{}_0 > 0$. The restriction to proper Lorentz transformations is not essential, whereas the restriction to orthochronous transformations is essential, since we want to study irreverible processes which are not invariant under time inversion transformations.

In the following we consider the set Σ of flat, spacelike hypersurfaces σ in Minkowski space. Each such hypersurface σ can be characterized by its unit normal vector n^μ satisfying

$$n^2 \equiv n^\mu n_\mu = 1, \quad n^0 \geq 1, \tag{4}$$

and by some fixed point y^μ on this surface. Having fixed such a pair (n, y) of 4–vectors the points $x \in \sigma$ are given by the equation

$$n(x - y) \equiv n^\mu (x_\mu - y_\mu) = 0. \tag{5}$$

In the following we shall write

$$\sigma = \sigma(n, y) \tag{6}$$

for the hypersurface given by the 4–vectors n and y.

Note that the above parametrization of the hypersurfaces by n and y is not unique, since $\sigma(n, y) = \sigma(n, y')$ if $n(y - y') = 0$. Note further that, since the hypersurfaces are spacelike, each point $x \in \sigma(n, y)$ has a unique representation of the form

$$x = (x^0(\sigma, \boldsymbol{x}), \boldsymbol{x}) \tag{7}$$

where $\boldsymbol{x} \in \mathbb{R}^3$ and the time coordinate is given by

$$x^0(\sigma, \boldsymbol{x}) = \frac{\boldsymbol{n} \cdot \boldsymbol{x} + ny}{n^0}. \tag{8}$$

2.2 The Transport Equation

With each hypersurface $\sigma(n, y)$ we can associate a Hilbert space $\mathcal{H}(\sigma)$ of Dirac wave functions in the following way. Given a specific hypersurface $\sigma(n, y)$ we can introduce a Dirac wave function ψ defined on this surface by writing

$$\psi(x) = \psi(\sigma, \boldsymbol{x}). \tag{9}$$

This equation means that $\psi(\sigma, \boldsymbol{x})$ is the value of the Dirac wave function ψ taken at the point $x \in \sigma(n, y)$ which is uniquely determined by $\boldsymbol{x}$.

In order to make the set of Dirac wave functions on a fixed hypersurface $\sigma(n, y)$ a Hilbert space we need to define a positive definite Hermitian scalar product in this space. To this end, we observe first that the Lorentz metric induces a positive, 3–dimensional volume element $\mathrm{d}\sigma$ on each hypersurface given by

$$\mathrm{d}\sigma = \frac{1}{n^0}\mathrm{d}^3x. \tag{10}$$

Since n is the unit vector normal to $\sigma(n, y)$ the 4–vector $n_\mu \mathrm{d}\sigma$ is the surface element of the hypersurface. On using the Dirac current

$$j^\mu = \bar{\psi}\gamma^\mu\psi, \tag{11}$$

where $\bar{\psi} = \psi^\dagger \gamma^0$ and the γ^μ denote the usual gamma matrices (we use the notation of [15]), we can form the Lorentz invariant bilinear expression

$$\langle\psi|\psi\rangle_\sigma \equiv \int_\sigma j^\mu n_\mu \mathrm{d}\sigma. \tag{12}$$

Expressing the probability flow through the hypersurface $\sigma(n, y)$, this equation suggests the following definition of a Hermitian scalar product:

$$\langle\psi|\phi\rangle_\sigma \equiv \int_\sigma \mathrm{d}\sigma \bar{\psi} \not{n} \phi = \int \frac{\mathrm{d}^3x}{n^0} \psi^\dagger(\sigma, \boldsymbol{x}) \gamma^0 \not{n} \phi(\sigma, \boldsymbol{x}), \tag{13}$$

where we define $\not{n} \equiv n_\mu \gamma^\mu$. As is easily verified, the expression (13) has the properties of a Hermitian scalar product. Most importantly, positivity of (13) is a direct consequence of the fact that the (4×4)–matrix $\gamma^0 \not{n}$ is positive definite. Note further that for the special case $n = (1, 0, 0, 0)$ definition (13) reduces to the usual scalar product [15] taken at a fixed time $x^0 = y^0$,

$$\langle\psi|\phi\rangle_\sigma = \int \mathrm{d}^3x \; \psi^\dagger(x^0, \boldsymbol{x}) \phi(x^0, \boldsymbol{x}). \tag{14}$$

Having defined a Hermitian scalar product on the set of Dirac spinors on $\sigma(n, y)$, we can now define an associated Hilbert space $\mathcal{H}(\sigma)$ consisting of the Dirac spinors with finite norm

$$||\psi||_\sigma \equiv \sqrt{\langle\psi|\psi\rangle_\sigma}. \tag{15}$$

Thus, we have associated with each hypersurface a corresponding Hilbert space of Dirac wave functions,

$$\sigma \mapsto \mathcal{H}(\sigma). \tag{16}$$

In a differential geometric language, this mapping defines a so-called fiber bundle. Such a bundle consists of a base manifold and fibers attached to each point of the base manifold. In our case the base manifold is the set Σ of hypersurfaces and the fibres are given by the Hilbert spaces $\mathcal{H}(\sigma)$. All these Hilbert spaces are isomorphic and the above construction yields an (infinite–dimensional) fiber bundle which will be called Hilbert bundle.

We now turn to the formulation of the dynamics in the above Hilbert bundle. To get a covariant dynamics we have to start, of course, from a covariant one–particle equation. In order to be specific we take the Dirac equation for an electron (charge e, mass m) in an external electromagnetic field described by the vector potential $A_\mu(x)$,

$$\left(\mathrm{i}\gamma^\mu \left[\partial_\mu + \mathrm{i}eA_\mu(x)\right] - m\right) \Phi(x) = 0. \tag{17}$$

Any solution of Eq. (17) defines a Dirac wave function $\Phi = \Phi(x)$ on the space–time continuum. Our aim is now to reformulate the Dirac equation as a dynamical equation for the Dirac spinors $\psi(\sigma, \boldsymbol{x})$ which are defined on the various hypersurfaces $\sigma(n, y)$. Since the resulting equation should be equivalent to the

Dirac equation, $\psi(\sigma, \boldsymbol{x})$ is obtained just by restricting the space–time function $\Phi(x)$ to the hypersurface $\sigma(n, y)$. Therefore we set

$$\psi(\sigma, \boldsymbol{x}) = \Phi(x^0(\sigma, \boldsymbol{x}), \boldsymbol{x}). \tag{18}$$

When differentiating this equation with respect to the space coordinates one has to distinguish carefully between partial and total derivatives of Φ. We denote by $(\partial_i \Phi)$ partial space derivatives and by $\partial_i \Phi$ total space derivatives. We then get from (18)

$$(\partial_i \Phi)(x^0, \boldsymbol{x}) = \partial_i \Phi(x^0, \boldsymbol{x}) - \frac{\partial x^0}{\partial x^i}(\partial_0 \Phi)(x^0, \boldsymbol{x}) = \partial_i \psi(\sigma, \boldsymbol{x}) - \frac{n^i}{n^0}(\partial_0 \Phi)(x^0, \boldsymbol{x}). \tag{19}$$

The Dirac equation can be written as

$$\partial_0 \Phi = -\mathrm{i}\gamma^0 \left(-\mathrm{i}\boldsymbol{\gamma} \cdot \boldsymbol{\nabla} + e\not{A}(x) + m\right) \Phi \equiv -\mathrm{i} H_D \Phi, \tag{20}$$

where H_D denotes the (possibly time–dependent) Dirac Hamiltonian. Thus we have on using (19)

$$\begin{aligned} H_D \Phi(x^0, \boldsymbol{x}) &= \gamma^0 \left(-\mathrm{i}\boldsymbol{\gamma} \cdot \boldsymbol{\nabla} + e\not{A}(x) + m\right) \Phi(x^0, \boldsymbol{x}) \\ &= \gamma^0 \left(-\mathrm{i}\boldsymbol{\gamma} \cdot \boldsymbol{\nabla} + e\not{A}(x) + m\right) \psi(\sigma, \boldsymbol{x}) + \gamma^0 \left(\mathrm{i}\boldsymbol{\gamma} \cdot \frac{\boldsymbol{n}}{n^0}(\partial_0 \Phi)(x^0, \boldsymbol{x})\right) \\ &= H_D \psi(\sigma, \boldsymbol{x}) + \mathrm{i}\gamma^0 \boldsymbol{\gamma} \cdot \frac{\boldsymbol{n}}{n^0}(\partial_0 \Phi)(x^0, \boldsymbol{x}) \\ &= \mathrm{i}(\partial_0 \Phi)(x^0, \boldsymbol{x}). \end{aligned} \tag{21}$$

From the last equation we get

$$\left(I - \gamma^0 \boldsymbol{\gamma} \cdot \frac{\boldsymbol{n}}{n^0}\right)(\partial_0 \Phi) = -\mathrm{i} H_D \psi(\sigma, \boldsymbol{x}), \tag{22}$$

where I denotes the unit matrix. On multiplying this equation with $n^0 \gamma^0$ we find

$$\not{n}(\partial_0 \Phi) = -\mathrm{i} n^0 \gamma^0 H_D \psi(\sigma, \boldsymbol{x}). \tag{23}$$

Now, we multiply both sides with $\not{n}$, and use the fact that $\not{n}\not{n} = n^2 = 1$. This yields the following equation for the time derivative of Φ,

$$(\partial_0 \Phi)(x^0, \boldsymbol{x}) = -\mathrm{i} n^0 H \psi(\sigma, \boldsymbol{x}), \tag{24}$$

where we have introduced the operator

$$H \equiv H(\sigma) = \not{n}\gamma^0 H_D. \tag{25}$$

As will be shown below, $H(\sigma)$ is in fact a Hermitian operator in the Hilbert space $\mathcal{H}(\sigma)$.

We now consider the total differential of ψ under variations of n and y,

$$\mathrm{d}\psi \equiv \psi(\sigma(n + \mathrm{d}n, y + \mathrm{d}y), \boldsymbol{x}) - \psi(\sigma(n, y), \boldsymbol{x}) = \frac{\partial \psi}{\partial n^\mu}\mathrm{d}n^\mu + \frac{\partial \psi}{\partial y^\mu}\mathrm{d}y^\mu. \tag{26}$$

Employing (18) and (24) we obtain

$$\begin{aligned} d\psi &= \left(\frac{\partial x^0}{\partial n^\mu} dn^\mu + \frac{\partial x^0}{\partial y^\mu} dy^\mu \right) (\partial_0 \Phi) \\ &= -i \left(n^0 \frac{\partial x^0}{\partial n^\mu} dn^\mu + n^0 \frac{\partial x^0}{\partial y^\mu} dy^\mu \right) H\psi. \end{aligned} \tag{27}$$

Since

$$n^0 \frac{\partial x^0}{\partial n^\mu} = y_\mu - x_\mu, \quad n^0 \frac{\partial x^0}{\partial y^\mu} = n_\mu, \tag{28}$$

we finally get the following differential equation

$$d\psi(\sigma) = -i \left(W_\mu dn^\mu + n_\mu dy^\mu \right) H\psi(\sigma), \tag{29}$$

where we have defined the 4–vector operator (acting on $\mathcal{H}(\sigma)$)

$$W_\mu = y_\mu - x_\mu. \tag{30}$$

For reasons that will become clear later we call (29) *transport equation* [16]. This equation is a reformulation of the Dirac equation (17) which is appropriate for our aim of formulating a covariant continuous measurement theory. Eq. (29) describes the change of the wave function ψ under variations of the hypersurface $\sigma(n, y)$ parametrized by n and y. Note that $d\psi$ vanishes if we consider variations of y that do not really change the hypersurface $\sigma(n, y)$, that is, we have $d\psi = 0$ if $dn = 0$ and $n\, dy = 0$. We remark that our formulation of the evolution in terms of an equation for the state vector as a function on the set of hypersurfaces is very similar to the Schwinger–Tomonaga equation [17] of relativistic quantum field theory. In a different physical context, a similar method has been used in [16,18,19].

As has been mentioned already, H and W_μ as defined in (25) and (30) have to be regarded as operators in $\mathcal{H}(\sigma)$. We now demonstrate that these operators are, in fact, Hermitian with respect to the scalar product (13). This is obviously true for the vector operator W_μ which is diagonal in spin space. To prove the Hermiticity of $H(\sigma)$ one uses the fact that the Dirac operator H_D is Hermitian in $\mathcal{H}(\sigma)$, where $n = (1, 0, 0, 0)$. Hence we have

$$\int d^3x\, \psi^\dagger H_D \phi = \int d^3x\, (H_D \psi)^\dagger \phi. \tag{31}$$

Therefore, we find on using (25) and $\gamma^0 \not{n} = (\gamma^0 \not{n})^\dagger$

$$\begin{aligned} \langle \psi | H\phi \rangle_\sigma &= \int \frac{d^3x}{n^0} \bar{\psi} \not{n} H\phi = \int \frac{d^3x}{n^0} \psi^\dagger H_D \phi \\ &= \int \frac{d^3x}{n^0} (H_D \psi)^\dagger \phi = \int \frac{d^3x}{n^0} (H\psi)^\dagger \gamma^0 \not{n} \phi \\ &= \langle H\psi | \phi \rangle_\sigma. \end{aligned}$$

This proves the Hermiticity of $H = H(\sigma)$.

For the considerations in this paper, in particular for the formulation of the various integrability conditions, it is very useful to employ a compact notation in terms of differential forms. To this end, we introduce the differential 1–form

$$\mathrm{d}G(\sigma) \equiv (W_\mu \mathrm{d}n^\mu + n_\mu \mathrm{d}y^\mu)\, H = (W_\mu H)\, \mathrm{d}n^\mu + (n_\mu H)\, \mathrm{d}y^\mu \tag{32}$$

which enables one to cast the transport equation into the compact form

$$\mathrm{d}\psi(\sigma) = -\mathrm{i}\mathrm{d}G(\sigma)\psi(\sigma). \tag{33}$$

More precisely, definition (32) introduces an operator–valued 1–form, that is, a 1–form in Σ whose coefficients (namely $W_\mu H$ and $n_\mu H$) at a certain point (n, y) are operators in the Hilbert space $\mathcal{H}(\sigma)$. Such differential forms naturally arise in the differential geometry of vector bundles.

2.3 Properties of the Transport Equation

Normalization. First we shall demonstrate that the transport equation is norm–conserving and thus defines a unitary evolution in the Hilbert bundle. By making use of the definition of the scalar product and taking into account its explicit dependence on n we obtain

$$\mathrm{d}\langle\psi|\psi\rangle_\sigma = \langle\psi|\mathrm{d}\psi\rangle_\sigma + \langle\mathrm{d}\psi|\psi\rangle_\sigma + \left\langle \psi \left| \not{n}\gamma_\mu - \frac{1}{n^0}\delta_{0,\mu} \right| \psi \right\rangle_\sigma \mathrm{d}n^\mu. \tag{34}$$

The last term on the right–hand side stems from the σ–dependence of the scalar product. Inserting the transport equation (33) we get

$$\mathrm{d}\langle\psi|\psi\rangle_\sigma = -\mathrm{i}\langle\psi|\mathrm{d}G - \mathrm{d}G^\dagger|\psi\rangle_\sigma + \left\langle \psi \left| \not{n}\gamma_\mu - \frac{1}{n^0}\delta_{0,\mu} \right| \psi \right\rangle_\sigma \mathrm{d}n^\mu. \tag{35}$$

Since

$$\mathrm{d}G - \mathrm{d}G^\dagger = [W_\mu, H]\mathrm{d}n^\mu \tag{36}$$

and since the commutator is given by

$$[W_\mu, H] = -\mathrm{i}\not{n}\gamma_\mu + \frac{\mathrm{i}}{n^0}\delta_{0,\mu} \tag{37}$$

we immediately find

$$\mathrm{d}\langle\psi|\psi\rangle_\sigma = -\mathrm{i}\langle\psi|[W_\mu, H]|\psi\rangle_\sigma \mathrm{d}n^\mu + \left\langle \psi \left| \not{n}\gamma_\mu - \frac{1}{n^0}\delta_{0,\mu} \right| \psi \right\rangle_\sigma \mathrm{d}n^\mu = 0. \tag{38}$$

Thus, the transport equation yields normalized solutions for any normalized initial value (given on a specific hypersurface).

Integrability Condition. Integrability of the transport equation (33) means the following. Consider a curve

$$\sigma(\tau) = \sigma(n(\tau), y(\tau)), \quad 0 \leq \tau \leq 1, \tag{39}$$

in the space Σ of hypersurfaces. Evaluated along such a curve, the transport equation gives rise to an ordinary differential equation,

$$\frac{\mathrm{d}\psi(\tau)}{\mathrm{d}\tau} = -\mathrm{i}\left(W_\mu \frac{\mathrm{d}n^\mu}{\mathrm{d}\tau} + n_\mu \frac{\mathrm{d}y^\mu}{\mathrm{d}\tau}\right) H\psi(\tau), \tag{40}$$

where $\psi(\tau) = \psi(\sigma(\tau))$. If we start from a normalized initial Dirac wave function $\psi(0)$ given on $\sigma(0)$ the corresponding solution $\psi(\tau)$ of (40) yields a normalized wave function along the curve $\sigma(\tau)$. In particular we get a normalized state $\psi(1)$ in $\mathcal{H}(\sigma(1))$ at the endpoint $\sigma(1)$ of the curve.

Now, if we have two such curves $\sigma_1(\tau)$ and $\sigma_2(\tau)$ with common initial point and common endpoint, integrability means that the two corresponding wave functions ψ_1 and ψ_2 thus obtained at the common endpoint coincide, that is

$$\psi_1(\tau = 1) = \psi_2(\tau = 1). \tag{41}$$

If this condition is true for all curves σ_1 and σ_2 of the above type and for all initial conditions, the transport equation (33) is said to be integrable. Equivalently, integrability means that for any initial condition on a specific hypersurface, the transport equation can be solved to give a *function* $\psi = \psi(\sigma)$ of σ, i. e., a function on the set Σ of hypersurfaces.

It should be obvious that the integrability condition is trivially satisfied for our transport equation, since we know by construction that its solutions $\psi(\sigma)$ are given by (18), that is, by restricting the corresponding solutions of the Dirac equation to the various hypersurfaces σ. In mathematical terms the necessary and sufficient condition for the integrability of the transport equation is

$$\mathrm{d} \wedge \mathrm{d}\psi = 0, \tag{42}$$

where $\wedge$ denotes the wedge product of differential forms. Now, on using the transport equation we get from this condition

$$\begin{aligned} \mathrm{d} \wedge \mathrm{d}G\psi &= (\mathrm{d} \wedge \mathrm{d}G)\psi - \mathrm{d}G \wedge \mathrm{d}\psi \\ &= (\mathrm{d} \wedge \mathrm{d}G + \mathrm{i}\mathrm{d}G \wedge \mathrm{d}G)\,\psi = 0. \end{aligned}$$

From the known integrability of the transport equation we thus conclude that the following relation holds

$$F \equiv \mathrm{d} \wedge \mathrm{d}G + \mathrm{i}\mathrm{d}G \wedge \mathrm{d}G = 0. \tag{43}$$

By making use of the explicit formula (32) this equation may be verified explicitly by a lengthy calculation which is omitted here since we know that it is true by the general argument.

We note that the above formalism has a nice interpretation in terms of differential geometric concepts. In a differential geometric language a wave function which is a solution of equation (40) is said to be parallel transported along the the curve $\sigma(\tau)$. This is the reason why (33) is called *transport equation.* The differential 1–form $\mathrm{d}G$ can then be interpreted as a so–called connection form or gauge potential in our Hilbert bundle. Integrability means that parallel transport along a closed curve does not change the wave function. The 2–form F defined in (43) is nothing but the curvature or field strength of the Hilbert bundle. Thus we may say that the integrability condition is equivalent to the vanishing of the curvature of the Hilbert bundle.

Density Matrix and von Neumann–Type Equation. As a preparation for the investigations in section 3 we consider the pure state density matrix defined in the position representation by

$$\rho(\sigma, \boldsymbol{x}, \boldsymbol{x}') \equiv \psi(\sigma, \boldsymbol{x})\psi^{\dagger}(\sigma, \boldsymbol{x}')\gamma^0 \not{n} \frac{1}{n^0}. \tag{44}$$

To understand this definition we first remark that we use the following definition for the trace of an operator $A(\sigma)$,

$$\mathrm{tr}A(\sigma) \equiv \int \mathrm{d}^3x \; \mathrm{tr}_{\mathrm{spin}} A(\sigma, \boldsymbol{x}, \boldsymbol{x}), \tag{45}$$

where $\mathrm{tr}_{\mathrm{spin}}$ denotes the trace over the spinor indices. On using this definition we obtain from (44)

$$\begin{aligned} \mathrm{tr}\rho(\sigma) &= \int \mathrm{d}^3x \; \mathrm{tr}_{\mathrm{spin}} \psi(\sigma, \boldsymbol{x})\psi^{\dagger}(\sigma, \boldsymbol{x})\gamma^0 \not{n} \frac{1}{n^0} \\ &= \langle \psi | \psi \rangle_{\sigma} = 1, \end{aligned}$$

and

$$\mathrm{tr}\left\{\rho(\sigma)A(\sigma)\right\} = \langle \psi | A | \psi \rangle_{\sigma}. \tag{46}$$

Definition (44) ensures that the density matrix $\rho(\sigma)$ is Hermitian with respect to the scalar product in $\mathcal{H}(\sigma)$. Thus, a straightforward calculation yields the following equation of motion for the density matrix,

$$\mathrm{d}\rho(\sigma) = -\mathrm{i}[\mathrm{d}G(\sigma), \rho(\sigma)]. \tag{47}$$

Note that, due to the integrability of the transport equation, also this von Neumann–type equation is integrable. This means that Eq. (47) can be solved to yield a density matrix which is a function on the set of hypersurfaces.

Lorentz Covariance. The covariance of the transport equation (33) under Lorentz transformations (3) is immediately clear from the following argument. The construction of the transport equation in Sect. 2.2 has been performed using the coordinates x of a certain inertial frame A, say. Consider now another inertial frame B whoose coordinates x' are connected to the coordinates of A by the Lorentz transformation (3). As is obvious from equation (5) the parameters of the hypersurfaces in frame B are connected to those of frame A by

$$n' = \Lambda n, \quad y' = \Lambda y + b, \tag{48}$$

which means that n transforms as a normal vector and y as a coordinate vector. Since the Dirac equation (17) is covariant, the Lorentz transformed wave function

$$\Phi'(x') = S(\Lambda)\Phi(x) \tag{49}$$

fulfills the Dirac equation in frame B with Lorentz transformed vector potential $A'_\mu(x')$, where $S(\Lambda)$ denotes the usual spinor representation of the Lorentz group [15].

In accordance with (18) the observer in frame B now defines

$$\psi'(\sigma', \boldsymbol{x}') = \Phi'(x') = S(\Lambda)\Phi(x) = S(\Lambda)\psi(\sigma, \boldsymbol{x}), \tag{50}$$

where the relation between the space coordinates x^i in frame A and x'^i in frame B is given by

$$x^i = x^i(\sigma', \boldsymbol{x}') = \left(\Lambda^{-1}\right)^i{}_0 \, x'^0(\sigma', \boldsymbol{x}') + \left(\Lambda^{-1}\right)^i{}_j \, x'^j. \tag{51}$$

The transformation law for the wave function thus reads explicitly

$$\psi'(\sigma', \boldsymbol{x}') = S(\Lambda)\psi(\sigma, \boldsymbol{x}(\sigma', \boldsymbol{x}')). \tag{52}$$

Now, if one repeats the derivation of the transport equation using the coordinates x' and the parameters n' and y' of frame B, one obtains, of course, an equation which takes on the same form as the transport equation in frame A. Thus, the covariance of (33) is an immediate consequence of the covariance of the Dirac equation and of the covariant transformation properties (48) of the the parameters n and y that define the hypersurfaces $\sigma(n, y)$.

3 Continuous Measurement Theory and Stochastic Dynamics in the Hilbert Bundle

We now employ the general framework developed in the preceding section to construct a relativistically covariant stochastic state vector dynamics for continuous measurements. Sect. 3.1 deals with a specific example from quantum optics, namely a direct photodetection scheme, corresponding to the case of a single Lindblad operator. Our result will be generalized in the Sect. 3.2 to include the most general case of a Markovian quantum dynamical semigroup which is described through an arbitrary number of Lindblad generators.

3.1 Piecewise Deterministic Jump Processes: Direct Photodetection

As a simple example for a stochastic state vector evolution we shall investigate in this subsection the output signal of a two-level atom in an optical cavity which is detected by a photocounter. The photodetector moves with velocity $\boldsymbol{v}$ relative to the cavity in the direction of the output signal, thereby following a world line $y(\tau)$ in Minkowski space $\mathbb{R}^4$. The parameter τ denotes the proper time of the detector, that is, the time of a clock attached to the detector. In the following we allow for an accelerated motion for which the 4–velocity

$$n(\tau) \equiv \frac{\mathrm{d}y}{\mathrm{d}\tau} = (\gamma, \gamma\boldsymbol{v}) \tag{53}$$

is not a constant, where $n^\mu(\tau)n_\mu(\tau) = 1$, and $\gamma \equiv (1 - \boldsymbol{v}^2)^{-1/2}$.

An observer O moving with the detector can then construct a co–moving coordinate frame (Fermi–Walker frame) as follows [20]. At each fixed proper time τ the time axis is given by the unit vector $n(\tau)$, whereas instantaneous 3–space at that time is given by the flat, spacelike hypersurface $\sigma = \sigma(n, y)$ which is orthogonal to $n(\tau)$ and contains the point $y(\tau)$, i.e., which is defined by (5). Thus, the hypersurface $\sigma(\tau) = \sigma(n(\tau), y(\tau))$ is the set of those spacetime points x to which observer O assigns one and the same time coordinate τ. The family of hypersurfaces $\sigma(\tau)$ determined by the detector path $y(\tau)$ will be called a foliation[1].

Our aim is now to construct a relativistically covariant, stochastic representation of the source dynamics induced by the continuous, selective measurement of the radiated photons through the moving photodetector. Each time a photon is detected the wave function of the source undergoes an instantaneous change. We assume, as in the non–relativistic theory [3], that this change is obtained through the application of some Lindblad operator L to the state vector. Due to the lack of an absolute time the term *instantaneous change of the wave function* is not a relativistically covariant one. However, by the very principles of quantum mechanics, source, radiation field, and detector have to be regarded as a whole. The important conclusion is that a Lorentz transformation affects the quantum object and the probe as well as the detector and the hypersurfaces $\sigma(\tau)$ of the foliation of observer O. Thus we do get a covariant prescription for the state vector reduction if we postulate that the state vector reduction occurs instantaneously in the detector's rest frame, that is, along a certain spacelike hypersurface $\sigma(\tau)$ of the foliation of the observer O associated to it.

[1] More precisely, the family of hypersurfaces $\sigma(\tau)$ represents a foliation of that space-time region in which different hypersurfaces do not intersect. The corresponding condition is $l(x) < g^{-1}$, where $l(x)$ is the distance of a point x from the world line $y(\tau)$ and g is the acceleration of the detector measured in its own rest frame [20]. The set of points x which fulfill this condition is uniquely parameterized by the Fermi–Walker frame given above. Beyond a distance g^{-1} from the detector's world line a complete, continuous measurement and, therefore, a stochastic representation of the source dynamics ceases to exist.

As a consequence of these considerations the jump operator becomes a function $L = L(\sigma)$ of the hypersurfaces which will be determined below. In direct analogy to the non–relativistic formulation we obtain on the basis of our state reduction postulate the following Markovian stochastic state vector equation describing a piecewise deterministic process:

$$\begin{aligned} \mathrm{d}\psi(\sigma) = & -\mathrm{i}\mathrm{d}G(\sigma)\psi(\sigma) \\ & -\frac{1}{2}\left(L^{\dagger}(\sigma)L(\sigma) - \langle L^{\dagger}(\sigma)L(\sigma)\rangle\right)\psi(\sigma)\mathrm{d}a \\ & +\left(\frac{L(\sigma)\psi(\sigma)}{\sqrt{\langle L^{\dagger}(\sigma)L(\sigma)\rangle}} - \psi(\sigma)\right)\mathrm{d}N(a), \end{aligned} \tag{54}$$

where we have introduced the abbreviation

$$\langle L^{\dagger}(\sigma)L(\sigma)\rangle \equiv \langle \psi | L^{\dagger}(\sigma)L(\sigma)|\psi\rangle_{\sigma}. \tag{55}$$

The first term on the right-hand side of (54) represents the unitary dynamics as it is described through the transport equation (29). The second and the third term yield the irreversible part of the dynamics induced by the continuous monitoring of the quantum object. The structure of these terms is similar to the one encountered in quantum optical applications. Here, the quantity $\mathrm{d}N(a)$ is the increment of a Poisson process which obeys

$$E\left[\mathrm{d}N(a)\right] = \langle L^{\dagger}(\sigma)L(\sigma)\rangle\mathrm{d}a, \quad \left[\mathrm{d}N(a)\right]^2 = \mathrm{d}N(a), \tag{56}$$

where E denotes the expectation value, and the quantity a plays the role of an invariant time parameter which will be determined below. The second relation in (56) tells us that $\mathrm{d}N(a)$ takes on the values 0 or 1. As long as no photon is detected we have $\mathrm{d}N(a) = 0$ and, thus, the second term of (54) gives the evolution of the state vector conditioned on the outcome that no photon is detected. If a photon is detected we have $\mathrm{d}N(a) = 1$, such that the third term of (54) yields the corresponding jump of the state vector given by

$$\psi(\sigma) \longrightarrow \frac{L(\sigma)\psi(\sigma)}{\sqrt{\langle L^{\dagger}(\sigma)L(\sigma)\rangle}}. \tag{57}$$

Thus, the Poisson process $N(a)$ simply counts the number of photon detection events.

According to (57) the stochastic jumps of the state vector occur along the hypersurfaces $\sigma(\tau)$ of the foliation given by the detector path. If a photon has been detected at a certain proper time τ the state vector reduction has to be performed at the corresponding retarded proper time τ_{ret} taking into account the time required by the light signal to propagate from the source to the detector. This follows directly from the fact that the detected signal yields information on the state of the source at the retarded time and is made explicit in the input/output–formalism of quantum optics [21]. Thus, the precise prescription for the state vector reduction takes the following form: The reduction of the state vector occurs along the spacelike hypersurface $\sigma(\tau_{\rm ret})$ at the retarded proper time $\tau_{\rm ret}$ which corresponds to the proper time τ of the actual measuring event. Thus, (54) leads to a stochastic equation of motion for the source wave function $\psi(\tau_{\rm ret}) = \psi(\sigma(\tau_{\rm ret}))$.

In order to determine the invariant time parameter a we first observe that, according to Eq. (56), the photocurrent as measured in the rest frame of the detector is given by

$$J = \langle L^\dagger(\sigma)L(\sigma)\rangle \frac{\mathrm{d}a}{\mathrm{d}\tau}. \tag{58}$$

J is the average number of photon counts per unit of the proper time interval $\mathrm{d}\tau$. Due to the Lorentz invariant nature of the scalar product we may simply set $\langle L^\dagger(\sigma)L(\sigma)\rangle = \Gamma_0$, where Γ_0 is an invariant emission rate characteristic of the source. Thus we have $J = \Gamma_0 \mathrm{d}a/\mathrm{d}\tau$. If the detector is at rest with respect to the source ($\boldsymbol{v} = 0$), the detected photocurrent must be $J_0 = \Gamma_0$. Thus, a must be equal to the proper time of the source, i.e., a is the time of a clock fixed at a position in the vicinity of the source. To see that this conclusion is correct we consider the case of a moving detector ($\boldsymbol{v} \neq 0$). It is easy to show with the help of the transformation laws for the electromagnetic field strength tensor that the photocurrrent J as measured in the rest frame of the detector is given by

$$J = \sqrt{\frac{1-v}{1+v}} J_0, \tag{59}$$

where $\boldsymbol{v} = v\boldsymbol{e}$ and $\boldsymbol{e}$ denotes the unit vector in the direction of the radiation (we do not assume that v is constant). Now, with the above choice for the quantity a the square root expression is identically equal to $\mathrm{d}a/\mathrm{d}\tau$ which proves our claim[2].

The covariance of our stochastic state vector equation (54) under Lorentz transformations is obvious. Since $\mathrm{d}a$ is an invariant, $\mathrm{d}a' = \mathrm{d}a$, the quantity $\mathrm{d}N(a)$ is an invariant stochastic process. Furthermore, since the scalar product transforms as a Lorentz scalar,

$$\langle \psi' | \psi' \rangle_{\sigma'} = \langle \psi | \psi \rangle_\sigma, \tag{60}$$

both the dissipative and the stochastic term of (54) transform covariantly. It is important to emphasize that the transformation laws also involve a transformation of the jump operator, namely

$$L'(\sigma') = U L(\sigma) U^\dagger, \tag{61}$$

where U denotes the unitary representation of the Lorentz transformation derived in the preceding section. Physically, this means that the quantum object

[2] As mentioned earlier the source must lie within a distance $l < g^{-1}$ from the world line of the detector. This condition implies that $\mathrm{d}a/\mathrm{d}\tau_{\rm ret} > 0$ and that, therefore, (54) represents a sensible stochastic state vector equation with a positive increment $\mathrm{d}a$. If this condition is violated, obviously no complete continuous monitoring by the detector is possible. As an example one might think of a detector in hyperbolic motion, in which case the observer can outrun the photons radiated by the source.

as well as the environment and the measuring apparatus have to be Lorentz transformed in order to obtain covariance of the stochastic process. In view of the physical meaning of the process as a continuous measurement, this is a plausible prescription which is in full agreement with both quantum mechanics and special relativity.

It remains to derive an expression for the Lindblad operator $L = L(\sigma)$. To this end we consider the density matrix which is given by the covariance of the stochastic process [16],

$$\rho(\sigma, \boldsymbol{x}, \boldsymbol{y}) = E\left[\psi(\sigma, \boldsymbol{x})\psi^{\dagger}(\sigma, \boldsymbol{y})\gamma^{0}\not{n}\frac{1}{n^{0}}\right], \tag{62}$$

and which describes the continuous measurement on the non-selective level. Its equation of motion is obtained from (54) as a closed equation in Lindblad form

$$\begin{aligned} \mathrm{d}\rho(\sigma) = & -\mathrm{i}[\mathrm{d}G(\sigma), \rho(\sigma)] \\ & + \left(L(\sigma)\rho(\sigma)L^{\dagger}(\sigma) - \frac{1}{2}L^{\dagger}(\sigma)L(\sigma)\rho(\sigma) - \frac{1}{2}\rho(\sigma)L^{\dagger}(\sigma)L(\sigma)\right)\mathrm{d}a. \end{aligned} \tag{63}$$

Given a certain detector wordline $y(\tau)$ the solution of the stochastic evolution equation (54) yields a stochastic process which corresponds to a specific foliation through the hypersurfaces $\sigma(\tau)$. In this sense, each detector path constitutes a specific measurement scheme, that is, a certain method to gain complete information on the source. Turning from a description in terms of a stochastic process $\psi(\sigma)$ to a description in terms of a density matrix $\rho(\sigma)$ corresponds to a transition from a selective to a non-selective measurement. Therefore, the resultant density matrix equation only depends upon the structure of the quantum object (source) and of the quantum probe (electromagnetic field) and *not* on the detector path $y(\tau)$: Once the field variables have been eliminated the dynamics of the source on the non-selective level does not depend on the way one detects its radiated field.

The conclusion from the foregoing is that the density matrix $\rho(\sigma)$ as determined by (63) must be path-independent, that is it must be integrable in the same sense as described in the preceding section. A straightforward calculation shows that integrability holds if the condition

$$\mathrm{d}L(\sigma) + \mathrm{i}[\mathrm{d}G(\sigma), L(\sigma)] = R(\sigma)\mathrm{d}a \tag{64}$$

is fulfilled, where $R(\sigma)$ may be any operator. It can be shown that (64) has a unique solution $L = L(\sigma)$ once the Lindblad operator has been specified in the rest frame of the source. This is an important conclusion to be drawn from our theory: The path-independence of the density matrix follows from the path-independence of the measurement on the non-selective level and leads to the integrability condition (64) which, in turn, allows the determination of the Lindblad operator $L(\sigma)$.

For the present example of a two-level atom we know from the non-relativistic theory that in the rest frame of the source the Lindblad operator takes on the form $L_0 = \sqrt{\Gamma_0}\Phi_g\Phi_e^{\dagger}$, where Γ_0 is the spontaneous emission rate (measured in the rest frame of the source) and Φ_g, Φ_e are ground and excited state, respectively (stationary states of the atomic Dirac equation). The corresponding solution of the integrability condition then reads

$$L(\sigma) = \sqrt{\Gamma_0}\psi_g(\sigma)\psi_e^\dagger(\sigma)\gamma^0 \not{n}\frac{1}{n^0}, \tag{65}$$

where $\psi_g(\sigma)$ and $\psi_e(\sigma)$ are obtained by restricting the spacetime functions $\Phi_g(x)$, $\Phi_e(x)$ to the hypersurface σ. This completes the construction of a relativistic stochastic state vector dynamics for the continuous measurement through direct photodetection.

3.2 Generalization to the Case of Several Lindblad Operators

The above stochastic evolution equation can easily be generalized to the case of an arbitrary number of Lindblad operators. The corresponding density matrix equation then takes on the general form for a Markovian quantum dynamical semigroup.

If we have any number of Lindblad generators $L_\lambda(\sigma)$, labeled by some index λ, we consider the corresponding piecewise deterministic process given by,

$$\begin{aligned} \mathrm{d}\psi(\sigma) &= -\mathrm{i}\mathrm{d}G(\sigma)\psi(\sigma) \\ &\quad -\frac{1}{2}\sum_\lambda \left(L_\lambda^\dagger(\sigma)L_\lambda(\sigma) - \langle L_\lambda^\dagger(\sigma)L_\lambda(\sigma)\rangle\right)\psi(\sigma)\mathrm{d}a_\lambda \\ &\quad +\sum_\lambda \left(\frac{L_\lambda(\sigma)\psi(\sigma)}{\sqrt{\langle L_\lambda^\dagger(\sigma)L_\lambda(\sigma)\rangle}} - \psi(\sigma)\right)\mathrm{d}N_\lambda(a_\lambda). \end{aligned} \tag{66}$$

The quantities $\mathrm{d}N_\lambda(a_\lambda)$ are independent Poisson increments which satisfy the relations,

$$E\left[\mathrm{d}N_\lambda(a_\lambda)\right] = \langle L_\lambda^\dagger(\sigma)L_\lambda(\sigma)\rangle \mathrm{d}a_\lambda, \quad \mathrm{d}N_\lambda(a_\lambda)\mathrm{d}N_{\lambda'}(a_{\lambda'}) = \mathrm{d}N_\lambda(a_\lambda)\delta_{\lambda\lambda'}. \tag{67}$$

It is then easy to verify that the associated density matrix is governed by an equation in Lindblad form,

$$\begin{aligned} \mathrm{d}\rho(\sigma) &= -\mathrm{i}[\mathrm{d}G(\sigma), \rho(\sigma)] \\ &+\sum_\lambda \left(L_\lambda(\sigma)\rho(\sigma)L_\lambda^\dagger(\sigma) - \frac{1}{2}L_\lambda^\dagger(\sigma)L_\lambda(\sigma)\rho(\sigma) - \frac{1}{2}\rho(\sigma)L_\lambda^\dagger(\sigma)L_\lambda(\sigma)\right)\mathrm{d}a_\lambda. \end{aligned} \tag{68}$$

The physical interpretation of this equation of motion is similar to the one given in the preceding subsection. In particular, the requirement of integrability of the density matrix equation leads to the following sufficient conditions,

$$\mathrm{d}L_\lambda(\sigma) + \mathrm{i}[\mathrm{d}G(\sigma), L_\lambda(\sigma)] = R_\lambda(\sigma)\mathrm{d}a_\lambda \tag{69}$$

and

$$[L_\lambda(\sigma), L_{\lambda'}(\sigma)] = 0. \tag{70}$$

The first integrability condition (69) is the analogue of condition (64) and must be satisfied for all λ. The second integrability condition (70) is required to hold for all pairs $\lambda \neq \lambda'$ of indices for which $\mathrm{d}a_\lambda \wedge \mathrm{d}a_{\lambda'} \neq 0$. This condition expresses the requirement of causality [22]. For example, we could have two different Lindblad operators $L_\lambda(\sigma)$ and $L_{\lambda'}(\sigma)$ which describe local measurements at two different space-time points. If these points are located on the same hypersurface σ they are separated by a spacelike distance and, consequently, the corresponding operators must commute in order to guarantee causality.

Throughout this paper we have assumed that the reduced system dynamics is Markovian, giving rise to an equation of motion for the density matrix in Lindblad form. We remark that this is, of course, an approximation the validity of which depends on the nature of the interaction and/or the measuring device. A non-Markovian dynamics emerges, for example, if the measuring device involves a quantum probe which is in an entangled quantum state [22].

4 Conclusions

It is sometimes claimed that the state reduction postulate of quantum mechanics is incompatible in some sense with the principles of special relativity. However, as has been shown in this paper, quite the contrary is true if the state reduction postulate formulated in section 3 is used. Regardless of any philosophical interpretation, the state reduction postulate gives a prescription for the change of probability amplitudes for future measurements *conditioned on the outcome of presently known measurements.* Relativistic quantum theory then leads to the conclusion that the wave function is not a function on the space-time continuum but must be regarded as a functional on the set of spacelike hypersurfaces. Consequently, the key point of our construction is that the Lorentz group acts on the Hilbert bundle when measurement interactions are taken into account and a reduced description in terms of the quantum object's wave function is adopted. Thus, if we agree to consider the state vector as a functional on the set of spacelike hypersurfaces no conflict with special relativity arises and a fully consistent description of continuous measurements in terms of a relativistically covariant piecewise deterministic process for the state vector can be developed.

Acknowledgment
We would like to thank the Istituto Italiano per gli Studi Filosofici in Naples for financial support of this research project.

References

1. V. B. Braginsky, F. Ya. Khalili: *Quantum measurement* (Cambridge University Press, Cambridge 1992)
2. M. O. Scully and M. S. Zubairy: *Quantum Optics* (Cambridge University Press, Cambridge 1997)

3. H. M. Wiseman and G. Milburn: Phys. Rev. **A47**, 642 (1993); Phys. Rev. **A47**, 1652 (1993)
4. H. P. Breuer and F. Petruccione: Fortschr. Phys. **45**, 39 (1997)
5. J. Dalibard, Y. Castin and K. Mølmer: Phys. Rev. Lett. **68**, 580 (1992)
6. R. Dum, P. Zoller and H. Ritsch: Phys. Rev. **A45**, 4879 (1992)
7. H. Carmichael: *An Open Systems Approach to Quantum Optics* (Springer, Berlin 1993)
8. H. P. Breuer and F. Petruccione: Phys. Rev. Lett. **74**, 3788 (1995)
9. Y. Castin, K. Mølmer: Phys. Rev. Lett. **74**, 3772 (1995)
10. F. B. de Jong, R. J. C. Spreeuw, H. B. van Linden van den Heuvell: Phys. Rev. **A55**, 3918 (1997)
11. B. Wolfseder and W. Domcke: Chem Phys. Lett. **235**, 370 (1995)
12. A. Imamoglu, Y. Yamamoto: Phys. Lett. **A191**, 425 (1994) A. Imamoglu: Phys. Rev. **A50**, 3650 (1994)
13. H. P. Breuer, B. Kappler, F. Petruccione: Phys. Rev. **A56**, 2334 (1997)
14. H. P. Breuer, B. Kappler, F. Petruccione: Phys. Rev. **A59**, 1633 (1999)
15. J. D. Bjorken and S. D. Drell: *Relativistic Quantum Mechanics* (McGraw-Hill, New York 1964)
16. H. P. Breuer and F. Petruccione: J. Phys. A: Math. Gen. **31**, 33 (1998)
17. S. S. Schweber: *An Introduction to Relativistic Quantum Field Theory* (Row, Peterson and Company, New York 1961)
18. Y. Aharonov, D. Z. Albert: Phys. Rev. **D29**, 228 (1984)
19. G. C. Ghirardi, R. Grassi, and P. Pearle: Foundations of Physics **20**, 1271 (1990)
20. C. W. Misner, K. S. Thorne, J. A. Wheeler: *Gravitation* (Freeman, San Francisco 1973)
21. C. W. Gardiner: *Quantum Noise* (Springer, Berlin 1991)
22. H. P. Breuer and F. Petruccione: 'Stochastic Unraveling of Relativistic Quantum Measurements'. In: *Open Systems and Measurement in Relativistic Quantum Theory, Proceedings of the Workshop Held at the Istituto Italiano per gli Studi Filosofici, Naples, 1998*, ed. by H. P. Breuer, F. Petruccione (Springer, Berlin 1999)

Probabilistic Results for Six Detectors in a Three-Particle GHZ Experiment

José Acacio de Barros[1,2] and Patrick Suppes[1]

[1] CSLI – Stanford University, Stanford, CA 94305-4115, USA
[2] Dep. de Física – ICE, UFJF, Juiz de Fora, MG, 36036-330, Brazil

Abstract. In this paper we show that the GHZ theorem can be reformulated as a probablistic theorem allowing for inefficiencies in the detectors. We show quantitatively that taking into account these inefficiencies, the published results of the Innsbruck experiment support the nonexistence of a joint probability distribution for the six correlated spin variables, and hence the nonexistence of hidden variables that explain the experimental results.

1 Introduction

One of the most influential papers on the foundations of quantum mechanics is undoubtedly that of Einstein, Podolski and Rosen (EPR), where they analyzed the correlations between observables of entangled two-particle states [1]. In their paper, EPR conclude that, because entangled systems can be perfectly correlated, Bohr's interpretation of quantum mechanics was incomplete, and a new theory using hidden variables was necessary. However, in 1963 John Bell showed that if we wanted a hidden-variable theory for quantum mechanics, as EPR did, this theory would be falsifiable, as it would yield correlations incompatible with the ones predicted by quantum mechanics [2]. In 1982, in a famous experiment, Aspect, Dalibard and Roger showed that quantum mechanical correlations were satisfied, in contradiction to the local hidden-variable predictions [3].

Recently, Greenberger, Horne and Zeilinger (GHZ) proposed another test for quantum mechanics against hidden-variable theories [4]. What makes the GHZ proposal distinct from Bell's inequalities is the fact that, by using the quantum mechanical predicted correlations for entangled states of more than two particles, they could obtain a mathematical contradiction if a local hidden-variable theory were assumed. The contradiction obtained for the case of three particles, as derived by Mermin in [5], goes as follows. Let $|\psi\rangle$ be a three-particle entangled spin state defined by

$$|\psi\rangle = \frac{1}{\sqrt{2}}(|+++\rangle + |---\rangle), \tag{1}$$

where we use the notation $|+++\rangle = |+\rangle_1 \otimes |+\rangle_2 \otimes |+\rangle_3$, and $\hat{\sigma}_{1z}|+\rangle_1 = |+\rangle_1$, $\hat{\sigma}_{1z}|-\rangle_1 = -|-\rangle_1$, $\hat{\sigma}_{1z}$ being the spin observable in the z direction acting on particle 1, and so on. The wavefunction (1) is an eigenstate of the following operators:

$$\hat{A}|\psi\rangle = \hat{\sigma}_{1x}\hat{\sigma}_{2y}\hat{\sigma}_{3y}|\psi\rangle = |\psi\rangle, \quad \hat{B}|\psi\rangle = \hat{\sigma}_{1y}\hat{\sigma}_{2x}\hat{\sigma}_{3y}|\psi\rangle = |\psi\rangle, \tag{2}$$

$$\hat{C}|\psi\rangle = \hat{\sigma}_{1y}\hat{\sigma}_{2y}\hat{\sigma}_{3x}|\psi\rangle = |\psi\rangle, \quad \hat{D}|\psi\rangle = \hat{\sigma}_{1x}\hat{\sigma}_{2x}\hat{\sigma}_{3x}|\psi\rangle = -|\psi\rangle. \tag{3}$$

From equations (2)–(3) we have at once that

$$E(\hat{A}) = E(\hat{B}) = E(\hat{C}) = 1 \tag{4}$$

and

$$E(\hat{D}) = -1. \tag{5}$$

To show the contradiction, we assume there is a hidden variable λ, that is a function of the source that generated the entangled state and that determines the value of each ot the components of spin for the particles. Let us denote by $m_{ix}(\lambda)$ the value ± 1 of the spin $\hat{\sigma}_{ix}$ in the x direction for particle i, $i = 1,2,3$, and by $m_{iy}(\lambda)$ the value of the spin $\hat{\sigma}_{iy}$ in the y direction. Since the expectations are 1, we can write, from (2)–(3), that

$$m_{1x}(\lambda)m_{2y}(\lambda)m_{3y}(\lambda) = 1, \tag{6}$$

$$m_{1y}(\lambda)m_{2x}(\lambda)m_{3y}(\lambda) = 1, \tag{7}$$

$$m_{1y}(\lambda)m_{2y}(\lambda)m_{3x}(\lambda) = 1. \tag{8}$$

But

$$\begin{aligned}\left[m_{1x}m_{2y}m_{3y}\right]\left[m_{1y}m_{2x}m_{3y}\right]\left[m_{1y}m_{2y}m_{3x}\right] &= \left[m_{1x}m_{2x}m_{3x}\right]\left[m_{1y}^2 m_{2x}^2 m_{3y}^2\right] \\ &= \left[m_{1x}m_{2x}m_{3x}\right],\end{aligned} \tag{9}$$

where we dropped the λ to simplify the notation. Equations (6)–(9) imply that

$$E(\hat{D}) = m_{1x}m_{2x}m_{3x} = 1,$$

in clear contradiction with the quantum mechanical prediction given by (5), thus showing that noncontextual hidden-variables are incompatible with the quantum mechanical predictions. Hence, quantum mechanics is not compatible with the completeness requirements of EPR. We must stress at this point that the proof of the GHZ theorem relies on the fact that, in (9), all three terms on the left-hand side of the equation have the same hidden variable. However, if we allow the hidden variable to be contextual, i.e., for its value to depend on the experimental setup given by the observable operators, then no contradiction can be derived, as can be shown by proving that a nonmonotonic upper probability distribution exists that is compatible with the quantum mechanical expectations [6].

The main characteristic of GHZ's proof, however, has a major problem. How can we verify experimentally predictions based on correlation-one statements, since experimentally one cannot obtain events perfectly correlated? This problem was also present in Bell's original paper, where he considered cases where the correlations were one. To "avoid Bell's experimentally unrealistic restrictions",

Clauser, Horne, Shimony and Holt [7] derived a new set of inequalities that would take into account imperfections in the measurement process. However, Bell's inequalities are quite different from the GHZ case, where in the latter it is *necessary* to have experimentally unrealistic perfect correlations to justify the proof leading to a contradiction. This can be seen from the following theorem (a simplified version of the theorem found in [8]), where the previous proof is reduced to pure probabilistic arguments.

Theorem 1 Let $\mathbf{X}_1$, $\mathbf{X}_2$, $\mathbf{X}_3$, $\mathbf{Y}_1$, $\mathbf{Y}_2$ and $\mathbf{Y}_3$ be six ± 1 random variables, representing the outcome of spin measurements in the x and y directions for particles $1, 2$, and 3, and let
(i) $E(\mathbf{X}_1\mathbf{Y}_2\mathbf{Y}_3) = E(\mathbf{Y}_1\mathbf{X}_2\mathbf{Y}_3) = E(\mathbf{Y}_1\mathbf{Y}_2\mathbf{X}_3) = 1$,
(ii) $E(\mathbf{X}_1\mathbf{X}_2\mathbf{X}_3) = -1$.
Then (i)–(ii) imply a contradiction.

In a previous paper [9] we analyzed the Innsbruck three-particle GHZ experiment [10], where a GHZ state was generated, and showed that their experiment was not compatible with a noncontextual hidden-variable theory. To do this, we showed that a set of inequalities for three-particle correlations were necessary and sufficient for the existence of a noncontextual hidden-variable. However, our proof did not use the full set of six detection random variables $\mathbf{X}_1$, $\mathbf{X}_2$, $\mathbf{X}_3$, $\mathbf{Y}_1$, $\mathbf{Y}_2$ and $\mathbf{Y}_3$, but instead used random variables defined as products of the $\mathbf{X}_i$'s and $\mathbf{Y}_i$'s, in a way similar to that used in the proof of Theorem 1. As a consequence, the set of inequalities presented in [9] is not symmetric in all random variables, as are Bell's inequalities. The present paper can be considered as a continuation of our early work [9], as we obtain conditions that, if verified experimentally, guarantee the nonexistence of a joint probability distribution for the particle observables $\mathbf{X}_1$, $\mathbf{X}_2$, $\mathbf{X}_3$, $\mathbf{Y}_1$, $\mathbf{Y}_2$ and $\mathbf{Y}_3$, and hence the nonexistence of a hidden-variable [11]. The main difference from the inequalities obtained here and the ones from [9] is that, in the form presented below in Theorem 2, the inequalities assume a completely symmetric form. Also, since the publication of [9] new data appeared from the group in Innsbruck [12], and in Section 3 we refine the analysis made previously in the light of these new data, concluding once again that no noncontextual hidden variables are compatible with the outcomes of the experiment.

2 Bell-Like Inequalities for the GHZ State

In this Section we want to derive inequalities that guarantee the nonexistence of hidden-variables for the observables used in GHZ's proof. Before we derive those inequalities, it is important to note that if we could measure all the random variables $\mathbf{X}_1$, $\mathbf{X}_2$, $\mathbf{X}_3$, $\mathbf{Y}_1$, $\mathbf{Y}_2$ and $\mathbf{Y}_3$ of Theorem 1 simultaneously, we would have a joint probability distribution. The existence of a joint probability distribution is a necessary and sufficient condition for the existence of a noncontextual hidden variable [11]. Hence, if the quantum mechanical GHZ correlations are obtained, then no such hidden variable exists. However, Theorem 1 still involves probability-one statements. On the other hand, the quantum mechanical correlations present in the GHZ state are so strong that even if we allow for experimental errors, the nonexistence of a joint distribution can still be verified, as we show in the following theorem, which, as we said above, extends the results in [9].

Theorem 2 Let $\mathbf{X}_i$ and $\mathbf{Y}_i$, $1 \leq i \leq 3$, be six ± 1 random variables. Then, there exists a joint probability distribution for all six random variables if and only if the following inequalities are satisfied:
$-2 \leq E(\mathbf{X}_1\mathbf{Y}_2\mathbf{Y}_3) + E(\mathbf{Y}_1\mathbf{X}_2\mathbf{Y}_3) + E(\mathbf{Y}_1\mathbf{Y}_2\mathbf{X}_3) - E(\mathbf{X}_1\mathbf{X}_2\mathbf{X}_3) \leq 2,$
$-2 \leq -E(\mathbf{X}_1\mathbf{Y}_2\mathbf{Y}_3) + E(\mathbf{Y}_1\mathbf{X}_2\mathbf{Y}_3) + E(\mathbf{Y}_1\mathbf{Y}_2\mathbf{X}_3) + E(\mathbf{X}_1\mathbf{X}_2\mathbf{X}_3) \leq 2,$
$-2 \leq E(\mathbf{X}_1\mathbf{Y}_2\mathbf{Y}_3) - E(\mathbf{Y}_1\mathbf{X}_2\mathbf{Y}_3) + E(\mathbf{Y}_1\mathbf{Y}_2\mathbf{X}_3) + E(\mathbf{X}_1\mathbf{X}_2\mathbf{X}_3) \leq 2,$
$-2 \leq E(\mathbf{X}_1\mathbf{Y}_2\mathbf{Y}_3) + E(\mathbf{Y}_1\mathbf{X}_2\mathbf{Y}_3) - E(\mathbf{Y}_1\mathbf{Y}_2\mathbf{X}_3) + E(\mathbf{X}_1\mathbf{X}_2\mathbf{X}_3) \leq 2.$

Proof: The argument is similar to the one found in [9]. To simplify, we use a notation where $x_1y_2y_3$ means $\mathbf{X}_1\mathbf{Y}_2\mathbf{Y}_3 = 1$, $\overline{x_1y_2y_3}$ means $\mathbf{X}_1\mathbf{Y}_2\mathbf{Y}_3 = -1$. We prove first that the existence of a joint probability distribution implies the four inequalities. Then, we have by an elementary probability computation that

$$P(x_1y_2y_3) = P(x_1y_2y_3, y_1x_2y_3, y_1y_2x_3) + P(x_1y_2y_3, \overline{y_1x_2y_3}, y_1y_2x_3) + P(x_1y_2y_3, y_1x_2y_3, \overline{y_1y_2x_3}) + P(x_1y_2y_3, \overline{y_1x_2y_3}, \overline{y_1y_2x_3})$$

and

$$P(\overline{x_1y_2y_3}) = P(\overline{x_1y_2y_3}, y_1x_2y_3, y_1y_2x_3) + P(\overline{x_1y_2y_3}, \overline{y_1x_2y_3}, y_1y_2x_3) + P(\overline{x_1y_2y_3}, y_1x_2y_3, \overline{y_1y_2x_3}) + P(\overline{x_1y_2y_3}, \overline{y_1x_2y_3}, \overline{y_1y_2x_3}),$$

with similar equations for $\mathbf{Y}_1\mathbf{X}_2\mathbf{Y}_3$ and $\mathbf{Y}_1\mathbf{Y}_2\mathbf{X}_3$. But

$$\mathbf{X}_1\mathbf{X}_2\mathbf{X}_3 = (\mathbf{X}_1\mathbf{Y}_2\mathbf{Y}_3)(\mathbf{Y}_1\mathbf{X}_2\mathbf{Y}_3)(\mathbf{Y}_1\mathbf{Y}_2\mathbf{X}_3),$$

and so we have that

$$P(x_1x_2x_3) = P(x_1y_2y_3, y_1x_2y_3, y_1y_2x_3) + P(\overline{x_1y_2y_3}, \overline{y_1x_2y_3}, y_1y_2x_3) + P(x_1y_2y_3, \overline{y_1x_2y_3}, \overline{y_1y_2x_3}) + P(\overline{x_1y_2y_3}, y_1x_2y_3, \overline{y_1y_2x_3})$$

and

$$P(\overline{x_1x_2x_3}) = P(\overline{x_1y_2y_3}, \overline{y_1x_2y_3}, \overline{y_1y_2x_3}) + P(\overline{x_1y_2y_3}, y_1x_2y_3, y_1y_2x_3) + P(x_1y_2y_3, \overline{y_1x_2y_3}, y_1y_2x_3) + P(x_1y_2y_3, y_1x_2y_3, \overline{y_1y_2x_3}).$$

A straightforward computation shows that

$$\begin{aligned} F = {} & 2[P(x_1y_2y_3, y_1x_2y_3, y_1y_2x_3) + P(\overline{x_1y_2y_3}, y_1x_2y_3, y_1y_2x_3) \\ & + P(x_1y_2y_3, \overline{y_1x_2y_3}, y_1y_2x_3) + P(x_1y_2y_3, y_1x_2y_3, \overline{y_1y_2x_3})] \\ & - 2[P(\overline{x_1y_2y_3}, \overline{y_1x_2y_3}, \overline{y_1y_2x_3}) + P(\overline{x_1y_2y_3}, \overline{y_1x_2y_3}, y_1y_2x_3) \\ & + P(\overline{x_1y_2y_3}, y_1x_2y_3, \overline{y_1y_2x_3}) + P(x_1y_2y_3, \overline{y_1x_2y_3}, \overline{y_1y_2x_3})], \end{aligned}$$

where F is defined by

$$F = E(\mathbf{X}_1\mathbf{Y}_2\mathbf{Y}_3) + E(\mathbf{Y}_1\mathbf{X}_2\mathbf{Y}_3) + E(\mathbf{Y}_1\mathbf{Y}_2\mathbf{X}_3) - E(\mathbf{X}_1\mathbf{X}_2\mathbf{X}_3).$$

Since all probabilities are non-negative and sum to ≤ 1, we infer the first inequality at once. The derivation of the other inequalities is similar.

Now for the sufficiency part. First, we assume the symmetric case where

$$E(\mathbf{X}_1\mathbf{Y}_2\mathbf{Y}_3) = E(\mathbf{Y}_1\mathbf{X}_2\mathbf{Y}_3) = E(\mathbf{Y}_1\mathbf{Y}_2\mathbf{X}_3) = 2p - 1, \tag{10}$$

and

$$E(\mathbf{X}_1\mathbf{X}_2\mathbf{X}_3) = -(2p - 1). \tag{11}$$

Then, the first inequality yields

$$\frac{1}{4} \leq p \leq \frac{3}{4}, \tag{12}$$

while the other ones yield

$$0 \leq p \leq 1. \tag{13}$$

Since $\mathbf{X}_i$ and $\mathbf{Y}_i$ are ± 1 random variables, p has to belong to the interval $[0, 1]$, and inequality (13) doesn't add anything new. We will prove the existence of a joint probability distribution for this symmetric case by showing that, given any p, $\frac{1}{4} \leq p \leq \frac{3}{4}$, we can assign values to the atoms that have the proper marginal distributions.

The probability space for $\mathbf{X}_i$ and $\mathbf{Y}_i$ has 64 atoms. It is difficult to handle a problem of this size, so we will assume some further symmetries to reduce the number of atoms. We will first introduce the following notation: if a group of symbols is between square brackets, all the possible permutations of the barred symbol will considered. For example, we assume $a_5 = P([\bar{x}_1x_2x_3]y_1y_2y_3)$, meaning that $P(\bar{x}_1x_2x_3y_1y_2y_3) = a_5$, $P(x_1\bar{x}_2x_3y_1y_2y_3) = a_5$, and $P(x_1x_2\bar{x}_3y_1y_2y_3) = a_5$. Then, the number of atoms in the problem is reduced to the following 16: $a_1 = P(x_1x_2x_3y_1y_2y_3)$, $a_2 = P(x_1x_2x_3\bar{y}_1\bar{y}_2\bar{y}_3)$, $a_3 = P(x_1x_2x_3[\bar{y}_1y_2y_3])$, $a_4 = P(x_1x_2x_3[\bar{y}_1\bar{y}_2y_3])$, $a_5 = P([\bar{x}_1x_2x_3]y_1y_2y_3)$, $a_6 = P([\bar{x}_1x_2x_3]\bar{y}_1\bar{y}_2\bar{y}_3)$, $a_7 = P([\bar{x}_1x_2x_3][\bar{y}_1y_2y_3])$, $a_8 = P([\bar{x}_1x_2x_3][\bar{y}_1\bar{y}_2y_3])$, $a_9 = P([\bar{x}_1\bar{x}_2x_3][\bar{y}_1y_2y_3])$, $a_{10} = P([\bar{x}_1\bar{x}_2x_3][\bar{y}_1\bar{y}_2y_3])$, $a_{11} = P([\bar{x}_1\bar{x}_2x_3]y_1y_2y_3)$, $a_{12} = P([\bar{x}_1\bar{x}_2x_3]\bar{y}_1\bar{y}_2\bar{y}_3)$, $a_{13} = P(\bar{x}_1\bar{x}_2\bar{x}_3[\bar{y}_1y_2y_3])$, $a_{14} = P(\bar{x}_1\bar{x}_2\bar{x}_3[\bar{y}_1\bar{y}_2y_3])$, $a_{15} = P(\bar{x}_1\bar{x}_2\bar{x}_3y_1y_2y_3)$, $a_{16} = P(\bar{x}_1\bar{x}_2\bar{x}_3\bar{y}_1\bar{y}_2\bar{y}_3)$.

These new added symmetries reduce the problem from 64 to 16 variables. The atoms have to satisfy various sets of equations. The first set comes just from the requirement that $E(\mathbf{X}_i) = E(\mathbf{Y}_i) = 0$, for $i = 1, 2, 3$, but two of the six equations are redundant, and so we are left with the following four.

$$\begin{aligned} &a_1 + a_2 + 3a_3 + 3a_4 + a_5 + a_6 + 3a_7 + 3a_8 - 3a_9 \\ &-3a_{10} - a_{11} - a_{12} - 3a_{13} - 3a_{14} - a_{15} - a_{16} \quad = 0, \end{aligned} \tag{14}$$

$$\begin{aligned} &a_1 - a_2 + a_3 - a_4 + 3a_5 - 3a_6 + 3a_7 - 3a_8 + 3a_9 \\ &-3a_{10} + 3a_{11} - 3a_{12} + a_{13} - a_{14} + a_{15} - a_{16} \quad = 0, \end{aligned} \tag{15}$$

$$\begin{aligned} &a_1 - a_2 + a_3 - a_4 + 3a_5 - 3a_6 + 3a_7 - 3a_8 + 3a_9 \\ &-3a_{10} + 3a_{11} - 3a_{12} - a_{13} + a_{14} + a_{15} - a_{16} \quad = 0, \end{aligned} \tag{16}$$

$$\begin{aligned} &a_1 + a_2 + 3a_3 + 3a_4 - a_5 - a_6 - 3a_7 - 3a_8 + 3a_9 \\ &+3a_{10} + a_{11} + a_{12} - 3a_{13} - 3a_{14} - a_{15} - a_{16} \quad = 0, \end{aligned} \tag{17}$$

where (14) comes from $E(\mathbf{X}_1) = 0$, (15) from $E(\mathbf{X}_2) = 0$, (16) from $E(\mathbf{Y}_1) = 0$, and (17) from $E(\mathbf{Y}_2) = 0$. The triple expectations also imply

$$\begin{aligned} &a_1 + a_2 + 3a_3 + 3a_4 - 3a_5 - 3a_6 - 9a_7 - 9a_8 + 9a_9 \\ &+9a_{10} + 3a_{11} + 3a_{12} - 3a_{13} - 3a_{14} - a_{15} - a_{16} \quad = -2p + 1, \end{aligned} \tag{18}$$

$$\begin{aligned} &a_1 + a_2 - a_4 - a_4 + a_5 + a_6 - a_7 - a_8 + a_9 \\ &+a_{10} - a_{11} - a_{12} + a_{13} + a_{14} - a_{15} - a_{16} \quad = 2p - 1, \end{aligned} \tag{19}$$

and

$$\begin{aligned} &a_1 - a_2 - 3a_3 + 3a_4 + 3a_5 - 3a_6 - 9a_7 + 9a_8 - 9a_9 \\ &+9a_{10} + 3a_{11} - 3a_{12} - 3a_{13} + 3a_{14} + a_{15} - a_{16} \quad = 2p - 1. \end{aligned} \tag{20}$$

Finally, the probabilities of all atoms have to sum to one, yielding the last equation

$$\begin{aligned} &a_1 + a_2 + 3a_3 + 3a_4 + 3a_5 + 3a_6 + 9a_7 + 9a_8 + 9a_9 \\ &+9a_{10} + 3a_{11} + 3a_{12} + 3a_{13} + 3a_{14} + a_{15} + a_{16} \quad = 1. \end{aligned} \tag{21}$$

Even with the symmetries reducing the problem to 16 variables, we still have an infinite number of solutions that satisfy Eqns. (14)–(21). Since it is very hard to exhibit a general solution for (14)–(21) with the constraints $0 \le a_i \le 1$, $i = 1 \ldots 16$, we will just show that particular solutions exist for an arbitrary p satisfying the inequality (12). To do so, we will divide the problem into two parts: one where we will exhibit an explicit solution for the atoms $a_1, \ldots, a_{16}$ that form a proper probability distribution for $p \in [\frac{1}{4}, \frac{1}{2}]$, and another explicit solution for $p \in [\frac{1}{2}, \frac{3}{4}]$.

It is easy to verify that, given an arbitrary p in $[\frac{1}{4}, \frac{1}{2}]$, the following set of values constitute a solution of Eqns. (14)–(21): $a_1 = 0$, $a_2 = -\frac{1}{2} + 2p$, $a_3 = \frac{1}{4} - \frac{1}{2}p$, $a_4 = 0$, $a_5 = 0$, $a_6 = 0$, $a_7 = 0$, $a_8 = 0$, $a_9 = 0$, $a_{10} = 0$, $a_{11} = 0$, $a_{12} = \frac{1}{4} - \frac{1}{2}p$, $a_{13} = 0$, $a_{14} = 0$, $a_{15} = p$, $a_{16} = 0$. For p in $[\frac{1}{2}, \frac{3}{4}]$ the following set of values constitute a solution of Eqns. (14)–(21): $a_1 = -\frac{1}{8} + \frac{1}{2}p$, $a_2 = 0$, $a_3 = \frac{3}{8} - \frac{1}{2}p$, $a_4 = 0$, $a_5 = -\frac{5}{24} + \frac{1}{3}p$, $a_6 = -\frac{1}{24} + \frac{1}{6}p$, $a_7 = 0$, $a_8 = 0$, $a_9 = 0$, $a_{10} = 0$, $a_{11} = 0$, $a_{12} = 0$, $a_{13} = 0$, $a_{14} = \frac{1}{8}$, $a_{15} = \frac{3}{8} - \frac{1}{2}p$, $a_{16} = 0$. So, for p satisfying the inequality $\frac{1}{4} \le p \le \frac{3}{4}$ we can always construct a probability distribution for the atoms consistent with the marginals, and this concludes the proof. ◇

We note that the form of the inequalities of Theorem 2 is actually that of the Clauser et al. inequalities [7] for the Bell case, when the Bell binary correlations are replaced by the GHZ triple correlations. The inequalities from Theorem 2 immediately yield the following.

Corollary Let $\mathbf{X}_i$ and $\mathbf{Y}_i$, $1 \le i \le 3$, be six ± 1 random variables, and let
(i) $E(\mathbf{X}_1\mathbf{Y}_2\mathbf{Y}_3) = E(\mathbf{Y}_1\mathbf{X}_2\mathbf{Y}_3) = E(\mathbf{Y}_1\mathbf{Y}_2\mathbf{X}_3) = 1 - \varepsilon$,
(ii) $E(\mathbf{X}_1\mathbf{X}_2\mathbf{X}_3) = -1 + \varepsilon$, with $\varepsilon \in [0, 1]$.
Then there cannot exist a joint probability distribution of $\mathbf{X}_i$ and $\mathbf{Y}_i$, $1 \le i \le 3$, satisfying (i) and (ii) if $\varepsilon < \frac{1}{2}$.

Proof. If a joint probability exists, then

$$-2 \le E(\mathbf{X}_1\mathbf{Y}_2\mathbf{Y}_3) + E(\mathbf{Y}_1\mathbf{X}_2\mathbf{Y}_3) + E(\mathbf{Y}_1\mathbf{Y}_2\mathbf{X}_3) - E(\mathbf{X}_1\mathbf{X}_2\mathbf{X}_3) \le 2.$$

But

$$E(\mathbf{X}_1\mathbf{Y}_2\mathbf{Y}_3) + E(\mathbf{Y}_1\mathbf{X}_2\mathbf{Y}_3) + E(\mathbf{Y}_1\mathbf{Y}_2\mathbf{X}_3) - E(\mathbf{X}_1\mathbf{X}_2\mathbf{X}_3) = 4 - 4\varepsilon,$$

and the inequality is satisfied only if $\varepsilon \ge \frac{1}{2}$. Hence, if $\varepsilon < \frac{1}{2}$ no joint probability exists. ♢

In the Corollary, ε may represent, for example, a deviation, due to experimental errors, from the predicted quantum mechanical correlations. So, we see that to prove the nonexistence of a joint probability distribution for the GHZ experiment, we do not need to have perfect measurements and 1 or -1 correlations. In fact, from the inequality obtained above, it is clear that any experiment that satisfies the strong symmetry of the Corollary and obtains a correlation for the triples stronger than 0.5 (and -0.5 for one of them) cannot have a joint probability distribution. It is worth mentioning at this point that the inequalities derived in Theorem 2 have a completely different origin than do Bell's inequalities. The inequalities of Theorem 2 are not satisfied by a particular model, but they just accommodate the theoretical conditions in *GHZ* to possible experimental deviations. Also, Theorem 2 does not rely on any "enhancement" hypothesis to reach its conclusion. Thus, with this reformulation of the *GHZ* theorem it is possible to use strong, yet imperfect, experimental correlations to prove that a hidden-variable theory is incompatible with the experimental results.

To complete this Section, we will now present a result that is similar to one obtained by Garg and Mermin for the case of Bell's inequalities [13]. Garg and Mermin constructed a set of eight ± 1 random variables that satisfy Bell's inequalities, but do not have a joint probability distribution. For our example, it is sufficient to increase the number of random variables from six to seven.

Theorem 3 Let $\mathbf{X}_i$, $\mathbf{Y}_i$, and $\mathbf{Y}_1'$, $1 \le i \le 3$, be seven ± 1 random variables with expectations equal to zero. The satisfaction of the inequalities of Theorem 2 by the subsets of random variables $\{\mathbf{X}_1, \mathbf{X}_2, \mathbf{X}_3, \mathbf{Y}_1, \mathbf{Y}_2, \mathbf{Y}_3\}$ and $\{\mathbf{X}_1, \mathbf{X}_2, \mathbf{X}_3, \mathbf{Y}_1', \mathbf{Y}_2, \mathbf{Y}_3\}$ does not guarantee the existence of a joint probability distribution for the full set of random variables.

Proof. We prove Theorem 3 by constructing an example of seven variables not having a joint probability distribution. Say we have the following expectations, all satisfying the inequalities of Theorem 2:

$$E(\mathbf{Y}_1\mathbf{X}_2\mathbf{X}_3) = E(\mathbf{X}_1\mathbf{Y}_2\mathbf{X}_3) = E(\mathbf{X}_1\mathbf{X}_2\mathbf{Y}_3) = E(\mathbf{Y}_1\mathbf{Y}_2\mathbf{Y}_3) = -1, \tag{22}$$

$$E(\mathbf{Y}_1'\mathbf{X}_2\mathbf{X}_3) = E(\mathbf{Y}_1'\mathbf{Y}_2\mathbf{Y}_3) = 0, \quad E(\mathbf{X}_1\mathbf{Y}_2\mathbf{X}_3) = E(\mathbf{X}_1\mathbf{X}_2\mathbf{Y}_3) = -1. \tag{23}$$

Since $E(\mathbf{Y}_1'\mathbf{X}_2\mathbf{X}_3) = E(\mathbf{Y}_1'\mathbf{Y}_2\mathbf{Y}_3) = 0$, a joint probability for the six variables of (23) exists with the correlations

$$E(\mathbf{Y}_1'\mathbf{Y}_2) = 1, \quad E(\mathbf{Y}_1'\mathbf{X}_2) = 1, \quad E(\mathbf{Y}_1'\mathbf{Y}_3) = 1, \quad E(\mathbf{Y}_1'\mathbf{X}_3) = -1.$$

From the correlations fixed above, we can obtain immediately that

$$E(\mathbf{X}_2\mathbf{Y}_2) = 1, \quad E(\mathbf{X}_3\mathbf{Y}_3) = -1.$$

But, we have also that

$$E(\mathbf{Y}_1\mathbf{Y}_2\mathbf{Y}_3) = E((\mathbf{Y}_1\mathbf{X}_2\mathbf{X}_3)(\mathbf{X}_2\mathbf{Y}_2)(\mathbf{X}_3\mathbf{Y}_3)) = -E(\mathbf{Y}_1\mathbf{X}_2\mathbf{X}_3),$$

thus showing that the assumption of a joint probability distribution leads to a contradiction with (22). ◇

The counterexample to the existence of a joint probability distribution just constructed, and also the one constructed by Garg and Mermin for Bell's inequalities, simply illustrate a well-known feature of probability theory that has nothing to do with quantum mechanics. The feature is this. Given marginal probability distributions for pairs, triples, or, in general, n-tuples, $n = 1, 2, \ldots,$ of random variables, there will not exist a joint probability distribution for m-tuples of random variables, $m > n$, without making special assumptions. Of course, distributions satisfying special assumptions are of great importance, e.g., the Gaussian or normal distribution of n random variables for which only means, variances and covariances are needed to fix the distribution uniquely. Consequently, for a Gaussian, all higher moments are already fixed by the first and second moments.

It is also to be emphasized that giving inequalities for larger numbers of random variables, either in the Bell or GHZ case, necessary and sufficient for the existence of a joint probability distribution is difficult, and very likely intractable for arbitrary n. For a sense of the difficulties in the Bell case, see Suppes and Zanotti [11].

3 Analysis of the Innsbruck Experiment

In this Section, our results from [9] are applied to the most recent experimental results of the Innsbruck group [10,12]. Before we proceed, we need to lay down the experimental setup that we consider. A schematic of the Innsbruck experiment is shown in Figure 1. In this experiment an UV laser pulse is sent into a nonlinear crystal, generating correlated pairs of photons, that can be detected in four detectors T, D_1, D_2, and D_3. In the Figure, "Pol." are polarizers, "BS" is a 50-50 beam splitter, and $\lambda/2$ is a half-wave plate. There is a small, yet nonzero, probability that two pairs of photons are generated simultaneously, where by simultaneous we mean within a window of observation Δt. It can be shown theoretically that, for the case of two-pair production, when one of the photons is detected in the trigger detector T, the other three photons detected in D_1,

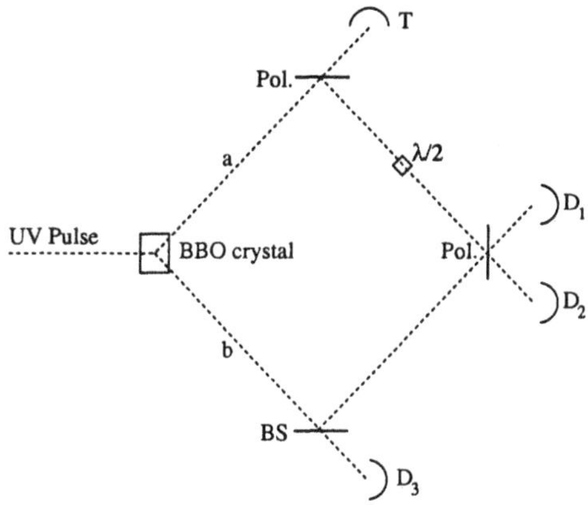

Fig. 1. Schematics of the GHZ experiment of Bouwmeester et al.

D_2, and D_3 are in a GHZ entangled state. In other words, if we conditionalize the experimental data to a detection at T, then D_1, D_2, and D_3 should show a GHZ correlation. There are several different sources of inefficiencies that we have taken into account in [9]: the detection efficiency, the dark-count rate, and the misfiring probability, just to name a few, but important, examples. We use the analysis from [9] of what happens to the GHZ correlations in a setup like Figure 1 when the detectors have efficiency $d \in [0,1]$ and a dark-count rate of probability γ. We assumed that the misfiring probabilities are negligible, as the polarizers used in the wavelengths in question are very efficient. The analysis also assumed that whenever a double pair is produced, this double pair has the expected GHZ correlation for D_1, D_2, and D_3 if T correctly detects a photon. The probability of no photon being generated and having a fourfold coincidence registered was considered negligible, as this would require a dark-count rate much bigger than found in modern photon detectors. Also, triple-pair creations were neglected because they are of extremely low probability compared to the other events. In [9] we obtained as the expression for the spin correlations the following.

$$E\left(\mathbf{S}_1\mathbf{S}_2\mathbf{S}_3 \mid td_1d_2d_3\right) = \frac{E\left(\mathbf{S}_1\mathbf{S}_2\mathbf{S}_3 \mid td_1d_2d_3 \;\&\; GHZ\right)}{\left[1+6\frac{P(p_1p_2)}{P(p_1\ldots p_4)}\frac{\gamma^2}{d^2}\right]}. \tag{24}$$

This value is the corrected expression for the correlations if we have detector efficiency taken into account.

We estimate the values of γ and d to see how much $E\left(\mathbf{S}_1\mathbf{S}_2\mathbf{S}_3 \mid td_1d_2d_3\right)$ would change due to experimental errors. For that purpose, we will use typical rates of detectors [14] for the frequency used at the Innsbruck experiment, as well as their reported data [10]. First, modern detectors usually have $d \approx 0.5$ for the wavelengths used at Innsbruck. We assume a dark-count rate of about 3×10^2 counts/s. With a time window of coincidence measurement of 2×10^{-9} s, we then have that the probability of a dark count in this window is $\gamma = 6\times10^{-7}$. From [10] we use that the ratio $P(p_1p_2)/P(p_1\ldots p_2)$ is on the order of 10^{10}. Substituting these numerical values in (24) we have $E(\mathbf{S}_1\mathbf{S}_2\mathbf{S}_3 \mid td_1d_2d_3) \cong 0.9$. From this expression it is clear that the change in correlation imposed by the dark-count rates is significant for the given parameters. However, it is also clear that the value of the correlation is quite sensitive to changes in the values of both γ and d.

We can now compare the values we obtained with the ones observed by Bouwmeester et al. for GHZ and $\overline{GHZ}$ states [12]. Using the notation from previous sections, they report that for the settings $\mathbf{Y}_1\mathbf{X}_2\mathbf{X}_3$, $\mathbf{X}_1\mathbf{Y}_2\mathbf{X}_3$, and $\mathbf{X}_1\mathbf{X}_2\mathbf{Y}_3$ a fraction of 0.85 ± 0.04 had the predicted quantum mechanical correlation, i.e.,

$$E(\mathbf{Y}_1\mathbf{X}_2\mathbf{X}_3) + E(\mathbf{X}_1\mathbf{Y}_2\mathbf{X}_3) + E(\mathbf{X}_1\mathbf{X}_2\mathbf{Y}_3) = 2.55.$$

However, for the experimental setting $\mathbf{Y}_1\mathbf{Y}_2\mathbf{Y}_3$ they report a correlation

$$E(\mathbf{Y}_1\mathbf{Y}_2\mathbf{Y}_3) = -0.74.$$

If we use the inequalities from Theorem 2, we see that the first inequality is clearly violated, as

$$E(\mathbf{Y}_1\mathbf{X}_2\mathbf{X}_3) + E(\mathbf{X}_1\mathbf{Y}_2\mathbf{X}_3) + E(\mathbf{X}_1\mathbf{X}_2\mathbf{Y}_3) - E(\mathbf{Y}_1\mathbf{Y}_2\mathbf{Y}_3) = 3.29 > 2.$$

Using Theorem 2, this result strongly supports the nonexistence of a joint probability distribution.

References

1. A. Einstein, B. Podolski, N. Rosen: Phys. Rev. **47,** 777 (1935)
2. J. S. Bell: Speakable and Unspeakable in Quantum Mechanics (Cambridge University Press, Cambridge, 1987)
3. A. Aspect, J. Dalibard, and G. Roger: Phys. Rev. Lett. **49,** 1804 (1982)
4. D. M. Greenberger, M. Horne, and A. Zeillinger: 'Going Beyond Bell's Theorem'. In: *Bell's Theorem, Quantum Theory, and Conceptions of the Universe,* ed. by M. Kafatos (Kluwer, Dordrecht, 1989) pp. 69–72
5. N. D. Mermin: Phys. Rev. Lett. **65**, 1838 (1990).
6. J. Acacio de Barros and P. Suppes: 'Some conceptual issues involving probabilities in quantum mechanics'. To appear in: *Reverberations of a Shaky Game,* ed. by Roger Jones and Philip Ehrlich, quant-ph/0001017.
7. J. F. Clauser, M. A. Horne, A. Shimony, R. A. Holt: Phys. Rev. Lett. **23,** 880 (1969)
8. P. Suppes, J. Acacio de Barros, G. Oas: 'A Collection of Hidden-variable Theorems and Counterexamples'. In: *Waves, Information and Foundations of Physics,* ed. by Riccardo Pratesi and Laura Ronchi (Italian Physics Society, Bologna, 1998) pp. 267–292.
9. J. Acacio de Barros, P. Suppes: Phys. Rev. Lett. **84,** 793 (2000)
10. D. Bouwmeester, J-W. Pan, M. Daniell, H. Weinfurter, A. Zeilinger: Phys. Rev. Lett. **82,** 1345 (1999).
11. P. Suppes, M. Zannoti: Synthese **48,** 191 (1981)

12. J-W. Pan, D. Bouwmeester, M. Daniell, H. Weinfurter, A. Zeilinger: Nature **403,** 515 (2000)
13. A. Garg, N. D. Mermin: Phys. Rev. Lett. **49,** 1220 (1982)
14. Single photon count module specifications for EG&G's SPCM-AG series were obtained from EG&G's web page at `http://www.egginc.com`.

Classical Versus Quantum Probabilities

Enrico G. Beltrametti

Department of Physics, University of Genoa
and Istituto Nazionale di Fisica Nucleare, Sezione di Genova
via Dodecaneso 33, I-16146 Genova, Italy

Abstract. Some features that accompany the branching between classical and quantum probabilities are reviewed. We first review some aspects of the so called logical approach in which the states are viewed as probability measures on ordered structures, the latter being Boolean algebras in the classical case and nondistributive orthomodular lattices in the quantum case. The problem of providing intrinsic characterizations of classical and of quantum probabilities is also summarized. Generalizations of the usual classical probability framework that are able to host typical features of quantum probabilities are discussed: we consider in particular an approach which rests on a generalization of the usual notion of random variables.

1 Introduction

It is generally agreed that the probabilistic structure underlying quantum theories has non classical aspects. Quantum phenomena often do not fit with the standard, or Kolmogorovian, probability theory: we might think, for instance, of interferometry phenomena, of empirical violations of Bell inequalities, and so on.

In Sect. 2 we briefly recall the standard mathematical frames accompanying the statistical aspects of classical and quantum mechanics. In both cases the states of the physical system under discussion appear as probability measures on the *events* and the branching between the two theories can be read out of the different ordered structure associated to the classical and the quantum events. This is an old chapter in the studies in the mathematical foundations of quantum theory, which is briefly summarized in Sect. 3.

The problem of having criteria to decide whether some given set of empirically observed probabilities fits with the classical or the quantum formalism is a nonsimple one: the Bell inequalities provide the most celebrated example of such a criterion. Some remarks on this problem are collected in Sect. 4.

There have been various proposal for "non Kolmogorovian" probability theories. A relevant attempt is provided by the so-called algebraic noncommutative approach [1]. In Sect. 5 we examine an alternative approach which is based on commutative algebras (thus retaining a classical aspects) but admits a class of observables wider than the usual family of random variables. The emerging new probability theory, to be called *operational probability theory*, embodies indeterministic features and hosts interesting extensions of the quantum description.

2 Standard Frames for Classical and Quantum Probabilities

Let us first focus attention on the general mathematical frames hosting the (kinematical) description of classical and quantum mechanics.

In the classical case a measurable space Ω is attached to the physical system under discussion and the states of this system are identified with the probability measures on the σ-Boolean algebra $\mathcal{B}(\Omega)$ of the measurable subsets of Ω: we write $M_1^+(\Omega)$ for the set of these probability measures. The Dirac measures, namely the probability measures concentrated at a point of Ω, are the extreme elements of the convex set $M_1^+(\Omega)$ and represent the pure states. Hence Ω itself can be viewed as the set of the pure states. Notice that, by definition, the convex set $M_1^+(\Omega)$ has the structure of a simplex: any nonpure state determines uniquely the pure states of which it is a convex combination. The real-valued observables are then defined as those affine maps of $M_1^+(\Omega)$ into $M_1^+(\mathbb{R})$ that map Dirac measures on Ω into Dirac measures on $\mathbb{R}$. Thus a real-valued observable A corresponds to a function $f_A : \Omega \to \mathbb{R}$, generally called a random variable. The probability of getting an outcome of the observable A in the set $X \in \mathcal{B}(\mathbb{R})$, given that the state is $\mu \in M_1^+(\Omega)$, is then provided by $\mu(f_A^{-1}(X))$.

In the quantum case a separable, complex Hilbert space $\mathcal{H}$ is attached to the physical system under discussion and the states of this system are identified with the probability measures on the set $\mathcal{P}(\mathcal{H})$ of the projectors (on the closed subspaces) of $\mathcal{H}$: the Gleason theorem ensures that they correspond to the density operators of $\mathcal{H}$ if the dimension of $\mathcal{H}$ is not smaller than 3. The probability measures concentrated at a one-dimensional projector of $\mathcal{H}$ are the extreme elements of the convex set of states and represent the pure states. The "rays" of $\mathcal{H}$ can thus be viewed as the pure states. Notice that the set of states is no longer a simplex: a nonpure state does not determine uniquely the pure states of which it is a convex combination. The real-valued observables are defined as those affine maps of the set of states into $M_1^+(\mathbb{R})$ which come from the self-adjoint operators of $\mathcal{H}$: if A is an observable (we will denote by the same tetter also the corresponding self-adjoint operator), then the probability of getting an outcome of A in $X \in \mathcal{B}(\mathbb{R})$, given that the state is the density operator D, is provided by $\mathrm{tr}(DP_X^A)$, where P_X^A is the projector specified by the pair (A, X) according to the spectral theorem for self-adjoint operators.

Notice that the pair formed by a real-valued observable A and a measurable subset $X \in \mathcal{B}(\mathbb{R})$ carries a notion of dichotomy since we can associate to such a pair the two-valued question "does the outcome of A fall in X?". As seen before, in the classical case the pair (A, X) determines the subset $f_A^{-1}(X) \in \mathcal{B}(\Omega)$ while in the quantum case it determines the projector $P_X^A \in \mathcal{P}(\mathcal{H})$. The pairs of the form (A, X) will be called the events pertaining to the physical system, and the probability of getting an outcome of A in X will be called the probability of occurrence of the event (A, X).

According to the above terminology, we can say that, both in the classical and in the quantum case, the states of the system are the probability measures on the events. Similarly, the real-valued observables are easily seen to be the event-valued measures on $\mathcal{B}(\mathbb{R})$. This points at the fact that the family of events is a plausible building block for the mathematical frames of classical and quantum theories and it should embody the information about the branching between the two theories.

3 Classical Versus Quantum Events

We write $a, b, c, \ldots$ for the events, and $\mathcal{L}$ for the set they form, no matter whether the classical or the quantum description applies. The set of states provides an order relation on the events: $a \leq b$ if each state gives to a a probability of occurrence not bigger than the one given to b. In the classical case, where the events are subsets of $\mathcal{B}(\Omega)$, and similarly in the quantum case where the events can be seen as closed subspaces of $\mathcal{H}$, this amounts to set-theoretic containment [2]. With respect to this order relation, and for any two events a, b, there exist in $\mathcal{L}$, their join $a \vee b$ (i.e., their minimal majorization) and their meet $a \wedge b$ (i.e., their maximal minorization): in other words $\mathcal{L}$ is a lattice. We write $\mathbf{0}$ for the minimal element of $\mathcal{L}$ and $\mathbf{1}$ for the maximal element. The minimal majorizations of $\mathbf{0}$ will be called atoms: every element a of $\mathcal{L}$ turns out to be the join of the atoms it contains, while the join of a and an atom not contained in a is a minimal majorization of a. This property is summarized by saying that $\mathcal{L}$ is an atomic lattice with the covering property.

$\mathcal{L}$ is also equipped with a unary relation $a \to a^{\perp}$ satisfying $a \vee a^{\perp} = \mathbf{1}$ and $a \wedge a^{\perp} = \mathbf{0}$: $a^{\perp}$ is said to be the orthocomplement of a and $\mathcal{L}$ is said to be orthocomplemented. In the classical case a is a subset of Ω and $a^{\perp}$ is its set complement; in the quantum case a is a projector of $\mathcal{H}$ and $a^{\perp}$ is the identity minus that projector [2].

As already outlined in the seminal paper of Birkhoff and von Neumann [3] the key difference between the classical and the quantum case comes out when we look at the distributivity conditions

$$a \vee (b \wedge c) = (a \vee b) \wedge (a \vee c), \quad a \wedge (b \vee c) = (a \wedge b) \vee (a \wedge c).$$

In the classical case these conditions hold true for any triple a, b, c and $\mathcal{L}$ is accordingly called distributive: it is indeed the Boolean algebra $\mathcal{B}(\Omega)$. In the quantum case the above conditions are ensured only for triples in which $a \leq b$ and $c = a^{\perp}$: this weaker property qualifies $\mathcal{L}$, that is $\mathcal{P}(\mathcal{H})$, as an orthomodular (nondistributive) lattice [2].

Classical states are thus probability measures on Boolean algebras while quantum states are probability measures on nondistributive orthomodular lattices.

Orthomodular lattices provide a basic weakening of the Boolean algebras: not only the projectors of a Hilbert space form an orthomodular lattice but the same do the projectors of any von Neumann algebra. Also the idempotents elements of any involutive ring form an orthomodular lattice and a significant correlation emerges: if this orthomodular lattice is nondistributive then the ring is noncommutative [2]. Thus we see that the nondistributivity of $\mathcal{P}(\mathcal{H})$ reflects the noncommutativity of the involutive ring of all bounded operators of $\mathcal{H}$. The occurrence of probability measures on orthomodular nondistributive lattices is typical of noncommutative algebraic frameworks.

While quantum mechanics naturally brings to orthomodularity, we can ask to what extent the orthomodularity of the set of events brings to Hilbert-space quantum mechanics. Let us recall that a number of mathematical result, relevant to this issue, came in the sixties and seventies: think, for instance, of the works of G.W. Mackey [4], V.S. Varadarajan [5], F. Maeda and S. Maeda [6], C. Piron [7]. In particular it emerged (see, e.g., [2]) that if the events (assumed as primitive elements) form an atomic orthomodular lattice $\mathcal{L}$ with the covering property, then there exist a vector space V over an involutive field K and an Hermitean form h such that $\mathcal{L}$ is isomorphic to the lattice of the closed subspaces of the Hermitean space $\{V, K, h\}$. For a while it was conjectured that orthomodularity might be enough to force K into the Archimedean completions $\mathbb{R}, \mathbb{C}, \mathbb{Q}$ of the rationals and $\{V, K, h\}$ into a Hilbert space, until H. Keller [8] put an end to this conjecture by exhibiting Hermitean spaces over nonclassical fields whose closed subspaces form orthomodular lattices. In 1995 M.P. Soler [9] found a bridge on this gap showing that if the Hermitean space $\{V, K, h\}$ has an infinite orthonormal sequence then $K = \mathbb{R}, \mathbb{C}, \mathbb{Q}$ and $\{V, K, h\}$ becomes a Hilbert space.

4 Intrinsic Characterization of Classical and Quantum Probabilities

A state is a probability measure on the events, both in the classical and in the quantum case. The following question naturally arises: given the observed probabilities of occurrence of a number of events are there rules to decide whether these empirical probabilities fit with the classical or the quantum context? As we shall briefly summarize, various results emerged which provide criteria to separate the classical from the nonclassical situation, but more complicated appears the purpose of giving tests to single out the quantum case in a general nonclassical situation.

A general method to determine necessary and sufficient conditions for the classical representability of a family of observed probabilities was already worked out by G. Boole in 1862 [10]: he spoke of *conditions of possible experience*, the quantum experience being not yet in view. Great attention to the question of classical vs quantum representability of empirical probabilities came after Bell's work on the hidden-variable interpretation of the Einstein-Podolsky-Rosen (EPR) correlation, Bell's inequalities being a particular case of Boole's *conditions of possible experience.* It should also be outlined Accardi's proposal [11] of *statistical invariants* to characterize and separate the classical and the quantum representability of probabilities in some simple situation. Let us further quote Gudder and Zanghi [12] for additional results. In 1989 Pitowski [13] formulated his remarkable *polytope approach* in which, for instance, the conditions for classical representability take the form of inequalities like $0 \leq l \leq 1$ where l is a real linear combination of the observed probabilities.

Both Boole's and Pitowski's approaches have exponential complexity since the number of necessary and sufficient conditions increases exponentially with the number of involved events [14]. For instance, in the usual EPR example involving a physical system decaying into two spin-$\frac{1}{2}$ particles, with the probabilities p_1, p_2 of one particle passing a polarization filter along the directions $1, 2$, the probabilities p_3, p_4 of the other particle passing a polarization filter along the directions $3, 4$, and the joint probabilities $p_{1,3}, p_{1,4}, p_{2,3}, p_{2,4}$ of both particles passing the filters, the number of necessary and sufficient conditions for classical representability is 88 (including some trivial conditions). But if we move to the example of a physical system decaying into three spin-$\frac{1}{2}$ particles, and consider the analogous family of probabilities of passing polarization filters, the number of necessary and sufficient conditions for classical representability becomes 51,676.

The exponential complexity of the problem of giving a complete set of necessary and sufficient conditions for classical representability makes interesting the search of simple rules to generate some necessary condition. Here we summarize one such rule. Consider n events and let $p_1, p_2, ..., p_n$ be the probabilities of their occurrence. Suppose that also the probabilities of some joint events are observed, so that the family of the observed probabilities will take the form $\{p_T \mid T \in F\}$ where F is some collection of subsets of $\{1, 2, ..., n\}$. As an example we can think of the usual EPR pattern recalled above, in which n is 4 and F is the family $\{1\}, \{2\}, \{3\}, \{4\}, \{1,3\}, \{1,4\}, \{2,3\}, \{2,4\}$. More generally we could think of a compound physical system in a given state and suppose that we detect the probabilities of occurrence of some events on each subsystem as well as the probabilities of joint occurrence of numbers of the events so generated. Then we have the following result [14,15]: the inequality

$$0 \leq \sum_{T \in F} c_T p_T \leq 1,$$

where the c_T's are real coefficients, is a necessary condition for the classical representability of the observed probabilities whenever,

$$0 \leq \sum_{T \subseteq S} c_T \leq 1 \quad \text{for every } S \subseteq \{1, 2, ..., n\}.$$

As an example, we come again to the quoted EPR correlation and observe that the inequality

$$0 \leq p_1 + p_3 - p_{1,3} - p_{1,4} - p_{2,3} + p_{2,4} \leq 1$$

involves coefficients that satisfy the above requirement so that it is a necessary condition for classical representability: it is indeed a Bell inequality. The result above provides a generalization of the Bell inequality issue.

5 Operational Probability Theory

The question we consider in this section is whether there are natural generalizations of the classical probability theory able to encompass the typical features of quantum probabilities.

The standard classical frame, recalled in Sec. 2, is characterized by choosing for the set of states the simplex $M_1^+(\Omega)$ and for the set of (real-valued) observables the affine functions $M_1^+(\Omega) \to M_1^+(\mathbb{R})$ that map Dirac measures on Ω into Dirac measures on $\mathbb{R}$. The last requirement represents the deterministic nature of the standard classical frame. The generalization here proposed [16,17] preserves the simplex structure $M_1^+(\Omega)$ of the states (thus ensuring the unique decomposition of mixed states into pure states) but drops out the deterministic requirement that observables have to send Dirac measures into Dirac measures. In other words, we take for the set of the (real-valued) observables all the affine functions $M_1^+(\Omega) \to M_1^+(\mathbb{R})$, thus accepting also the "indetermistic", or "fuzzy", ones that map a measure δ_ω concentrated at a point ω of Ω into a diffused measure on $\mathbb{R}$. Therefore, if A is an observable and $X \in \mathcal{B}(\mathbb{R})$, we have $(A\delta_\omega)(X) \in [0,1]$, rather than $(A\delta_\omega)(X) \in \{0,1\}$ as it holds true in the standard classical case. The statistical framework so emerging will be called *operational probability theory*, according to a terminology proposed by Davies [18]: the expression it fuzzy probability theory has also been used in the literature [19,20]. Thus we see that in this new framework the pair (A,X) defines a function $\Omega \to [0,1]$, hence a fuzzy subset of Ω. This generalizes the pattern of the standard classical case, where the pair (A,X) defines a function $\Omega \to \{0,1\}$, hence a crisp subset of Ω. For the pairs (A,X) the word *effects* will be preferred, in place of events.

The effects form an algebraic structure more general than a Boolean algebra, actually even more general that an orthomodular lattice: it is usual to call *effect algebra* the structure they form, and we outline that such structures have been the object of many studies in the recent years: we refer to [21] also for further bibliography.

The operational probability theory (OPT for short) preserves relevant classical features, like the simplex structure of the convex set of states or the fact that any two observables admit a joint observable. Nevertheless OPT can host typical quantum features. A first instance [16] is the fact that uncertainty relations can occur in OPT; indeed there are pairs of observables, say A, B, such that the product of their variances $Var(A,\mu)$, $Var(B,\mu)$ in a same state μ of the physical system has a positive lower bound h:

$$Var(A,\mu) \cdot Var(B,\mu) \geq h \quad \text{for every } \mu \in M_1^+(\Omega).$$

As another instance [22] we can think of possible violations of Bell inequalities inside OPT.

Actually, quantum mechanics admits a "canonical extension" into the OPT frame which goes along the following steps [16]:

- take the measurable space Ω formed by the set of all quantum pure states (the one-dimensional projectors of $\mathcal{H}$);

- notice that there is an affine surjection R which maps $M_1^+(\Omega)$ into the set of density operators of $\mathcal{H}$, namely, into the set of quantum states: if D is a density operator, then all the convex combinations of one-dimensional projectors into which D can be decomposed are mapped by R into D itself;
- build up the OPT framework on the measurable space Ω said above;
- notice that if A is a real-valued quantum observable, that is a map of the set of density operators into $M_1^+(\mathbb{R})$, then the map composition $\widetilde{A} := A \circ R$ is an observable of the OPT framework, since it is an affine mapping of $M_1^+(\Omega)$ into $M_1^+(\mathbb{R})$: we say that $\widetilde{A}$ is the OPT-representative of A;
- notice that all the statistical properties of the quantum observable are shared by its OPT-representative: the quantum features of the observables are thus hosted in the OPT extension.

The family of the observables of the OPT framework described above contains the representatives of the usual quantum observables but it is not exhausted by them. Actually, this OPT frame can host even embeddings of generalzations of standard quantum mechanics.

One such generalization is the so-called "unsharp quantum mechanics" which differs from the standard framework summarized in Sec. 2 in the sense that a more general family of observables is adopted. Still considering the real-valued observables, as we did in the previous sections, all the affine maps $M_1^+(\mathcal{H}) \to M_1^+(\mathbb{R})$ are now accepted, dropping out the requirement of correspondence to the self-adjoint operators. It is known that the probability of getting the outcome of one such observable in $X \in \mathcal{B}(\mathbb{R})$, given that the state is the density operator D will take the form $\mathrm{tr}(DP)$ where P is some positive operator. In other words, the pair formed by an observable A and the subset $X \in \mathcal{B}(\mathbb{R})$ no longer defines a projector but it defines a positive operator. This more general set of observables can thus be identified with the so-called *positive operator measures* on $\mathbb{R}$ (in short POV-measures).

Motivations for the above generalization can be found in the fact that the standard formulation, based on observables that are projection-valued measures, is unable to qualify as "observables" a number of experimental devices. As an elementary example we can think of a beam splitter and notice that there is no projector able to describe the "event" which asserts that the particle is detected in one of the splitted beams. For a discussion of other physically relevant examples we refer to [23].

In unsharp quantum mechanics the pair (A, X) formed by a real-valued observable A and $X \in \mathcal{B}(\mathbb{R})$ corresponds to an affine function from the set of states (density operators) into the segment $[0, 1]$. There is an embedding of the observables of unsharp quantum mechanics into the observables of the so called canonical classical extension of quantum mechanics, and this embedding preserves all the statistical properties of the quantum observables [16]. Thus we see that the OPT framework not only hosts an extension of the standard quantum mechanics but also an analogous extension of the unsharp quantum mechanics.

There are even strong arguments to argue that OPT is the natural framework hosting nonlinear generalizations of quantum mechanics. It is known that the inclusion of nonlinear terms in the dynamics of a quantum system can lead to the separation of the convex combinations of pure states that correspond to a same density operator [24]: this is indeed exactly what occurs in the canonical classical extension of quantum mechanics. The possibility of describing the observables associated with nonlinear operators as observables of the OPT framework has been explicitly discussed in [25].

References

1. L. Accardi, A. Frigerio, J.T. Lewis, *Quantum Stochastic Processes*, Publ. RIMS, Kyoto Univ. **18**, 97 (1992)
2. E.G. Beltrametti and G. Cassinelli, *The Logic of Quantum Mechanics* (Addison-Wesley, Reading 1981)
3. G. Birkhoff and J. von Neumann, Ann. Math. **37**, 823 (1936)
4. G.W. Mackey, *Mathematical Foundations of Quantum Mechanics* (Benjamin, Reading 1963)
5. V.S. Varadarajan, *Geometry of Quantum Theory* (Springer, New York 1985)
6. F. Maeda and S. Maeda, *Theory of Symmetric Lattices* (Springer, New York 1970)
7. C. Piron, *Foundations of Quantum Physics* (Benjamin, Reading 1976)
8. H. Keller, Math. Z. **172**, 41 (1980)
9. M.P. Soler, Comm. in Algebra **23**, 219 (1995)
10. G. Boole, Phil. Trans. R. Soc. London **152**, 225 (1862)
11. L. Accardi, "The probabilistic roots of the quantum mechanical paradoxes" in *The Wave-Particle Dualism* ed. by S. Diner, G. Lochat and F. Selleri (North-Holland, Amsterdam 1976)
12. S. Gudder and N. Zanghi, Nuovo Cimento **79 B**, 291 (1984)
13. I. Pitowski, *Quantum Probability-Quantum Logic*, Lect. Notes in Phys. **321**, (Springer, Berlin 1989)
14. E.G. Beltrametti and M. Maczynski, Found. of Phys. **24**, 1153 (1994)
15. E.G. Beltrametti and M. Maczynski, J. Math. Phys. **34**, 4919 (1993)
16. E.G. Beltrametti, S. Bugajski, J. Phys. A: Math. Gen. **28**, 3329 (1995)
17. E.G. Beltrametti, S. Bugajski, Int. J. Theor. Phys. **34**, 1221 (1995)
18. E.B. Davies, J.T. Lewis, Comm. Math. Phys. **17**, 239 (1970)
19. S. Bugajski, Int. J. Theor. Phys. **35**, 2229 (1996)
20. S. Gudder, Demonstratio Mathematica **31**, 235 (1998)
21. E.G. Beltrametti, S. Bugajski, J. Math. Phys. **38**, 3020 (1997)
22. E. G. Beltrametti, S. Bugajski, J. Phys. A: Math. Gen. **29**, 247 (1996)
23. P. Busch, M. Grabowski, P. J. Lahti, *Operational Quantum Physics*, (Springer-Verlag, Berlin 1995)
24. R. Haag and U. Bannier, Comm. Math. Phys. **60**, 1 (1978)
25. S. Bugajski, Int. J. Theor. Phys. **30**, 961 (1992)

Part IV

Chaotic Systems

Does Quantum Chaos Exist?

Andreas Knauf

Mathematisches Institut der Universität Erlangen-Nürnberg, Bismarckstr. 1 1/2, 91054 Erlangen.

Abstract. The usual operational definition of the term 'Quantum Chaos', meaning a quantum system whose classical counterpart is non-integrable, is not self-contained. However it is argued that there cannot be any intrinsic definition of chaoticity of a finite quantum system which is not based on some kind of semiclassical limit.

Unlike for finite systems, the quantum dynamical entropy of infinite systems may be strictly positive. However, an example shows that this quantity may be lowered by interactions which lead to an increase of classical dynamical entropy.

1 Introduction

The popularity as well as the weakness of the vogue word 'chaos' is based on the richness of its connotations.

In this talk I shall exclusively use it as a name of an *intrinsic* property shared by many dynamical systems, namely that one cannot predict their long term time evolution by a measurement of the initial state.

There are several *operational definitions* which allow us to decide whether a given system is chaotic or not, such as

- its Lyapunov exponents
- dynamical entropy (measure theoretical resp. topological)
- ergodic properties (ergodicity, mixing etc.)

All of them involve a limit of infinite time, and in a sense such a limit is unavoidable if one intends to catch a qualitative phenomenon such as chaos.

Even integrable classical systems of *infinitely* many degrees of freedom like an ideal gas are chaotic in the above sense, because previously unobserved particles may move into the region under observation and thus introduce an element of unpredictability. Thus in order to compare classical and quantum dynamics, I shall first concentrate on systems with *finitely* many degrees of freedom.

The general result of this comparison is that quantum dynamics is much less chaotic than its classical counterparts. Even more, contrary to the classical case, there does not seem to exist a *qualitative* distinction between chaotic and regular motion in Quantum Mechanics.

Such a distinction would be possible with the help of an *order parameter*, a real-valued quantity, depending on the dynamical system (and possibly its state), which is zero for the regular systems and strictly positive for the chaotic systems.

The Lyapunov exponents and the dynamical entropies mentioned above are examples of order parameters for classical dynamical systems. Their quantum generalizations, however, are typically *zero* in the case of finitely many particles.

There are other quantities based on the spectral representation (diagonalization) of the Schrödinger operator $H_\hbar$ of the quantum system:

- One studies the statistics of the eigenvalues of $H_\hbar$
- One studies the probability distribution of its eigenfunctions in configuration space, or the corresponding probability distribution in phase space (the so-called Husimi function).

If one only considers the eigenvalues and corresponding eigenfunctions in a *bounded* energy interval, then one does not find a *qualitative* distinction between regular and chaotic motion.

If, however, an unbounded interval is considered (as has been done, e.g. for Laplace operators with Dirichlet boundary conditions – the so-called Quantum Billiards), or the semi-classical limit $\hbar \searrow 0$ for Planck's Constant $\hbar$ is performed for the Hamiltonian $H_\hbar$, such qualitative distinctions *are* found (see [6] for a general introduction)

- If the principal symbol H of the operator $H_\hbar$ (the associated Hamiltonian function) leads to an integrable classical flow, then the eigenvalue statistics tends to resemble the Poisson distribution. If, instead, the classical dynamics is fully chaotic (say, mixing), then eigenvalues tend to repel each other in a pattern predicted by Random Matrix Theory.
  Somewhat ironically, these numerical observations do not extend to the simplest systems, which can be analyzed by arithmetic techniques, like the multi-dimensional harmonic oscillator or motion in the modular domain [13]. There the eigenvalue statistics are quite different from the ones mentioned above.
- In the classically ergodic case the Husimi functions are proven to converge weakly to Liouville measure of the energy shell, up to a set of density zero as $\hbar \searrow 0$ (Schnirelman Theorem, [14], see also [10]).

In any case, quantum chaos is seen as a phenomenon emergent only in the semi-classical limit.

But even in the limit $\hbar \searrow 0$ one does not simply recover classical dynamics, since it is known not to commute with the limit of infinite time.

2 An Example

I shall exemplify that state of affairs by the example of a Schrödinger operator

$$H_\hbar = -\frac{\hbar^2}{2}\Delta + V(q) \qquad \text{on } L^2(\mathbb{R}^d),$$

where the potential $V : \mathbb{R}^d \to \mathbb{R}$ is invariant under translations in the regular lattice $\mathcal{L} \subset \mathbb{R}^d$.

This operator, which arises in solid state physics, is unitarily equivalent to the Bloch direct integral

$$\int_{\mathbb{T}^*} H_\hbar(k)\mathrm{d}k \quad \text{of operators} \quad H_\hbar(k) = \tfrac{1}{2}(-i\hbar\nabla + k)^2 + V(q) \quad \text{on } L^2(\mathbb{T})$$

with configuration space torus $\mathbb{T} := \mathbb{R}^d/\mathcal{L}$ and dual torus $\mathbb{T}^*$ (Brillouin zone). Denoting the n-th eigenvalue of $H_\hbar(k)$ by $E_n^\hbar(k)$, the group velocity is given by

$$\overline{v}^\hbar : \mathbb{N} \times \mathbb{T}^* \to \mathbb{R}^d \quad , \quad \overline{v}^\hbar(n,k) := \hbar^{-1}\nabla_k E_n^\hbar(k).$$

The corresponding classical quantity is the *mean velocity* $\overline{v}$, which depends on the initial point in the phase space $T^*\mathbb{T}$ over the configuration torus $\mathbb{T}$:

$$\overline{v} : T^*\mathbb{T} \to \mathbb{R}^d \qquad \overline{v}(x) := \lim_{T\to\infty} \frac{1}{T}\int_0^T p(t,x)\mathrm{d}t.$$

Birkhoff's Theorem implies that this time average w.r.t. the flow generated by the Hamiltonian function

$$H(p,q) := \tfrac{1}{2}p^2 + V(q) \quad \text{on} \quad T^*\mathbb{T}$$

exists almost everywhere. Now we fix some small energy interval and compare the distribution $\nu_\hbar$ of group velocities with the distribution ν of classical asymptotic velocities. $\nu_\hbar$ is the image measure on $\mathbb{R}^d$ of

$$(2\pi\hbar)^d \times \text{counting measure on } \mathbb{N} \times \text{Haar measure on } \mathbb{T}^*$$

w.r.t. $\overline{v}^\hbar$, whereas ν is the image measure on $\mathbb{R}^d$ of Liouville measure w.r.t. $\overline{v}$.

Conjecture $\quad w - \lim_{\hbar\searrow 0} \nu^\hbar = \nu.$

This conjecture looks quite natural and innocent. But the hard problem is to exclude that quantum tunneling in phase space reduces the speed in the semiclassical limit. Tunneling occurs on the time scale $\mathrm{e}^{c/\hbar}$ which is much longer than the time scale $\ln(1/\hbar)$ for which the quantal dynamics resembles the classical one (Egorov's Theorem, see e.g. [7]).

The conjecture has been proven in [1] for classically integrable and for classically ergodic flows. Recently it has been shown for the phase space region filled by KAM tori [2]. In the last case one uses quasimode techniques due to Lazutkin [11].

In general, however, the conjecture has not been proven or falsified.

Consider now the special case of a classically ergodic motion, where ν is localized at asymptotic velocity zero. To be concrete, I shall concentrate on a periodic potential V on $\mathbb{R}^2$ which is the sum of attracting coulombic potentials, representing the nuclei of our planar crystal. Then for energies above a threshold the following holds true [8,9]:

- *Classically* the motion is a *deterministic diffusion* process, that is the squared position diverges linearly in time

$$\langle q^2(t)\rangle \sim c \cdot t \qquad (t \to \infty),$$

the brackets denoting averaging w.r.t. an arbitrary a.c. probability measure on the space of initial conditions.

- *Quantum mechanically* the motion is *ballistic*, that is

$$\langle q^2(t)\rangle \sim c_\hbar \cdot t^2 \qquad (t \to \infty).$$

- The *speed* however goes to zero in the semi-classical limit

$$\lim_{\hbar \searrow 0} c_\hbar = 0.$$

Thus the velocity autocorrection function of the quantum particle does not converge to zero like the corresponding classical quantity and it never loses memory of its direction through the crystal.

3 Dynamical Entropy

There are several possibilities to generalize the Kolmogorov-Sinai dynamical entropy to non-abelian situations, one of them being the *CNT* entropy introduced in 1987 by Connes, Narnhofer and Thirring [5] (see also [3,12]). I shall shortly indicate its definition.

In order to compare classical and quantum dynamics, it is useful to choose an algebraic formulation. Thus classically one considers

- the algebra $L^\infty(\mathcal{X})$ of essentially bounded functions on the phase space $\mathcal{X}$,
- the state $\omega : L^\infty(\mathcal{X}) \to \mathbb{C}$, $\omega(f) := \int_{\mathcal{X}} f \, d\mu$,
- the subalgebra $B \subset L^\infty(\mathcal{X})$ generated by the characteristic functions $\mathbb{1}_{A_i}$ of a partition $A_1, \ldots, A_d \subset \mathcal{X}$ of $\mathcal{X}$.

Definition The entropy of the subalgebra B is given by

$$H_\omega(B) := \sum_{i=1}^{d} \eta(\omega(\mathbb{1}_{A_i})) \quad , \text{with} \quad \eta(x) := -x \ln x.$$

Now the idea is to generalize $H_\omega(B)$ to nonabelian algebras B. However, the choice $S(\omega\restriction_B)$ for $H_\omega(B)$, $S(\rho) := \mathrm{Tr}(\eta(\rho))$ being von Neumann entropy, turns out to be bad. One problem with S is that it lacks monotonicity, that is

$$B_1 \subset B_2 \not\Rightarrow S(\omega\restriction_{B_1}) \leq S(\omega\restriction_{B_2})$$

for two subalgebras B_i. Thinking of a subalgebra as a set of measurements, von Neumann entropy does not measure the information gain coming with these experiments, but the uncertainty about their outcome. Whereas classically these two notions coincide, quantum mechanically the latter is the larger quantity.

Therefore one has to substract the loss of information caused by the measurement itself. This is done by decomposing the state ω into a convex combination $\Sigma_i \mu_i \omega_i$ of states ω_i:

Definition The entropy of a subalgebra B is given by

$$H_\omega(B) := S(\omega\restriction_B) - \inf_{\omega=\Sigma_i \mu_i \omega_i} \mu_i S(\omega_i \restriction_B).$$

If one wants to define the entropy $H_\omega(B_1, \ldots, B_n)$ of n subalgebras $B_1, \ldots, B_n$, corresponding to n measurements, one has to guarantee that $H_\omega(B, \ldots, B) = H_\omega(B)$, since repeated measurements of the same quantity do not lead to additional information. This problem was solved in [5].

Armed with these definitions, the introduction of the quantum dynamical entropy proceeds along the lines known from the classical dynamical entropy:

Definition The CNT entropy of an automorphism $\sigma^t : \mathcal{A} \to \mathcal{A}(t \in \mathbb{R})$ of the algebra $\mathcal{A}$ is given by

$$h_\omega(\sigma) := \sup_{B \subset \mathcal{A}} h_\omega(\sigma, B),$$

with

$$h_\omega(\sigma, B) := \lim_{n\to\infty} \frac{1}{n} H_\omega\left(B, \sigma(B), \ldots, \sigma^{n-1}(B)\right).$$

Now consider the standard situation encountered in the quantum theory of finitely many particles, that is

- the algebra $\mathcal{A} := B(\mathcal{H})$ of bounded operators on a Hilbert space $\mathcal{H}$,
- the time evolution $\sigma^t(x) := U_t^* x U_t \qquad (x \in B(\mathcal{H}))$ induced by a one parameter family of unitary operators U_t,
- the state $\omega(x) := \mathrm{Tr}(\rho x)$ given by a density matrix ρ.

Under these circumstances one can easily show [4] that the quantum dynamical entropy vanishes:

$$h_\omega(\sigma) = 0.$$

So CNT entropy cannot serve as an order parameter for quantum chaos (and no such order is expected to exist).

4 Infinite Particle Systems

As mentioned in the introductory section, infinite particle systems tend to look more chaotic than their finite counterparts, and indeed many infinite quantum systems have positive CNT entropy.

As for classical infinite particle systems, this kind of randomness is not necessarily caused by a complicated single-particle dynamics.

What may be astonishing, however, is the fact that an increase of the classical single particle entropy may lead to a *decrease* in entropy of the corresponding quantum gas.

In [4] we consider such a gas of non-interacting particles in a crystal, described by a periodic potential $V : \mathbb{R}^d \to \mathbb{R}$.

We assume observations to be performed in a bounded configuration space region $B \subset \mathbb{R}^d$. This situation is modelled by

- The *CAR* (canonical anticommutation relations) algebra $\mathcal{A}(B)$ over the Hilbert space $L^2(B)$
- The quasifree state ω for a density $\rho : \mathbb{N} \times \mathbb{T}^* \to \mathbb{R}_0^+$ on the Brillouin zones
- The Bogoliubov automorphism $\alpha_H : \mathcal{A} \to \mathcal{A}$ on the *CAR* algebra $\mathcal{A} := \mathcal{A}(\mathbb{R}^d)$ induced by the one particle Hamiltonian $H_\hbar$.

Defining the dynamical entropy w.r.t. observations in B by

$$h_{\omega,B}(\alpha_H) := \sup_{A \subset \mathcal{A}(B)} H_\omega(\alpha_H, A),$$

one first notices that this quantity is not extensive, that is, does not increase proportionally to the volume of the observation region $B \subset \mathbb{R}^d$.

Instead it is a *surface effect.* Namely if we rescale the region by a factor $\lambda > 0$, we obtain (modulo a convergence argument, see [4]) the formula

$$\lim_{\lambda \to \infty} \frac{h_{\omega,\lambda B}(\alpha_H)}{\lambda^{d-1}} = \frac{\hbar^{-1}}{2} \sum_{m=1}^{\infty} \int_{\mathbb{T}^*} \int_{\partial B} |\nabla_k E_m^\hbar \cdot \mathrm{d}n| \, \eta_F(\rho(m,k)) \, \mathrm{d}k.$$

Here $n(x)$ for $x \in \partial B$ is the unit normal vector to the boundary ∂B of B. So the integrand $\hbar^{-1}|\nabla_k E_m^\hbar \cdot dn|$ measures the incoming flux of particles with band index m and quasimomentum k. Finally, $\eta_F(x) := \eta(x) + \eta(1-x)$ is the fermionic entropy function.

Now we have seen above that a more complicated classical dynamics arising from a more complicated periodic potential V tends to *decrease* the absolute value of the group velocity $\hbar^{-1}\nabla_k E_m^\hbar(k)$ and thus to lead to a *smaller* quantum dynamical entropy! The reason for this seemingly paradoxical statement is the fact that less previously unobserved particles enter the region of observation.

5 Open Questions

As we have seen, dynamical entropy behaves discontinuously in the semiclassical limit. From a conceptional point of view this discontinuity is quite irritating.

After all, the world we live in is quantal, and one could be led to the conclusion that classical dynamical entropy is a totally unphysical concept.

Still that conclusion need not be inevitable, for the following reason. We never observe totally isolated physical systems, but there is always an interaction with the environment. Now denote the coupling constant for this interaction by ϵ, and first perform the semiclassical limit for the dynamical entropy of the coupled quantum system, followed by the decoupling limit $\epsilon \to 0$. It could well be that for physically realistic situations one thereby recovers the classical dynamical entropy of the uncoupled system.

If this is the case, then the environment only plays the role of an obstetrician.

In order to answer this and related questions, one needs phase space microlocalization techniques for *CNT* entropy, as opposed to the configuration space localization techniques used in [4].

Another question related to quantum entropy is the one of quantum information processing. Unfortunately, I can only mention in passing this fascinating new subject.

References

1. Asch, J., Knauf, A.: Motion in Periodic Potentials. Nonlinearity **11**, 175–200 (1998)
2. Asch, J., Knauf, A.: Quantum Transport on KAM Tori. Commun. Math. Phys. **205**, 113–128 (1999)
3. Benatti, F.: Deterministic Chaos in Infinite Quantum Systems. Trieste Notes in Physics. Berlin: Springer 1993
4. Benatti F., Hudetz, T., Knauf, A.: Quantum Chaos and Dynamical Entropy. Commun. Math. Phys. **198**, 607–688 (1998)
5. Connes, A., Narnhofer, H., Thirring, W.: Dynamical Entropy for C* algebras and von Neumann Algebras. Commun. Math. Phys. **112**, 691 (1987).
6. Gutzwiller, M.: Chaos in Classical and Quantum Mechanics. Berlin, Heidelberg, New York: Springer; 1990
7. Helffer, B.: h–pseudodifferential operators and applications: an introduction. Rauch, Jeffrey (ed.) et al., Quasiclassical methods. Proceedings based on talks given at the IMA workshop, Minneapolis, MN, USA, May 22–26, 1995. New York, NY: Springer. IMA Vol. Math. Appl. 95, 1-49 (1997)
8. Knauf, A.: Ergodic and Topological Properties of Coulombic Periodic Potentials. Commun. Math. Phys. **110**, 89–112 (1987)
9. Knauf, A.: Coulombic Periodic Potentials: The Quantum Case. Annals of Physics **191**, 205–240 (1989)
10. Knauf, A., Sinai, Ya.: Classical Nonintegrability, Quantum Chaos. DMV–Seminar Band 27. Basel: Birkhäuser 1997
11. Lazutkin, V. F.: KAM theory and semiclassical approximations to eigenfunctions. Ergebnisse der Mathematik und ihrer Grenzgebiete, Vol. **24**. Berlin; New York: Springer 1993.
12. Ohya, M., Petz, D.: Quantum Entropy and its Use. Berlin: Springer 1993.
13. Sarnak, P.: Arithmetic Quantum Chaos. Israel Math. Conf. Proc. **8**, 183–236 (1995)
14. Schnirelman, A.I.: Ergodic Properties of Eigenfunctions. Usp. Math. Nauk. **29**, 181–182 (1974)

Time-Scales for the Approach to Thermal Equilibrium

Stefano Ruffo

Dipartimento di Energetica "S. Stecco", Università di Firenze, via S. Marta 3, INFN and INFM I-50139 Firenze (Italy).

Abstract. The approach to equilibrium of an isolated system constitutes the basic principle of thermodynamics. Fermi, Pasta and Ulam (FPU) studied numerically this process for a chain of coupled oscillators. The FPU "experiment" has been an amazingly rich source of problems in modern dynamical system theory and has played in this context a role analogue to the Ising model for statistical mechanics. Recent results have shown the presence of increasingly long time-scales of the relaxation process as the energy is lowered. States which were previously classified as "frozen" have been instead discovered to "diffuse" very slowly to the equipartition state. The dependence of the diffusive time-scale τ_D on energy E and number of degrees of freedom N has been found both analytically and numerically for some classes of initial conditions. For "thermodynamical" initial states τ_D is found to scale at small ϵ with an inverse power-law of $\epsilon = E/N$.

1 Introduction

The numerical "experiment" by Fermi, Pasta, Ulam (FPU) and Tsingou[1] [1] was the first historical attempt to check the predictions of classical statistical mechanics concerning the long-time behavior of a nonlinear Hamiltonian system with a large number N of degrees of freedom.

This experiment plays also a relevant role for the basic understanding of the so-called *Zeroth law* of thermodynamics, which states that

> *An isolated system will, in the course of time, approach a state of "thermal" equilibrium in which all macroscopic variables have reached steady values.*

Thermal equilibrium was detected in the FPU experiment by looking at energy equipartition among the quadratic modes (phonons) of the oscillator chain (the equipartition "principle" of Boltzmann); indeed a rather refined macroscopic variable. Other observables could have been chosen, like temperature, or specific heat (this has been done in [2]). However, one should first remark that the number of "physically interesting" thermodynamical observables (the parameters defining a thermal state, or macroscopic variables) is much smaller than the number of degrees of freedom, making the task of detecting thermal equilibrium less arduous.

[1] M. Tsingou contributed to the numerical work and then did not participate in the writing of the report.

The result of this experiment was a big surprise for the authors: the expected relaxation to energy equipartition was not revealed, during the time of observation and with low energy initial excitations. This also implies that *ergodicity* or *mixing* are not an obvious consequence of the non-existence of analytic first integrals of the motion besides the total energy. After an initial growth of the energy in the neighbouring modes, the authors observed that the energy sharing was restricted only to the first few modes, which showed a quite regular dynamics. They did not detect, as expected, a gradual and continuous energy flow from the first excited mode to the higher ones. Even more surprisingly, at later times, almost all the energy was flowing back into the initially excited mode, so that the system displayed a sort of quasiperiodic behavior.

Chirikov et al. [3] showed later that, at sufficiently high energy, the FPU model did relax to the equipartition state. It then became clear that the system had qualitatively different behaviors as the energy E, fixed by the initial condition, was varied.

These results stimulated numerical studies aiming at the determination of the dependence of the different observed behaviors of the FPU system on the number N of degrees of freedom (see [4] and references cited therein).

The qualitative picture emerging from all these studies is the following. The transition between a *quasi-integrable* behavior to a *mixing* one, is controlled by the energy E. At small energies the motion is weakly chaotic with positive but small Lyapunov exponents, revealing the presence of thin "stochastic" layers in the phase-space, which appears to be mostly filled with Kolmogorov-Arnold-Moser (KAM) tori. On the contrary, at higher energies the maximum Lyapunov exponent and the Kolmogorov-Sinai entropy rise considerably, revealing the growth of "stochastic" regions. This happens in an energy region which is extensive with N, i.e. one can define a "critical" energy density $\epsilon_c = E/N$ above which chaos is well developed in phase space [5] (this is also known under the name "strong stochasticity threshold" [6]). Well above the transition region $\epsilon \gg \epsilon_c$ all the signatures of large-scale chaos are present: the number of positive Lyapunov exponent increases with N, the orbit shows a fast diffusion on the constant energy hypersurface, spatio-temporal correlations rapidly decay.

In this paper we briefly review the most recent results concerning the scaling, with energy E and number of degrees of freedom N, of the "diffusive" (in a loose sense) time-scale τ_D for the relaxation to thermodynamical equilibrium (detected by energy equipartition among linear normal modes) in the β-FPU model for small energy densities ϵ.

2 Transition to Equipartition

The FPU system is an approximate model for analyzing the behavior of a classical solid at low temperatures. The reduction of complexity in comparison to the real physical situation is considerable. Only one spatial dimension is considered and the interaction (typically of the Lennard-Jones type) is expanded for small displacements around the equilibrium positions of the molecules, i.e. the weakly anharmonic case is considered (in practice the case of rather low temperatures).

Fermi, Pasta and Ulam then considered a one-dimensional chain of oscillators with unit mass and weakly nonlinear nearest-neighbour interaction (the lattice spacing is also taken of unitary length). Calling q_i and p_i the coordinates and, respectively, the momenta of the oscillators, the model is defined by the following Hamiltonian:

$$H = \sum_{i=1}^{N} \frac{p_i^2}{2} + \sum_{i=1}^{N} \left[\frac{1}{2}(q_{i+1} - q_i)^2 + \frac{\beta}{r}(q_{i+1} - q_i)^r \right] , \tag{1}$$

where $r = 3$ for the so-called FPU-α model, while $r = 4$ for the FPU-β model. Periodic or fixed boundary conditions have been considered. At fixed energy, the coupling constant β determines the amount of nonlinearity in the model. Conversely, for a fixed β the increasing departure from the harmonic behavior is controlled by increasing the energy. It can be easily shown that the dynamics depends only on the parameter $\beta E^{(r/2-1)}$.

The Hamiltonian (1), written in linear normal coordinates (Q_k, P_k) (phonons) becomes

$$H = \frac{1}{2} \sum_k \left(P_k^2 + \omega_k^2 Q_k^2 \right) + \beta V(\boldsymbol{Q}) \ . \tag{2}$$

with frequencies $\omega_k = 2\sin(\pi k/N)$ in the case of periodic boundary conditions, $\omega_k = 2\sin(\pi k/2(N+1))$ for fixed boundary conditions. The harmonic energy of mode k is defined by $E_k = (P_k^2 + \omega_k^2 Q_k^2)/2$.

The FPU experiment aimed at showing the progressive decorrelation of the system during its temporal evolution. To this end the authors chose a far from equilibrium initial condition, giving all the energy to the lowest ($k = 1$) normal mode only, and then calculating the instantaneous energies $E_k(t)$ of all modes. They expected to see a progressively uniform redistribution of energy between all modes, caused by the small anharmonic coupling among them. On the contrary they observed the well known FPU recurrent behavior: energy was flowing back regularly to mode $k = 1$ after an initial share. Return to the initial condition was not exact, but the possibility that relaxation was present on longer times was ruled out by a following numerical experiment, which revealed the "superperiod" phenomenon [7].

At higher energies, the equipartition state is instead reached in a relatively short time. A transition is present from a low energy region where the system appears not to be approaching equipartition, showing recurrent behavior in time, to a higher energy region where, on the contrary, equipartition is quickly reached.

A useful tool to characterize the approach to equilibrium is the so-called "spectral entropy" [5]. Let us define a weight as the ratio $p_k = E_k(t)/\sum_k E_k(t)$ between the energy of phonon k and the total harmonic energy (as it should be, this is non-negative and $\sum_k p_k = 1$). Define then a Shannon entropy using this weight $S = -\sum_k p_k \ln p_k$. This is a function of time which measures the effective number of excited normal modes $n_{\text{eff}}(t) = \exp(S(t))$.

The first clear numerical evidence of the existence of a transition region (later called *"strong stochasticity threshold"*) and of its stability with N was obtained in Ref. [5] using "spectral entropy". Above ϵ_c, the "spectral entropy" was shown to increase in time, reaching asymptotically its maximal value. Below ϵ_c the spectral entropy remained instead close to the value of the initial state. After a convenient normalization, the points showed a tendency to accumulate on a N independent curve. While the behavior above ϵ_c is confirmed also by the most recent numerical simulations, below ϵ_c a slower relaxation processes to equipartition is still present, which is ruled by much longer time-scales. Its origin will be briefly described in the next Section.

The transition to equipartition had been indeed suggested by Chirikov and Izrailev [3] using the "resonance overlap" criterion. Let us give a brief sketch of the application of this powerful criterion. The Hamiltonian of the β-FPU model can be rewritten in action-angle variable, considering as an approximation just one Fourier mode (this is justified when most of the energy is still residing in this mode, e.g. at the beginning of the time evolution)

$$H = H_0 + \beta H_1 \approx \omega_k J_k + \frac{\beta}{2N} \left(\omega_k J_k\right)^2 , \tag{3}$$

where $J_k = \omega_k Q_k^2$ is the action variable (in practice only the nonlinear self-energy of a mode is considered in this approximation) and H_0, H_1 are the unperturbed (integrable) Hamiltonian and the perturbation, respectively. $\omega_k J_k \approx H_0 \approx E$ if the energy is placed initially in mode k. It is then easy to compute the nonlinear correction to the linear frequency ω_k, giving the renormalized frequency ω_k^r

$$\omega_k^r = \frac{\partial H}{\partial J_k} = \omega_k + \frac{\beta}{N}\omega_k^2 J_k = \omega_k + \Omega_k . \tag{4}$$

When $N \gg k$

$$\Omega_k \approx \frac{\beta H_0 k}{N^2} . \tag{5}$$

If the frequency shift is of the order of the distance between two resonances

$$\Delta\omega_k = \omega_{k+1} - \omega_k \approx N^{-1} , \tag{6}$$

(the last approximation being again valid when $N \gg k$), i.e.

$$\Omega_k \approx \Delta\omega_k \tag{7}$$

(the last equation is the synthetic expression of the resonance overlap criterion), one obtains an estimate of the critical energy density multiplied by β, which is the control parameter for the development of sizeable chaotic regions,

$$\beta\epsilon_c \approx k^{-1} , \tag{8}$$

with $k = O(1) \ll N$. In other words, a critical energy density exists below which primary resonances are weakly coupled inducing an extremely slow relaxation process to equipartition (in the beginning the belief was that no relaxation was present and that the states were "frozen" out of equilibrium); above ϵ_c, on the contrary, a fast relaxation to equipartition is observed.

3 Time-Scales

A rapid increase of the relaxation time to equipartition at ϵ_c was first revealed in [8]. Later, a rapid decrease of the maximal Lyapunov exponent $\lambda_{\max}$ was found at ϵ_c [6], in correspondence of the transition region. No strong dependence on N of the $\lambda_{\max}$ *vs.* ϵ curve was detected, at sufficiently large N. At low energies $\lambda_{\max} \sim \epsilon^2$, implying that the "Lyapunov time" $\tau_\lambda = \lambda_{\max}^{-1}$, which measures the rate at which microscopic chaotic instabilities develop in time, increases as $\tau_\lambda \sim \epsilon^{-2}$ as ϵ is decreased. This was the first signature of the presence of a power-law for a "typical" time.

Concerning perturbation theory results, the findings obtained using Nekhoroshev estimates have been summarized in [9]. Nekhoroshev theory allows to evaluate lower bounds for the time variation of the unperturbed actions on times that, though being finite, increase exponentially as the perturbation parameter is decreased. It is possible, using this approach, to find results valid for initial conditions on open sets in the phase space, as opposed to methods based on the Kolmogorov-Arnold-Moser theorem (on the other hand the latter has the advantage to give statements valid for all times). The stability time τ_S of the single unperturbed actions (or action "freezing" time) is found to scale as

$$\tau_S = \tau_* \exp(\frac{\beta_*}{\beta})^d \tag{9}$$

where, in general, both τ_*, β_* and d depend on N. The most important dependence on N is that of d: the best estimated so far obtained for FPU gives $d \simeq N^{-1}$, a result confirmed also by numerical simulations (the estimate seems to be optimal). This result suggests that in the thermodynamic limit the freezing times might become short, or even vanishing, and the region of violation of energy equipartition could disappear. We must, however, remember that such estimates are valid in an energy region shrinking to zero as N is increased.

Normal form theory has been used in [10] to find an effective Hamiltonian describing the interaction among a reduced number of long wavelength modes. The main result of their theory is that above a critical energy E_c the system reaches equipartition on a time proportional to N^2; below this critical energy the time needed increases even faster with N (perhaps exponentially). This holds when the initial excitation is given to a subset of low modes whose center k and packet size Δk do not increase with N; this type of initial conditions has been recently defined to be in the "mechanical" class [11] (the meaning of this definition will become clear in the following). If instead $k \propto N$ (so-called "thermodynamical" class), the typical time-scale to equipartition increases like N. These predictions were also supported by numerical simulations. The idea to construct the effective Hamiltonian is to perform a large N expansion of the dispersion relation

$$\omega_k = 2\sin\left(\frac{\pi k}{2(N+1)}\right) \approx \frac{k}{N} - \left(\frac{k}{N}\right)^3 , \tag{10}$$

we are treating now the system with fixed ends and neglecting all factors $O(1)$ or $O(\pi)$ in the approximation. A four-wave resonance relations ($k_1+k_2+k_3+k_4=0$)

is then considered producing in the resonant normal form those angles which are slowly (adiabatically) varying; these latter are found to be $\theta_s = \theta_1 + \theta_3 - 2\theta_2$ and $\theta_{sp} = \theta_2 + \theta_4 - 2\theta_3$. This corresponds to the presence of a modified linear frequency

$$\omega_k^l = \omega_1 + \omega_3 - 2\omega_2 \approx \frac{k}{N^3} \; . \tag{11}$$

Then the frequency shift (5) can be written as

$$\Omega_k \approx R\omega_k^l \; , \tag{12}$$

with $R = \beta N H_0$ the new resonance overlap parameter, $R \approx 1$ corresponding to the resonance overlap condition. (This parameter controls the deformation of the actions, monotonic in the energy). The angles θ_s and θ_{sp} are slowly evolving with the frequency $\Omega_k \approx \beta k H_0/N^2$; the latter determines the characteristic evolution time for the actions $\tau \sim N^2/H_0$ for $k \simeq$ const., while $\tau \sim N/H_0$ if $k \propto N$ (these results are consistent with numerical results [12] obtained by looking at the time evolution of the "spectral entropy"). Moreover, since the resonance overlap parameter is proportional to $H_0 N$, chaos is present at very small energy if N is big enough [13]. This last result is consistent with the behavior of the maximal Lyapunov exponent [6].

Energy transfer to higher modes is present, but takes place on much longer times. Actually, it is also known that the energy fraction transferred to the highest modes is exponentially small in mode number [14]. A nonlinear frequency shift can be also estimated for high modes $\Omega_k^h \approx k/N^2$ [10], and when it becomes of the order of Ω_k a Melnikov-Arnold type of argument gives an estimate for the critical energy $E_c \approx 1/\beta$ below which no transfer to high modes should be present. This critical energy is however irrelevant in the thermodynamic limit, in this limit the transfer to high modes should be always present. Moreover, since the truncated Hamiltonian, studied in [13], does not evolve to equipartition, maintaining an exponential Fourier spectrum for all times, the coupling to high modes is really the crucial effect for the slow evolution towards equipartition below ϵ_c. This is also confirmed by a different approach based on the derivation of the breakdown (shock) time τ_{shock} for the non-dispersive limit of the mKdV equation (to which the FPU-β model reduces in the continuum limit) [15], which turns out to be $\tau_{\text{shock}} \approx \Omega_k^{-1}$. The development of the shock on the lattice produces the formation of fast spatial oscillations at the shock border, which corresponds to a growth of the short-wavelength Fourier components, a phenomenon which is well known to happen in the integration schemes of the Burgers equation in the zero dissipation limit [16]. It is also reasonable to conjecture that this is the reason why models of the electromagnetic field in a cavity or string models do not show evolution to equipartition, because the linear dispersion relation prevents from coupling high frequency modes [17].

This new approach to the study of the time-scale to equipartition below the "strong stochasticity threshold", where primary resonances do not overlap and hence chaos is "weak" can be summarized as follows for the β-FPU model. One can define a "diffusive" time scale to equipartition τ_D looking at the "typical" evolution time of the "spectral entropy", or better n_{eff}/N. For "mechanical" initial conditions, k and Δk fixed, that is not increasing with N, one finds that

$$\tau_D \sim \frac{\sqrt{N}}{\epsilon}. \tag{13}$$

This is why these initial conditions are called "mechanical", they are initial conditions of the "mechanical" finite-N system, which do not scale properly to the thermodynamic limit. As $N \to \infty$ they tend to live forever, never reaching the thermal state of equipartition (to this class did belong the initial condition used in the original FPU experiment). These initial condition do not respect the *Zeroth law* of thermodynamics and they must be excluded by hand in the construction of thermodynamics from mechanics. These are "special" (how much special is hard to say) initial states that, in the Boltzmann approach to the foundations of thermodynamics, live in the "less probable" part of the phase-space and never (as $N \to \infty$) flow to the "most probable" larger part. On the contrary for "thermodynamical" initial conditions, $k \propto N$ and Δk fixed or growing with N

$$\tau_D \sim \frac{1}{\epsilon^3}, \tag{14}$$

i.e. the diffusive time scale is *intensive*, it is finite in the thermodynamic limit. It increases quite sharply as the energy density is decreases, this is why previous numerical studies led to the belief that such initial states where "frozen".

This result is based on a *model* [11] where two crucial assumptions are made: *i)* that a low-mode set creates where the random phase approximation is valid allowing for the calculation of $\Omega_k \sim \beta E k/N^2$ as for the single mode initial condition; *ii)* that selective transfer to high-modes (denoted by h) happens only if Ω_k is bigger than $k\Delta h/N^2$ (Melnikov-Arnold argument). After a short but tricky calculation one gets an effective equation for modal energies

$$\frac{\mathrm{d}E_k}{\mathrm{d}t} = -\left(\frac{2\beta}{N}\right)\omega_k \frac{\beta E}{2\pi} E_k E_h \ , \tag{15}$$

which gives the diffusive time-scale

$$\tau_D = \frac{2\pi}{(\beta\epsilon)^3} \ln\left(\frac{\pi}{2\beta\epsilon}\right) \ . \tag{16}$$

Numerically, the logarithmic correction is not detected and the result is consistent with a $1/\epsilon^3$ divergence of the time-scale. This time-scale was first suggested in [18] and is in sharp contrast with the Lyapunov time-scale, which diverges at small energy as $1/\epsilon^2$. Hence, we can conclude that the process of relaxation to equipartition in the FPU model is not regulated by the microscopic chaotic instability, but by the typical time in which an orbit diffuses in phase-space, which is determined by the interaction among the phonons.

4 Concluding Remarks

All the above concerns the relaxation to equilibrium in anharmonic oscillator chains starting from long wavelength far from equilibrium initial states.

Very recently, the case of short wavelength initial conditions has been considered [19], revealing an unexpected process which leads to equipartition through the intermediate formation of a spatially localized chaotic state, which finally dies out into the thermal condition. The divergence at small ϵ of the time-scale of such initial states has been found to be ϵ^{-2}, in agreement with the Lyapunov time-scale. This new phenomenon is currently under study by many groups [20–22].

Approach to equilibrium starting from states close to equilibrium has been also considered [23], finding a ϵ^{-2} scaling. This process has closer relations with fluctuation-dissipation time-scales near equilibrium.

One should remark that the study of the approach to equilibrium has been mostly performed for coupled oscillator systems, *i.e.* for models of solids. Moreover, just a few studies exist of the relaxation of general thermodynamic functions (like pressure or specific heat), which would certainly be important and useful.

Approach to equilibrium in liquids should be better studied, it could reveal interesting FPU-like processes.

Time-scales of the so-called "aging" phenomenon in glasses have been recently thoroughly studied [24] and the connection of such phenomena with the FPU relaxation process has been hypothesized [25].

Isolated N body systems with long range forces show the presence of metastable states whose lifetime grows with N [26]. Indeed, equilibration in gravitational systems and in plasmas is also an active field of research [27].

Time-scales for establishing thermal equilibrium in finite systems is also a subject of current interest, since it appears that "small" systems like nuclei and atomic clusters of a few atoms already display at short time-scales thermodynamic behavior like phase-transitions [28,29].

The dynamical behavior of granular media reveals the presence of metastable states, that are not predicted by the hydrodynamical treatment [30], their lifetime might be interesting to study.

References

1. E. Fermi, J. Pasta, and S. Ulam, Los Alamos Report LA-1940 (1955), later published in *Collected Papers of Enrico Fermi*, E. Segré ed., University of Chicago Press, Chicago (1965) (Vol. II, p. 978); also reprinted in *Nonlinear Wave Motion*, A. C. Newell ed., Lect. Appl. Math. **15**, AMS, Providence, Rhode Island (1974); also in *The Many-Body Problem*, D. C. Mattis ed., World Scientific, Singapore (1993).
2. D. Escande, H. Kantz, R. Livi and S. Ruffo, J. Stat. Phys., **76**, 605 (1994).
3. F. M. Izrailev and B. V. Chirikov, Dokl. Akad. Nauk SSSR **166** (1966) 57 [Sov. Phys. Dokl. **11** (1966) 30].
4. G. Benettin, *Ordered and Chaotic Motions in Dynamical Systems with Many Degrees of Freedom*, in *Molecular Dynamics Simulation of Statistical Mechanical Systems*, Varenna XCVII Course, G. Ciccotti and W. G. Hoover eds., North-Holland, Amsterdam (1986).

5. R. Livi, M. Pettini, S. Ruffo, M. Sparpaglione, and A. Vulpiani, Phys. Rev. A **31** (1985) 1039; R. Livi, M. Pettini, S. Ruffo, and A. Vulpiani, Phys. Rev. A **31** (1985) 2740.
6. M. Pettini and M. Landolfi, Phys. Rev. A **41** (1990) 768.
7. J. L. Tuck, Los Alamos Report LA-3990 (1968); J. L. Tuck and M. T. Menzel, Adv. Math. **9** (1972) 399.
8. H. Kantz, Physica D **39** (1989) 322.
9. L. Galgani, A. Giorgilli, A. Martinoli, and S. Vanzini, Physica D **59** (1992) 334.
10. J. De Luca, A. J. Lichtenberg, and M. A.Lieberman, Chaos **5** (1995) 283.
11. J. De Luca, A. J. Lichtenberg, and S. Ruffo, Phys. Rev. E **60** (1999) 3781.
12. H. Kantz, R. Livi and S. Ruffo, J. Stat. Phys. **76** (1994) 627.
13. D. L. Shepelyansky, Nonlinearity **10** (1997) 1331.
14. F. Fucito, F. Marchesoni, E. Marinari, G. Parisi, L. Peliti, S. Ruffo, and A. Vulpiani, J. de Physique **43** (1982) 707; R. Livi, M. Pettini, S. Ruffo, M. Sparpaglione, and A. Vulpiani, Phys. Rev. A **28** (1983) 3544.
15. P. Poggi, S. Ruffo, and H. Kantz, Phys. Rev. E **52** (1995) 307.
16. C.D. Levermore and J-G Liu, Physica D **99** (1996) 191.
17. G. Benenti, G. Casati and I. Guarneri, *Chaotic dynamics of a classical radiant cavity*, preprint (1998).
18. L. Casetti, M. Cerruti-Sola, M. Pettini and E.G.D. Cohen, Phys. Rev. E, **55** (1997) 6566
19. T. Dauxois, S. Ruffo, A. Torcini and T. Cretegny, Physica D, **121** (1998) 109.
20. Yu. A. Kosevich, S. Lepri, Phys. Rev. B, **61** (2000) 299 .
21. K. Ullmann, A.J. Lichtenberg and G. Corso, Phys. Rev. E, **61** (2000) 2471.
22. K.O. Rasmussen, S. Aubry, A.R. Bishop and G.P. Tsironis, *Discrete nonlinear Schroedinger breathers in a phonon bath*, cond-mat/9901002.
23. S. Lepri, Phys. Rev. E, **58** (1998) 7165.
24. L.F. Cugliandolo, J. Kurchan and L. Peliti, Phys. Rev. E, **55** (1997) 3898.
25. A. Carati and L. Galgani, *On the specific heat of Fermi-Pasta-Ulam systems*, preprint (1999).
26. V. Latora, A. Rapisarda and S. Ruffo, Physica D, **131** (1999) 38.
27. D. Lynden-Bell, *Negative specific heat in astronomy, physics and chemistry*, cond-mat/9812172.
28. D.H. Gross, Rep. Progr. Phys., **53** (1990) 605.
29. M. Schmidt *et al.*, Phys. Rev. Lett., **79** (1997) 99.
30. S. McNamara and W.R. Young, Phys. Rev. E, **53** (1996) 5089.

Einstein's Nonconventional Conception of the Photon and the Modern Theory of Dynamical Systems

Andrea Carati and Luigi Galgani

Università di Milano,Dipartimento di Matematica, Via Saldini 50, 20133 Milano, Italy

Abstract. Everyone knows how Einstein introduced in the year 1905 the concept of the photon, by giving some conctreteness to the discretization of energy previously introduced by Planck at a formal level. Here we point out how, till the end of his life, Einstein considered such a conception just a "provisional way out", to be substituted by a conception involving continuous variations of energy. We explain how such a conception is understood by taking into account Einstein's contribution to the first Solvay conference. Finally we show how such a conception can be at least partially implemented in classical mechanics, through results from the modern theory of dynamical systems.

1 Introduction

The critical position of Einstein towards the standard interpretation of quantum mechanics is very well known, and is vividly witnessed by the famous paper he wrote with Podolsky and Rosen, which stimulated in more recent times so many discussions and controversies. There is however another specific point where Einstein manifested his uneasyness with respect to the standard interpretation of quantum mechanics; we refer to the starting point itself of quantum mechanics, namely the dilemma continuity–discontinuity (i.e. the problem of the very existence of energy levels), which came about in connection with Planck's law.

Apparently this fact remained unnoticed, or at least we were unable to find any reference to it in the literature (see [1]); and this might be a sufficient reason for discussing it in the present paper. Another element of interest is the fact that the nonconventional Einstein's conception of the photon we are referring to turns out to be strictly related to some of the most recent advances in the theory of dynamical systems. This was for the first time pointed out in [2], where it was shown how a relevant fluctuation formula of Einstein is a statistical counterpart of a purely dynamical formula that we like to call the Benettin–Jeans formula. So let us pass to illustrate what we mean by Einstein's nonconventional conception of the photon, and how we came to interpret it in terms of concepts from the theory of dynamical systems.

2 Einstein's Nonconventional Conception of the Photon

First of all, a nonconventional Einstein's conception of the photon indeed exists, at least potentially, in our opinion. A hint for this can be found in a famous page of Einstein's scientific autobiography, which was written a few years before his death. Indeed, he first recalls how, by inventing the photon, he had given some concreteness to the discretization of energy, previously introduced by Planck at a purely formal level. In his very words (see [3]): "*This way of considering the problem showed in a definitive and direct way that it is necessary to attribute a certain immediate concreteness to Planck's quanta and that, under the energetic aspect, radiation possesses a sort of molecular structure*". But after a few lines he adds: "*This interpretation, that almost all contemporary physicists consider as essentially definitive, to me appears instead as a simple provisional way out*". These words are actually so sharp that no doubts should be left; and this is indeed the first fact we are referring to. But the problem of understanding what Einstein actually had in mind, as a positive concrete proposal, when referring to a provisional way out, is a quite a different one. We try now to disclose this point.

In our opinion the clue is given by what Einstein wrote in two papers, in 1909 and 1911 (see [4] and [5]); the second of such papers constitutes in fact his contribution to the first Solvay conference, and is the one to which we will mostly make reference. It such a paper Einstein points out the relevance of fluctuations, and shows that formally Planck's formula is equivalent to assuming that there exists a certain functional relation between energy fluctuations and mean energy of a system of identical oscillators, namely

$$\sigma_E^2 = \epsilon U + U^2/N \ , \tag{1}$$

where U and σ_E^2 denote the mean energy and the fluctuation (precisely, the mean square deviation, or variance) respectively of the energy E of a system of N harmonic oscillators of the same frequency ω, while ϵ is the quantum of energy expressed by $\epsilon = \hbar\omega$ in terms of the reduced Planck's constant $\hbar$. The sense in which such a fluctuation formula is equivalent to Planck's formula will be illustrated in the next section; in the subsequent one we will instead show how an analytical formula of Einstein's type arises in classical mechanics.

For the moment however we just recall how Einstein interpreted his fluctuation formula in connection with the dilemma continuity–discontinuity. Indeed he makes reference to the corresponding formula for the relative fluctuations, namely

$$\frac{\sigma_E^2}{U^2} = \frac{\epsilon}{U} + \frac{1}{N} \ .$$

Then, by considering the limit of a large number N of oscillators or small energies, in which the formula takes the simpler form

$$\frac{\sigma_E^2}{U^2} \simeq \frac{\epsilon}{U} \ ,$$

he remarks (see [5]): "*If U becomes of the order of $\hbar\omega$ (namely of ϵ), the relative fluctuation becomes of the order of unity; in other terms, the fluctuation of energy is of the order of magnitude of energy itself, i.e. the total energy is alternatively present or absent, and consequently behaves as if it were not indefinitely divisible.*

It is not necessary to make the hypothesis that distinct energy elements of a definite magnitude exist".

This is actually what we mean by Einstein's nonconventional conception of the photon: one can conceive of the harmonic oscillator in a classical sense, as possessing at each time a well defined energy, ranging in the familiar domain $E \geq 0$, and if one finds a mechanics that produces Einstein's fluctuation formula for the energy, then the "level" ϵ turns out to be just that particular value of energy having the property that, correspondingly, "*the fluctuation of energy is of the order of magnitude of the energy itself*", i.e. that "*energy is alternatively present or absent*". In another passage of [5] Einstein even reinforces such an argument by saying that, in virtue of such a formula, "*the statistical properties of the phenomenon are the same as if energy were transferred through integer numbers of quanta* $\hbar\omega$". With this we presume we have given sufficient support to our claim that a nonconventional Einstein's conception of the photon indeed exists: it describes the energy of the oscillator in classical continuous terms, and the apparent quantum discontinuity just corresponds to a concise description of classical processes obeying a certain fluctuation law for energy, namely (1), giving a suitable functional relation between variance and mean. In our opinion, this is exactly what Einstein had in mind when he wrote the passage from his scientific autobiography quoted in the introduction. So apparently we are concerned here with a conception that he nurtured from at least the year 1909 till his death.

3 Einstein's Interpretation of Planck's Formula in Terms of Fluctuations

We now illustrate how Einstein came to conceive of his fluctuation formula (1). What he did was to provide a physical substantiation for the original deduction Planck had given of his law on October 19, 1900 (see [6]). To this end, let us recall preliminarily that it was only in later communications (starting from that of december 1900) that Planck gave his familiar deduction involving the standard statistical arguments with a discretization of energy, while in his first communication he was instead proceeding at a phenomenological level, without invoking any discretization at all.

Planck was concerned with the problem of finding a formula for the mean energy U of a system of N oscillators of the same frequency ω in equilibrium with a heat reservoir at absolute temperature T, or inverse temperature $\beta = 1/kT$, where k is Boltzmann's constant, and made the following remark (we are using here a contamination of the notations of Planck and of Einstein). He knew that Wien's law

$$U = C \exp(-\beta\epsilon) \ ,$$

with ϵ proportional to ω and a suitable constant C, fits well the experimental data for large frequencies, while the most recent experimental data available to him, which were referring to lower frequencies, turned out to rather fit the equipartition law

$$U = N/\beta = NkT \ .$$

On the other hand, Wien's law is obviously obtained as a solution of the differential equation

$$\frac{\mathrm{d}U}{\mathrm{d}\beta} = -\epsilon U \ ,$$

while the equipartition law obvioulsy satisfies the differential equation

$$\frac{\mathrm{d}U}{\mathrm{d}\beta} = -\frac{U^2}{N} \ ,$$

with a suitable choice of the integration constant. Just by virtue of imagination, through an interpolation he was then led to conceive of the differential equation

$$\frac{\mathrm{d}U}{\mathrm{d}\beta} = -(\epsilon U + \frac{U^2}{N}) \ , \tag{2}$$

which by integration, and a suitable choice of the integration constant, indeed gives Planck's formula, namely

$$U(\omega, T) = N\,(\frac{\epsilon}{e^{\beta\epsilon} - 1}) \ . \tag{3}$$

Planck's constant $\hbar$ was then introduced by fit with the experimental data through the relation $\epsilon = \hbar\omega$, because it was already known, by a general argument of Wien, that ϵ had to be taken proportional to frequency.

So much for what concerns Planck's first communication. The contribution of Einstein was the following one. Already in the year 1903 (see [7]) he had remarked that in the canonical ensemble the fluctuations of energy, described by the corresponding variance σ_E^2, is expressed in terms of the mean energy U through a relation having a kind of universal thermodynamic character, namely $\frac{\mathrm{d}U}{\mathrm{d}\beta} = -\sigma_E^2$. Thus Einstein was led to split Planck's differential equation (2) into two relations, namely

$$\frac{\mathrm{d}U}{\mathrm{d}\beta} = -\sigma_E^2 \tag{4}$$

and

$$\sigma_E^2 = \epsilon U + U^2/N \ ; \tag{5}$$

the former was conceived to be just a kind of general thermodynamic relation, while the latter should rather have a dynamical character, and might in principle be deducible from a microscopic dynamics. In his very words (see [5]): these two relations "*exhaust the thermodynamic content of Planck's*" formula; and: "*a mechanics compatible with the energy fluctuation* $\sigma_E^2 = \epsilon U + U^2/N$ *must then necessasily lead to Planck's*" formula. It was pointed out in [2] that the second of the above equations has indeed a mechanical character, coinciding essentially with what we call the Benettin–Jeans formula with a suitable ϵ (this is a delicate point in our result). So we might say that the "mechanics" conceived by Einstein as leading to Planck's formula perhaps is nothing but the dear old classical mechanics of Newton.

In this connection, however, we address preliminarly an important question of a general character, namely how can one obtain in classical mechanics something quantitatively comparable with quantum mechanics, as the latter involves a quantity, i.e. Planck's constant $\hbar$, which is completely extraneous to classical mechanics. A first answer, to which we limit oureselves in the present paper, is that Planck's constant can be introduced in classical mechanics simply through the molecular parameters. Indeed, consider for example a system of equal particles of the same mass m interacting through a typical interatomic potential, such as the familiar one of Lennard-Jones. Now, this potential contains two parameters, say V_0 and σ, with the dimensions of an energy and a length respectively, and from them and the mass m one constructs an action, namely $\sigma\sqrt{mV_0}$, which a priori can take any value. But if one takes for the parameters m, V_0, σ entering the model just the ones corresponding to actual atoms, as reported in the standard textbooks, one finds that the relation $\sigma\sqrt{mV_0} \simeq 2Z\hbar$ holds, where Z is the atomic number of the considered atoms. This is one way in which Planck's constant can be made to enter classical physics at the level of pure mechanics (see [8]). A more fundamental way would require considering the role of the electromagnetic field, but this interesting point will not be discussed here (see for example [9]).

4 A Dynamical Implementation of Einstein's Fluctuation Formula

The road that led us to provide a partial implementation of Einstein's fluctuation formula (1) in classical terms is a long one. It started from a serious attention given, since the early years 70's, to the paradoxical result obtained in the year 1954 by Fermi, Pasta and Ulam (FPU; see [10]). Such authors had shown, by numerical computations of the equations of motion of a one–dimensional model of a crystal, that at low energies classical dynamics seems to give results in contradiction with the law of equipartition, predicted by the Maxwell–Boltzmann distribution of classical equilibrium statistical mechanics. The first scientists that took up the problem, namely Izrailev and Chirikov (see [11]), put forward the very natural conjecture that the FPU paradox should disappear in the limit in which the number N of oscillators tends to infinity. Instead, in the paper [12] it was suggested that the lack of equipartition could persist in the limit of infinitely many oscillators, and in the paper [8] it was even found (still by numerical integration of the FPU model) that in classical mechanics Planck–like distributions seemed to occur. In fact, in the latter paper it was also realized, for the first time in a foundational context, that Planck's constant does in fact show up in classical mechanics through the molecular parameters in the way mentioned above.

There was then an intricate road passing through an appreciation of the many new possibilities offered by the modern results in the theory of dynamical systems (especially the stability results provided by KAM theorem and Nekhoroshev's theorem, see [13]). But it was finally realized (see [14]) that the simplest model describing the essence of the problem is that of a system of equal diatomic molecules on a line, where there are "internal" degrees of freedom (the oscillations of each of the molecules about its center of mass) and "external" ones (the center of mass of each molecule). The two subsystems (internal and external degrees of freedom, respectively) were found to go very rapidly to separate equilibria, and the problem remained of how would they go to a mutual equilibrium. It was then found numerically that the relaxation time to mutual equilibrium between the two subsystems increases as a stretched exponential with the frequency, and an analytical proof was provided in [15]. Eventually, the problem was then reduced to its very core, namely: the exchange of energy between a single spring (of frequency ω) and a colliding particle, the system moving on a line and the interaction being given by a potential between the particle and one extreme of the spring. This, by the way, is essentially equivalent to a model first discussed by Kelvin and Poincaré (see [16]) just in connection with the dynamical foundations of classical statistical mechanics.

So let us consider the problem of the exchange of energy δe of a spring on a single collision of one of its extremes with a point particle. An elementary calculation of a few lines made on a simplified version of the model gives the result (see [2])

$$\delta e = \eta^2 + 2\eta\sqrt{e_0}\cos\varphi_0 \ , \tag{6}$$

where e_0 is the initial energy of the spring, φ_0 its initial phase, while η is a quantity which tends to zero as a stretched exponential when the frequency ω of the spring increases and the velocity v of the particles decreases. We like to call formula (6) the Benettin–Jeans formula (see [17]). Now, one expects that the formula (6) should be correct if the frequency of the spring is sufficiently large and its energy sufficiently small, and this was proven in [18] by a quite delicate mathematical analysis, similar to the ones used in order to prove the exponentially small splitting of the separatrices in Melnikov's theorem.

Let us look now at the Benettin–Jeans formula (6), thinking of the impinging particle as mimicking a heat reservoir at a given temperature T, and of the spring as mimicking a crystal at a much lower temperature. Due to the exponential smallness of η, the formula implies that the echange of energy which should lead to equipartition with the reservoir is exponentially small with the characteristic internal frequency ω, so that the number of collisions required to go to equilibrium (i.e. the time required for it) is highly nonuniform in the frequency, being exponentially large with ω. Examples can be given in which there exists a frequency $\bar{\omega}$ that relaxes in 1 second, while the frequency $\bar{\omega}/2$ relaxes in 10^{-8} seconds and the frequency $2\bar{\omega}$ in 10^5 years. This fact is very important, because it explains the most relevant feature of Planck's formula, namely the circumstance that the high frequencies have a very small energy with respect to that of the reservoir: in dynamical terms, this is due to the fact that the high frequencies require an exponentially long time in order to go to equilibrium if they start from a negligible initial energy.

But the fact remains that the coefficients entering the exponentially small quantity η of the Benettin–Jeans formula (6) turn out to depend on the molecular parameters characterizing the particular interatomic potential considered. So one is lackng a formula possessing a sort of thermodynamic character. This fact was, for the whole group of people involved in the research described here, a great conceptual difficulty.

The clue was found by taking into consideration the second term appearing in the Benettin–Jeans formula (6), which produces a fluctuation of energy (depending on the phase φ_0) much larger than the drift term η^2. The relation between the Benettin–Jeans formula (6) and Einstein's fluctuation formula (1) was found in the following way (see [2]). Consider a sequence of k collisions and take the average over the phases (which as usual are assumed to be uniformly distributed). Denoting by u_k and σ_k^2 the mean energy and variance of energy respectively after k collisions, a completely elementary calculation gives the formulæ

$$u_k = e_0 + k\eta^2 \ , \quad \sigma_k^2 = 2e_0 k\eta^2 + (k\eta^2)^2 \ ,$$

which depend on time (i.e. on the number k of collisions). But one immediately sees that the "time" k can be eliminated, so that a functiomal relation exists between variance and mean, namely

$$\sigma_k^2 = 2e_0(u_k - e_0) + (u_k - e_0)^2 \ .$$

A similar relation also holds if one considers a system of N identical independent oscillators of frequency ω. Indeed, the quantities of interest are now the total energy $E_k = \sum_{i=1}^N e_k^{(i)}$ (where $e_k^{(i)}$ denotes the energy of the i–th oscillator after k collisions) and the corresponding exchanged energy $\tilde{E}_k = E_k - E_0$, where E_0 is the initial energy. By the central limit theorem, $\tilde{E}_k$ is normally distributed with a mean $\tilde{U}_k$ and a variance which are obtained by adding up the corresponding quantities for each oscillator, So, denoting by $\tilde{U}$ and $\sigma_{\tilde{E}}^2$ expectation and variance of the exchanged energy at any "time" k, one gets between $\sigma_{\tilde{E}}^2$ and $\tilde{U}$ a functional relation which is independent of "time" k, namely

$$\sigma_{\tilde{E}}^2 = 2a_0\omega\tilde{U} + \tilde{U}^2/N \ , \tag{7}$$

where a_0 denotes the initial action per oscillator, $a_0 := E_0/(\omega N)$. Notice that the quantity η, which contains the molecular parameters characterizing the particular system considered, has now completely disappeared, and formula (7) has some kind of thermodynamic universality.

In order to have something comparable to Einstein's fluctuation formula (1), we have however to get rid of the parameter still appearing in formula (7), namely the the quantity $2a_0$, twice the initial action per oscillator, which takes the place of Planck's constant $\hbar$. This is a delicate point that should deserve a deep investigation At the moment we are unable to say anything rigorous, and only present here some heuristic considerations. The point is that, as was recalled above, the Benettin–Jeans dynamical formula was established for small initial energies of the oscillators. So formula (7) should hold only for low enough energy or action per oscillator, say for $a_0 < a_*$, with a certain critical or threshold action a_*. This naturally leads to think of a situation with the N oscillators having random initial actions all smaller than a_*, so that, on averaging over the initial actions, uniformly distributed over the interval $(0, a_*)$, one would get a formula as (7) with $2a_*/2$, namely a_*. in place of Planck's constant. The actual mean energy would correspondingly be given by Planck's formula with the addition of an analog of the zero–point energy, namely $a_*\omega/2$, playing here the role of the initial energy. Notice that in such a fluctuation formula the molecular parameters enter only through the critical action a_* and so Planck's constant $\hbar$ finally appears in the way described above (see [8] and [19] [20]).

5 Conclusions

So we hope we were able to show that a nonconventional Einstein's conception of the photon, involving continuous variations of energy, indeed exists, and how it might be implemented in classical mechanics. We are well aware of the fact that we are still faced with many deep problems, but we like nevertheless to sketch here, in a few words, a perspective that seems now to be opened, in which Planck's law appears just as a first order approximation.

Namely, the law of equipartition can be considered just as a zeroth order approximation, in which the high nonuniformity of the relaxation times with respect to frequency is altogether neglected. Planck's law instead appears as a first order approximation, describing a kind of metaequilibrium state, similar to those occuring in glasses (an analogy first pointed out in [21]). Quantum mechanics would, in this sense, just be a first order approximation within classical mechanics. If our point of view is correct, deviations from Planck's law should show up, especially in the region of low frequencies, where equipartition would actually be present, with an "equipartition front" advancing with time, at an extremely slow pace. Such an effect was indeed predicted already by Jeans (a quotation can be found in [22]). For a review of the experimental data on laboratory black body up to some years ago, see the second part of the work [23], and also [24].

Finally, one also has a critical historical problem. Indeed, the point of view of metastability described above was advocated by Jeans at the beginning of the last century, but such an author then made a retractation (vividly documented in [25] and [26]), after Poincaré had proven (see [27]) that Planck's law seems to imply quantization, namely the existence of energy levels. So the problem is whether Poincaré's argument is really compulsory, but up to now we were unable to settle the question.

Acknowledgment

The work of one of us (A.C.) was made possible by a grant from Fondazione Cariplo per la Ricerca Scientifica.

References

1. T.S. Kuhn, Black–Body theory and the quantum discontinuity, 1894–1912, (Oxford U.P., Oxford, 1978); J.M. Klein, Max Planck and the beginning of quantum theory, Arch. Hist. Exact Sc. 1, 459–479 (1962).
2. A. Carati and L. Galgani, Analog of Planck's formula and effective temperature in classical statistical mechanics far from equilibrium, Phys. Rev. E, 61, (2000) 4791.
3. A. Einstein, Scientific autobiography, in P.A. Schilpp, Albert Einstein: philosopher–scientist, (Tudor P.C., New York, 1949).
4. A. Einstein, Phys. Zeits. 10 (1909) 185.
5. A. Einstein, Contribution to the 1911 Solvay Conference, in The collected papers of A. Einstein, (Princeton U.P., Princeton, 1993), Vol. 3, n. 26.
6. M. Planck, Verh. D. Phys. Ges. 2 (1900) ; reprinted in H. Kangro, Planck's original papers in quantum physics, (Taylor and Francis, London, 1972).
7. A. Einstein, Ann. Phys. 11 (1903) 170.
8. L. Galgani and A. Scotti, Phys. Rev. Lett. 28 (1972) 1173.
9. A. Carati, L. Galgani, The theory of dynamical systems and the relations between classical and quantum mechanics, in preparation.
10. E. Fermi, J. Pasta and S. Ulam, *Los Alamos Report* No. LA-1940 (1955), later published in *E. Fermi, Collected Papers*, University of Chicago Press (Chicago, 1965), and *Lect. Appl. Math.* 15(1974) 143 .
11. F.M. Izrailev and B.V. Chirikov, Sov. Phys. Dokl. 11 (1966) 30.
12. P. Bocchieri, A. Scotti, B. Bearzi and A. Loinger, Phys. Rev. A 2 (1970) 2013.
13. G. Benettin, L. Galgani and A. Giorgilli, Nature 311 (1984) 444.
14. G. Benettin, L. Galgani and A. Giorgilli, Phys. Lett. A 120 (1987) 23.
15. G. Benettin, L. Galgani and A. Giorgilli, Comm. Math. Phys. 121 (1989) 557.
16. H. Poincaré, Revue Générale des Sciences Pures et Appliquées, 5 (1894) 513–521, in *Oeuvres* X, 246–263.
17. O. Baldan and G.Benettin, J. Stat. Phys. 62 (1991) 201; G. Benettin, A. Carati and P. Sempio, J. Stat. Phys. 73 (1993) 175.
18. G. Benettin, A. Carati and G. Gallavotti, Nonlinearity 10 (1997) 479.
19. L. Galgani and A. Scotti, Recent progress in classical nonlinear dynamics, Rivista Nuovo Cim. 2 (1972) 189.
20. C. Cercignani, L. Galgani and A. Scotti, Phys. Lett. A38 (1972) 403.
21. A. Carati and L. Galgani, J. Stat. Phys. 94 (1999) 859.
22. L. Galgani, Relaxation Times and the Foundations of Classical Statistical Mechanics in the Light of Modern Perturbation Theory, in G. Gallavotti and P.F. Zweifel eds., Non-Linear Evolution and Chaotic Phenomena, NATO ASI Series R71B: Vol. 176, (Plenum Press, New York, 1988).
23. L. Galgani, Annales de la Fondation Louis de Broglie, 8 (1983) 19.
24. L. Crovini and L. Galgani, Lett. Nuovo Cimento, 39 (1984) 210.
25. *Physics at the British Association, Nature* **92**, 304–309 (1913).
26. P.P. Ewald, *Bericht über die Tagung der British Association in Birmingham (10 bis 17 September), Phys. Zeits.* **14**, 1297 (1913); see especially page 1298.
27. H. Poincaré, *J. Phys. Théor. Appl.* **5**, 5–34 (1912), in *Oeuvres* IX, 626–653.

Part V

Philosophy of Probability

What Interpretation for Probability in Physics?

Maria Carla Galavotti

University of Bologna
Dipartimento di Filosofia
Via Zamboni 38 – 40126 Bologna, Italy

Probability enters into modern science in three ways. (1) Measurement errors. This regards all of science; in physics, probability associated with errors of measurement enters – implicitly or explicitly – into all of its branches. In this sense, one can say that a probabilistic aspect is already present in classical mechanics. Indeed, the theory of errors came to completion around the end of the 18th Century, when Newtonian mechanics was at its peak. (2) The analysis of mass phenomena, like gas particles. In physics, this use of probability is linked to statistical mechanics. Here probability makes it possible to describe through mean values the behaviour of phenomena that are too complex to allow finer descriptions. In principle, these probabilities are not essential to the theory, they are needed because of the complexity of the phenomena under study, which makes a fully detailed analysis impossible. (3) Quantum mechanics. Here probability acquires a peculiar character in view of the fact of being 'primary' (to use an expression of Hermann Weyl), or 'intrinsic' to the theory (as Harold Jeffreys used to say). This is because the measurements realized on physical systems in quantum mechanics are genuinely random. On the assumption that the theory is complete, there is no way, not even in principle, of getting rid of probability.

The first sense in which probability is connected to physics, namely that relating to measurement errors, is linked with the classical interpretation of probability developed by Pierre Simon de Laplace. This is an epistemic interpretation, according to which probability expresses our ignorance about the true course of events. It says that probability can be determined as the ratio of favourable to all the equally possible alternatives, on the basis of the so called 'principle of insufficient reason', also called 'principle of indifference' (see [1,2]). Within Laplace's perspective, this notion of probability goes hand in hand with a deterministic world view, reflecting the mechanistic character of classical mechanics. Laplace's idea of an 'omniscient being' who would be able to predict all future events with certainty has been considered by Richard von Mises 'the most extreme formulation of Newton's determinism' ([3], p. 176).

The Laplacean, or 'classical' interpretation of probability has been no less influential than classical mechanics. In the second half on 19th Century, however, it was shown to be unsatisfactory in many ways. First of all, when applied to continuous properties it leads to paradoxical conclusions, as shown by the French probabilist Joseph Bertrand [4]. In addition, its applicability to a vast range of situations looks problematic; as observed by von Mises, 'equally likely cases... do not always exist' ([3], pp. 68–69). For instance, how do we compute the equally likely and the favourable cases relative to the probability of death of a person?

By 1900 the classical notion of probability had gradually become outdated, and a statistical view, according to which probability is to be defined in terms of frequencies, became prominent. The rise of the frequentist interpretation is strictly linked to the new trend taken by physics with the advent of statistical mechanics. The study of microphenomena gave rise to a number of theories of which probability is an indispensable ingredient, like the kinetic theory of gases, Brownian motion and the theory of radioactivity. According to von Mises, who can be considered the most outstanding representative of this interpretation of probability, the turning point was reached when Boltzmann 'conceived the remarkable idea of giving a statistical interpretation to one of the most important propositions of theoretical physics, the Second law of Thermodynamics' ([3], p. 176). In the hands of Richard von Mises frequentism became the 'official' interpretation of probability, to be applied to phenomena described by physical theories. The frequentist notion of probability, or statistical probability, corresponded to a widespread scientific practice, and became for this reason most popular with scientists, who accepted it quite unproblematically. In the period 1900–1930 the frequentist notion of probability also pervaded the mathematical theory of probability and theoretical statistics: several limiting theorems on the behaviour of frequencies were proved and various methods for testing statistical models and estimating statistical parameters were developed. With very few exceptions – notably that of Harold Jeffreys – frequentism was widely accepted and became the 'received view' of probability.

It is therefore not surprising that when quantum mechanics took shape, namely around 1925–27 with the work of Werner Heisenberg, Niels Bohr and Max Born, it did not destroy the frequentist interpretation, despite the fact that it contained elements that appear to be in contrast with frequentism. Von Mises claims that the frequentist interpretation can be readily extended to quantum mechanics and physicists like Heisenberg and Born refer to von Mises in their works. One might speculate that the empirical character of von Mises' conception and its operational attitude made it quite palatable to people like Born and Heisenberg.

Yet the extension of frequentism from the analysis of mass phenomena to quantum mechanics raises serious problems, like that of the single case. Quantum mechanics requires that probability be referred to the single case. Among others, Born claims that the statistical interpretation of quantum mechanics is meant to apply 'in any individual case' ([5], p. 27), and Heisenberg wants to talk of 'the observation of 'the electrons in a single atom" ([6], p. 64; see also [7]). On the contrary, within von Mises' approach it does not make sense to talk of the probability of a single event: one can speak of probability only with reference to infinite sequences of repeatable events. Therefore frequentism does not seem to allow for a straightforward application to quantum mechanics. The problem of interpreting quantum mechanical probabilities is left open.

In an attempt to answer this problem, around the end of the Fifties Karl R. Popper proposed the 'propensity' interpretation of probability [8–11]. This has undergone various changes in Popper's writings. To put things very simply, the underlying idea is that probability can be interpreted as a theoretical (unobservable) property of an experimental set-up – the propensity – expressing its tendency to exhibit certain frequencies that can be observed. The propensity interpretation is meant as 'purely objective', inasmuch as propensities are taken to be 'physically real'. For Popper propensity 'is a new physical (or perhaps metaphysical) hypothesis' analogous to Newtonian forces ([10], p. 360). Popper's move amounts to postulating the existence of propensities underpinning frequencies and to referring probabilities to the single experiments produced by the experimental set-up. Plainly, the term 'postulate' is mine, not Popper's. Popper uses the term 'conjecture' and claims that a propensity statement is a conjecture that has to be tested against observed frequencies. In order to gain applicability, Popper's propensity theory has to depend on frequencies. In view of this, the problems raised by frequentism are not removed, but only displaced.

If the empirical interpretations of probability – frequentism and propensionism – do not offer a satisfactory account of quantum mechanical probabilities, why not embrace an epistemic interpretation? The viability of this alternative has been maintained by Harold Jeffreys, an upholder of the logical interpretation of probability[1]. According to logicism, probability is a logical relation between a proposition describing certain experimental evidence and a proposition describing a hypothesis. Such a logical relation is objective in the sense that there is only one way of assigning it correctly. Given two people facing the same experimental evidence, if they disagree as to an assignment of probability to a certain hypothesis, they cannot both be right. Logicism is an objective interpretation of probability, even though it is epistemic. The opposition which is often made between 'epistemic' and 'objective' interpretations of probability does not capture the real features of the debate on the nature of probability, which is more faithfully reflected by the opposition between 'epistemic' and 'empirical', or 'physical'. This leaves room for a conception of probability that is both epistemic and objective, like logicism.

The most popular trend within the epistemic approach seems to be subjectivism. On the subjectivist side, we encounter different opinions. Bruno de Finetti, who, together with Frank Ramsey, can be rightly considered the father of modern subjectivism, took an intransigent attitude towards the notion of 'objective probability'. For him probability is utterly subjective and expresses the degree of belief of a person in the happening of an event[2]. If two people, on the basis of the same evidence, assign different probabilities to a hypothesis, this does not mean that one of them is wrong. It only means that they evaluate

[1] Jeffreys does not give details on how this idea can be worked out, but his work contains hints in this sense. See especially [12]. See also [13–15] for Jeffrey's theory of probability, which is framed in a very original epistemology.

[2] See [16–18]. On de Finetti's philosophy of probability see [19]. The application of de Finetti's subjectivism to quantum mechanics is discussed in [20].

probabilities in different ways. A probability judgment is a complex procedure, involving a number of factors in addition to the experimental evidence. In his extreme subjectivism, de Finetti refused to make a distinction between the assignments of probability made in everyday life and those suggested by physical theories. Only in the volume Filosofia della probabilità, containing the text of a series of lectures given by de Finetti in 1979, does he seem inclined to concede that scientific laws can give a 'more solid foundation' to personal opinions. But he adds that opinions are subjective, and that the only meaning that can be ascribed to laws is instrumental ([21], p. 117).

A different attitude is taken by Ramsey, who admits of an objective notion of 'probability in physics' within the framework of subjective probability[3]. The peculiarity of this notion, which marks its difference with subjective probabilities encountered in everyday life, is the fact that it makes reference to systems of beliefs rather than to the beliefs of certain people in particular situations. A distinctive feature of such systems is that they contain law statements. In case such laws are commonly accepted physical theories, the system is such that everybody agrees on it. Probability assignments made with reference to such systems are suggested directly by theories, and this confers on them a character of objectivity that is lacking from common assignments. Ramsey does not refer to quantum mechanics, having rather in mind statistical mechanics, but there is no reason why what he says could not be referred to quantum mechanics. In point of fact, he might have thought of quantum mechanics, since he was acquainted with the work of Jeffreys and Weyl, and might have gotten from them the idea that objective probability can be accommodated in an epistemic framework[4]. Ramsey's hint needs to be developed in more detail, but it points to the only meaningful way of talking about such things as 'physical probability', or 'chance', within subjectivism.

References

1. P.S. Laplace: Théorie analytique des probabilités (Paris 1812)
2. P.S. Laplace: Essai philosophique sur les probabilités (Paris 1814)
3. R. von Mises: Probability, Statistics and Truth (Allen and Unwin, London-New York 1939, second edition 1957)
4. J. Bertrand: Calcul des probabilités (Gauthier-Villars, Paris 1888)
5. M. Born: Physics in my Generation (Springer, New York 1969)
6. W. Heisenberg: The Physical Principles of Quantum Theory (Dover, New York 1949)
7. W. Heisenberg: Physics and Philosophy (Penguin Books, London 1958)
8. K.R. Popper: 'The Propensity Interpretation of the Calculus of Probability and the Quantum Theory', in Observation and Interpretation ed. by S. Körner (Butterworth, London 1957) pp. 65-70

[3] See [22,23]. For a comparison between de Finetti's and Ramsey's subjectivism see [24].

[4] On this point see [25,26].

9. K.R. Popper: 'The Propensity Interpretation of Probability', British Journal for the Philosophy of Science. 10, 25-42 (1959)
10. K.R. Popper: Realism and the Aim of Science (Hutchinson, London 1982)
11. K.R. Popper: A World of Propensities (Thoemmes, Bristol 1990).
12. H. Jeffreys: 'The Present Position in Probability Theory', British Journal for the Philosophy of Science. 5, 275-289 (1955)
13. H. Jeffreys: Scientific Inference 2nd edition (Cambridge University Press, Cambridge 1973)
14. H. Jeffreys: 'Scientific Method, Causality, and Reality', Proceedings of the Aristotelian Society, New Series. 37, 61-70 (1937)
15. H. Jeffreys: Theory of Probability 3rd edition (Clarendon Press, Oxford 1967)
16. B. de Finetti: 'Probabilismo', Logos (1931). English translation: 'Probabilism', Erkenntnis. 31, 169-233 (1989)
17. B. de Finetti: 'Sul significato soggettivo della probabilità', Fundamenta matematicae (1931). English translation: 'On the Subjective meaning of Probability', in Induction and Probability ed. by P. Monari and D. Cocchi (CLUEB, Bologna 1993) pp. 291-321
18. B. de Finetti: 'La prévision: ses lois logiques, ses lois subjectives', Annales de l'Institut Poincaré. 7, 1-68 (1937). English translation: 'Foresight: its Logical Laws, its subjective Sources', in Studies in Subjective Probability ed. by H.E. Smokler (Wiley, New York 1964) pp. 95-158.
19. M.C. Galavotti: 'Antirealism in the Philosophy of Probability: Bruno de Finetti's Subjectivism', Erkenntnis. 31, 239-261 (1989)
20. M.C. Galavotti: 'Operationism, probability and Quantum mechanics', Foundations of Science. 1, 99-118 (1995/96)
21. B. de Finetti, Filosofia della probabilità (Il Saggiatore, Milano 1995)
22. F.P. Ramsey: The Foundations of Mathematics and Other Logical Essays ed. by R. B. Braithwaite (Routledge, London 1931)
23. F.P. Ramsey: Notes on Philosophy, Probability and Mathematics ed. by M.C. Galavotti (Bibliopolis, Naples 1991)
24. M.C. Galavotti: 'The Notion of Subjective probability in the Work of Ramsey and de Finetti', Theoria. 57 239-259 (1991)
25. M.C. Galavotti: 'F.P. Ramsey and the Notion of 'Chance'', in The British Tradition in the 20th Century Philosophy, Proceedings of the 17th IWS 1994 ed. by J. Hintikka and K. Puhl (Hölder-Pichler-Tempsky Vienna 1995) pp. 330-340
26. M.C. Galavotti: 'Some Remarks on Objective Chance (F.P. Ramsey, K.R. Popper and N.R. Campbell)', in Language, Quantum, Music ed. by M.L. Dalla Chiara et al. (Kluwer, Dordrecht-Boston, 1999) pp. 73-82

Statistical Mechanics and the Propensity Interpretation of Probability*

Peter J. Clark

The University of St Andrews,
Dept of Logic and Metaphysics,
St Andrews, Fife Scotland KY16 9AL, UK

1 The Over-Determination Problem

One of the most fascinating problems for the philosopher posed by the corpus of statistical physics is the issue of the consistency problem which arises in both the classical and the quantum contexts. It arises starkly in classical statistical physics, there it is the issue of precisely how it is possible to add probabilistic assumptions to treat of an aggregate motion, the component submotions of which, being governed by the laws of mechanics, are entirely deterministic. Essentially the problem occurs because one of the two theories we want to employ viz. mechanics is a completely general theory, that is it ought to give a complete description of any physical situation to which it applies, hence if we put them together the suspicion must be that they will overdetermine the history of the physical system under consideration and inconsistency will result.[1]

To explain the macroscopic phenomena associated with thermodynamics we need to introduce stochastic postulates 'on top of' Hamiltonian mechanics, but the latter mechanics gives a complete state description of the physical situation and all the values of the observables are thereby determined as functions of the variables describing the instantaneous state.[2] The situation is of course compounded in the case of non-equilibrium statistical mechanics. The reason is that in theories of this kind we want to use deterministic and time-symmetric

* The author gratefully acknowledges the support provided by a Leverhulme Research Fellowship, during which this article was produced.

[1] The history of the kinetic theory of gases provides examples where the suspicion was more than well-founded, two of the most notorious examples are Maxwell's original assumption in his 1860 paper that the components of velocity of a single moving molecule are statistically independent thereby contradicting the law of the conservation of energy and Boltzmann's original claim that what is now called the hypothesis of molecular chaos holds all of the time in his first derivation of the H-theorem, contradicting the time-symmetry of classical mechanics. [1,2]

[2] However one views the problem of consistency, it is perfectly clear that statistical mechanics cannot be reduced to pure mechanics. Equally it is also perfectly clear that the theory cannot be reduced to probability theory alone, the equations of motion, Liouville's theorem and the existence of integrals of the equations of motion cannot be derived from probability theory for example. It is interesting however just how far probability can be taken and how little mechanics is actually required in some developments of the theory. See for example [28].

theories (once again classical dynamics) to obtain an explanation of the well-known phenomenon that the behaviour of physical systems that are far from equilibrium is not symmetric under time reversal. That is we want to obtain explanations of facts of the type subsumed by the second law of thermodynamics. Since there is no possibility whatever of combining time-symmetric deterministic dynamics to obtain, even in the limit, non-time-symmetric theories; some postulate of a stochastic kind (e.g. that the hypothesis of molecular chaos must hold for some part of the time in the approach to equilibrium) must replace the time-symmetrical dynamics.

The transition from classical mechanics to quantum mechanics will not resolve this issue despite the lack of a functional connection between state and observable in QM, for the problem immediately reappears in its standard form for the temporal evolution of density operators representing the instantaneous state of an isolated non-interacting quantum mechanical system evolves in a way described by the Schrödinger equation deterministically with the time (the evolution being given by a one parameter semi-group of unitary operators defined on the appropriate Hilbert Space). So in this form of evolution the situation is structurally directly analogous to the classical case. It is of course with the second form of evolution found in QM that real disanalogies with the classical case occur, for there we have discontinuous and indeterministic evolution. But the appearance of indeterminism here should not be taken as making the interpretation of probability ascriptions in QM easier than in the classical case, because of the way it appears in the theory it notoriously makes the matter very much harder.

As is very well known despite the enormous empirical success of statistical mechanics the conceptual problems surrounding it have persisted.[3] They are particularly acute if one further wishes to claim that the stochastic assumptions deployed in statistical mechanics describe physical facts, just as much as does the underlying dynamics. That is to say the stochastic postulates of the theory describe objective physical properties of the systems under consideration just as much as the underlying dynamics does, no more and no less. One very influential general account of objective physical probability, itself made up of a number of rather diffuse proposals is the 'propensity' interpretation of probability. Since the proposals falling under this heading are varied I will concentrate on one undoubtedly seminal approach developed over the last fifty years by Karl Popper and more recently by David Miller.[4]

[3] For a superb overview of the central conceptual problems of statistical mechanics see [3].

[4] Extensive treatments of the propensity theory of probability by Popper can be found in [4–9] and by Miller in [10,11,23]. An excellent treatment of the various versions of the propensity theory can be found in [14] and in [25–27].

2 Physical Probability and Propensity

In 1934, as a result of his work on the foundations of quantum mechanics, Popper became convinced of the need to be able to introduce well-defined probabilities for singular events.[5] He considered a very simple, natural extension of Von Mises' frequency theory. Von Mises had suggested that the probability of an event was to be identified with its limiting relative frequency in a collective, that is an infinite sequence of trials the outcomes of which satisfy the two postulates (i) of the existence of the limiting frequency and (ii) of the exclusion of a gambling system (see [13]). Popper added as an extension to this theory a claim explicitly denied by Von Mises that, if a single event is a member of a collective then its singular probability might be taken as equal to its probability in the collective as a whole. But as Popper realised any such suggestion would run up against the reference class problem. That is that the probability assigned to the single event would depend upon to which collective the single event had been assigned. But then the probability of the event would depend upon how we chose to describe it, and would no longer be an 'objective property' of the event. His solution to this difficulty was to try to locate the physically appropriate collective to which to assign the event. Popper's idea was to look at the physical set up which 'brings about' the event as a member of a set of outcomes of trials. It was in effect to appeal to the generating conditions of the collective. This apparently small change in the definition of a collective produced a very radical shift in the interpretation of the probability relation. Popper wrote ([6], p. 34):

> All this means that the frequency theorist is forced to introduce a modification of his theory – apparently a very slight one. He will now say that an admissible sequence of events (a reference sequence, a "collective") must always be a sequence of repeated experiments. Or more generally he will say that admissible sequences must be either virtual or actual sequences which are characterised by a set of generating conditions – by a set of conditions whose repeated realisations produces the elements of the sequences.

Thus in the revised theory it is the generating conditions which characterise the collective objectively, and Popper wrote of them as the origin of the propensity to produce the observed frequencies. Thus he argued ([6], p. 35):

> we have to visualise the conditions as endowed with a tendency or disposition, or propensity, to produce sequences whose frequencies are equal to the probabilities; which is precisely what the propensity interpretation asserts.

So we now have a theory which asserts (i) all probabilities are conditional probabilities (the conditioning event being the realisation of the generating conditions), (ii) probability is to be understood as the propensity of the experimental

[5] A full treatment of Popper's views on the foundations of Quantum mechanics can be found in [7,12].

set up – the generating conditions of the sequence of trials – to produce a relative frequency on repetition of the trial (iii) probability is well-defined for the single case (it is the disposition of the generating conditions to yield just that outcome in that particular trial), (iv) this account of objective physical probability as a disposition of generating conditions can only hold in an indeterministic universe, otherwise the generating conditions will fix the outcomes trivially and all probabilities will be zero or one. Condition (ii) gives the relation between the probability and the relative frequency over a long run, condition (iii) is precisely the constraint that the theory is designed to meet, while Popper regarded condition (i) as a particular benefit of the theory since by stressing the fundamental nature of conditional probability, he thought that his own theory 'corresponds to the transition from the frequency theory to the measure theoretical approach of Kolmogorov'.[6]

Condition (iv), which is really a consequence of condition (ii), however renders the application of the theory to statistical physics, its intended interpretation, impossible. Popper insists that 'propensities only obtain where exactly the same situation or 'state' may yield, different subsequent states; but that is incompatible with determinism' ([7], p. 105) and again David Miller repeats the claim 'for there to exist probabilities other than zero and one it is therefore essential that the class of future possibilities contains more than a single element' ([10], p. 186). However the immediate consequence is that propensities can't be present in statistical mechanics, because the underlying mechanics is deterministic. Popper was certainly aware of this point, but he simply chose to deny it. He wrote ([7], p. 105):

> Today I can see why so many determinists, and even ex-determinists who believe in the deterministic character of classical physics, seriously believe in a subjectivist interpretation of probability: it is in a way, the only reasonable possibility which they can accept; for objective physical probabilities are incompatible with determinism; and if classical physics is deterministic, it must be incompatible with an objective interpretation of classical statistical mechanics.

Clearly the reader is being invited to contrapose. The objective interpretation of statistical mechanics is the right one (with this I agree) so classical physics must be indeterministic. The argument then is as follows: since the statistical assertions of statistical mechanics are just as objective as the 'mechanical' assertions of that theory and since Popper argues those assertions cannot be treated as such if the underlying dynamical theory is deterministic, then the underlying dynamical theory must after all be indeterministic despite its *prima facie* deterministic character. Popper uses an example from non-equilibrium statistical mechanics thus: in order to explain irreversible approach to equilibrium, we

[6] [6], p. 40. Gillies points out that Kolomogorov had already made the suggestion in 1933 [15] that probabilities should be related to the outcomes of sets of repeatable conditions and not collectives. [14], p. 5

have to assume that the measure of the set of possible initial states which produce 'pathological' (i.e. non-approach to equilibrium) behaviour is zero and the measure of the set of possible initial dynamical states which do eventually yield equilibrium is one. Now this probabilistic hypothesis, that the set of 'pathological' states has measure zero, has to be physically interpreted as a claim about *propensities*, but propensities only obtain where exactly the same situation or 'state' may yield, in the time evolution of a system, different subsequent states; but that is incompatible with *determinism*. That is, Popper says: 'propensities can be accepted as physical realities (analogous to forces) only when determinism has been given up' (Ibid. and cf. [16], pp. 93–104). But the indeterminism of observational states is trivial here. Whatever propensities are present must be grounded in the *dynamics* (that which in complete exactness controls the behaviour of the system). So, to have dynamical propensities of a non-trivial kind (not always zero or one), we must have dynamical indeterminism.

However, in the context of classical statistical mechanics there is no other place to appeal to but classical mechanics, so giving up determinism must be the same as saying, classical mechanics is after all indeterministic. But in the context discussed this claim is manifestly false. Paradigmatically, as employed in standard formulations of statistical mechanics, the underlying mechanics is deterministic. This is because the equations of motion for the dynamical system (the canonical Hamiltonian equations, where the Hamiltonian is independent of the time) are such that we have a system of first order differential equations whose solutions *always exist* and are *unique*, given an arbitrary initial point in phase space at some instant t. Since both existence and uniqueness of solution of the differential equations of motion obtain, the dynamical states of the system are determined. It is thus extraordinary to see Popper claim: 'the propensity interpretation is I believe, that of classical statistical mechanics. Boltzmann spoke of a tendency. But I think that it describes best what classical authors really had in mind when speaking of the quotient of the equally likely to the possible cases; they had in mind that this quotient was a measure ... of the propensity, characteristic of certain specified conditions, to produce a given event' ([8], p. 398). Since one wants to agree with Popper that the physical probabilities occurring in classical statistical mechanics are as objective as the dynamical constraints, but disagree completely with him that the underlying mechanics in *this context* is indeterministic, one must, accepting the validity of his argument, deny that 'objective physical probabilities are incompatible with determinism'. And indeed so one must.

It seems to me that the issue of determinism versus indeterminism really ought to be (is) irrelevant to an interpretation of probability theory. As Kolmogorov put it in 1931: 'the applicability of probabilistic schemes stands in no relation to questions of chance in Nature' [17]. It just seems hopeless to do as Miller suggests and address the question 'Whether the requisite indeterministic loop-holes really exist in games of chance, mortality tables, and other macroscopic phenomena to which probabilistic ideas are customarily applied'

([10], p. 186): the epistemology and metaphysics of probability theory simply does not wait on the resolution of such questions.

As to the issue actually raised by Popper concerning why subjective theories of probability seemed the only reasonable alternative in respect of statistical mechanics it does seem that real issue concerns axiom two of the frequency interpretation, that is the notorious randomness postulate. Frequentists insist that probability exists only in sets of outcomes called collectives. A collective is an infinite set of outcomes which satisfies two axioms: first the axiom of convergence to the effect that limiting relative frequencies exist and second the axiom of randomness to the effect that the relative frequency remains unaltered in any subsequence selected by an "appropriately disinterested" rule. Clearly the thought must be that in classical mechanics the dynamics of evolution, the phase flow, itself allows for the specification of selection rules which will defeat the invariance of relative frequency whether that is given in terms of occupancy numbers or times of sojourn or of distributions over hypersurfaces in phase space. The real problem is thus posed by tying probability to the axiom of randomness and one of the great achievements of the Kolmogorov interpretation is to rid probability of its dependency upon this notion.

3 Statistical Stability

We are used now to the idea that it is very difficult indeed to give a probabilistic theory of causation. There are just too many counterexamples of the form in which causal factors lower rather than raise the probability of their effects ([18], p. 208–247). What about the other way round? What about generalised causation explaining probability? The idea is apparently clear enough. We are to see the probability of an outcome not as a measure of any kind of ratio or frequency, but as a measure of the inclination of the current state of affairs to realise that outcome. Popper sees propensities as a disposition of the current sate of affairs to produce outcomes of a certain sort. Writing of propensities he says: 'speaking of a propensity I wish to suggest an intuitive idea akin to that of a Newtonian force, yet distinct from a force in that it produces frequencies rather than accelerations' ([8], p. 397). What I want to look at more closely is the notion of the 'inclination or disposition of the current state of affairs'. One of the problems the propensity theory purports to solve is the problem of statistical stability. Popper says 'the tendency of statistical averages to remain stable if the conditions remain stable is one of the most remarkable characteristics of our Universe' ([9], p. 12). Miller sees the problem similarly, he says of it: 'For indeterminists – who may be tempted to regard such sequences as sequences made up of wholly undetermined events – it is the problem of explaining why an order at all should appear in the midst of disorder; truly disordered sequences might be expected to be just that, without sign of constancy' ([11], p. 137). Now the explanation of statistical stability is according to Miller to proceed like this: 'If propensities are postulated to satisfy the axioms of probability , then we may conclude, through the laws of large numbers , that in stable circumstances, in which the propensities of the events in question do not change, there is an overwhelming propensity for a longish sequence of throws to be statistically stable.' He comments: 'this explanation is not a magically deep one...' (Ibid., p. 138). It is not its depth I object to but its magical nature. Surely there is a real difficulty here. That is if the Universe is genuinely indeterministic in some parameter there is literally no inclination nor tendency, no generalised force or partial cause to produce any undetermined value for that parameter. What can it mean to say as Miller does that 'the World has a propensity, which is neither cast iron necessity nor cast iron impossibility, to develop in the way described' in a single case if the evolution of the parameter is indeterministic? Two ideas are being brought into play which are not compatible. Let all the constraints on the evolution of the state variables of a system be satisfied, let one of its variables evolve indeterministically: then nothing further fixes the evolution of that parameter, that is what we have just said but now we say that there is a propensity in this single case for that parameter to evolve in a certain way, but that means in the single case there is another constraint another cause on its evolution but that contradicts the claim that all the constraints on the system were originally satisfied. In the single case the propensity theory is having its cake and eating it.

Let us remind ourselves for a moment as to how we should think of determinism and indeterminism following Earman and Montague's analysis ([19] and [20]). To cut a quite long story short, *a theory* T is said to be *deterministic in the state variables* (say $\phi_1, \ldots, \phi_n$) if any two standard models, or *histories*, as they are called, of the theory (i.e. any two relational structures which are models of the theory) which agree at some given time, agree at all other times. In short the constraint entails that for a deterministic theory if two histories (or models) have identical states at one time then they have identical states at all times. A physical system may be said to be deterministic in the state variables $(\phi_1, \ldots, \phi_n)$ when its history *realises* (or satisfies) a *theory* deterministic in the state variables $(\phi_1, \ldots, \phi_n)$. This characterisation of a deterministic theory and deterministic system fits very well with classical, relativistic and quantum mechanics. It has the further advantage of not relying on the essentially extraneous notion of predictability at all.

So for a system which is indeterministic in say a state variable ϕ_n, literally nothing will serve to fix how ϕ_n will evolve. Many possible paths will lie open to it, all consistent with the past history of the system of which it is a state variable and the laws governing that system's evolution. What goes for an isolated physical system here will go for the Universe. But then this talk of disposition as partial cause, inclination, partial forcing, partial determining is quite out of place. What is really going on here is this, the notion of partial determinism is not well understood. Partial determinism means real indeterminism and indeterminism means just that, nothing fixes or produces, induces or forces the outcome. I want to try and be as clear as to the point I am making small though it is. I am not objecting to propensity understood as an intrinsic property of a chance set up which characterises stable frequencies generated by those conditions if such exists, what I am objecting to is regarding them as partial or generalised causes applicable to single cases, where it is argued to be explanatory of an exhibited statistical stability in the indeterministic case. If the system really is indeterministic then there is nothing inducing or partially forcing the outcome. It is the later theory of Popper, the combination of generalised causes and indeterminism that I find difficult to accept.

There is however another very deep difficulty in thinking of propensity as partial causation. This problem was pointed out by Wesley Salmon and Paul Humphreys [21,22]. The problem is that understood as generalised causation the propensity theory cannot be an interpretation of the probability calculus, because it can give no account of inverse probabilities which always exist when the conditional probability is defined, that is if $P(A|B)$ is defined so is $P(B|A)$. It may very well be the case that given that condition B is realised, this induces or produces a propensity or partial causation for the event A. But what in general can it mean to say that the event A (the outcome) induces a propensity or partial causation of the generating condition B? The issue is straightforward, if one thinks of propensity as partial causation, one has made it an asymmetric relation, but conditional probabilities, as opposed to propensities are symmetric. It is evident that something has to give. Since conditions (i), (ii) and (iv) above are central to the theory, dropping condition (iii), the single case requirement would seem the natural thing to do and this would leave us with a 'long run' propensity theory. But this is not the route that the later Popper or David Miller took. They did replace a condition, but condition (i). In the new theory they conditionalise on the entire state of the Universe at a given instant. In effect a probability measure is treated as a function of two variables $P(A; U(t))$ where A is an event type and $U(t)$ represents the entire complete specification of the instantaneous state of the Universe at time t. So $P(A; U(t))$ is then some suitably normalised measure of the number of possible states into which U at t could evolve in which A is satisfied in the class of all possible states into which U at t could evolve. The notion of 'could' being given by a notion of physical possibility. The idea of conditional probability is then given by $P(A|B; U(t))$ understood as the measure of the possible states into which U can evolve in which A is satisfied, given that it actually evolves to satisfy B. That is $P(A|B; U(t)) = P(A\&B; U(t))/P(B; U(t))$ where $P(B; U(t))$ is non-zero. So the inverse probability, $P(B|A; U(t))$ has a natural interpretation as the relative measure of the possible states into which U can evolve in which the condition B is satisfied among those states which evolve in such a way as to satisfy A. Of course in this case there is no causal link presupposed in the definition of the conditional probability, one is simply looking at the fraction of distinct histories evolving from the condition of the Universe at t which satisfy certain conditions. In discussing the relative probability $P(A|B; U(t))$ as it occurs in Popper's later theory, however Miller still insists that it is 'a genuine propensity, conditional on B, of the contemporary state of the world, not a propensity of the condition B' ([10], pp. 189–90). One is left however wondering why it is thought to be a propensity interpretation at all, all that is actually involved is a measure or weighting of possible histories of the Universe. A similar device is used by McCurdy when he explicitly severs the connection between theories of propensity and the propensity interpretation. He writes: 'Conditional propensities do not provide a measure of causal dependence between the singular conditioned event and the singular conditioning event. If one maintains that conditional propensities do provide a measure of causal dependence between the singular conditioned event and the (singular) conditioning event then one is no longer dealing with a propensity interpretation of probability, one is dealing with a theory of propensities' ([24], pp. 117–8).

4 Kolmogorov's Theory and the Application Problem

As is well known in Kolmogorov's account probability is characterised by an axiomatic theory in which either in the finite or infinitistic case the probability of an event is a unique real number associated with a member of a field of sets. The event being characterised as a subset of the set of elementary events. The function P which assigns real numbers to members of the field of subsets is characterised by the role it plays in the axioms. In other words it is an uninterpreted formal axiomatic theory. The question is how is the application of the theory to take place? To apply the theory (taking the finite case for simplicity) we need a certain complex of unlimitedly repeatable conditions, the outcomes of the repetitions of the conditions form the collection of elementary events. So applications of the theory are limited to repeatable events, there is no way of applying the theory to single events. The empirical probability of the single case can't be defined. What holds for the application of finite probability also holds for the infinite case, where the application of the strengthened axioms for the infinite theory require the existence of an *ensemble* to provide the elementary events, F, and to provide the field of sets to whose members a unique real number called the probability can be assigned (we take the familiar sigma-additive field of Borel subsets of F). Again there is clearly no hope of defining the notion of a single case probability. But now a familiar doubt must emerge: it is that in any application of the theory we are really bound to deploy a single case interpretation.

Consider the familiar case of explaining why my morning bath so unequally heated and far from equilibrium eventually attains equilibrium. Well one no doubt would want to refer the particular case to a typical ensemble of macroscopically identically prepared systems and claim that the particular bath was indeed a typical member of that typical ensemble. In such ensembles it is overwhelmingly probable that in a relatively short time the system will have relaxed to equilibrium . So being a typical member of the ensemble we regard the behaviour of the actual bath as explained. But of course we know that there is a probability zero set of members of our ensemble the dynamical initial conditions of which will exhibit anti-thermodynamic behaviour. So we have in fact claimed that we have overwhelming moral certainty this particular case isn't sitting with its initial conditions in that pathological set. But that is a judgement of practical reasoning, it is a subjective judgement to the effect that probability zero conditions, which are of course physically possible, won't occur. Of course the point generalises. In the finite case if $P(A)$ is very small one argues that it is practically certain that A will not appear in a very large repetition of the generating conditions whose outcomes form the ensemble and that if the conditions are repeated a great number of times one can be morally certain that the observed relative frequency m/n of the occurrence of say property A will differ only very little from $P(A)$. But none of this is justified by the theory rather it is a precondition for the application of the theory. But that admission seems to me at least to cast considerable doubt on the claim that we have given an objective interpretation of probability in these applications. That is why I am still left very uneasy about objective theories of physical probability: their ultimate reliance on postulates of moral or practical certainty.

References

1. J.C. Maxwell: 'Illustrations of the Dynamical theory of gases'. In: The Scientific papers of James Clerk Maxwell, ed. W.D. Niven (Dover Publications, New York 1965) pp. 377–409
2. L. Boltzmann: 'Weitere Studien über das Wärmegleichgewicht unter Gasmolekülen' Wien.Ber. **66**, 275 (1872)
3. L. Sklar: Physics and Chance (Cambridge University Press, Cambridge 1993)
4. K.R. Popper: The Logic of Scientific Discovery, sixth revised impression of the 1959 English translation (Hutchinson, London 1972)
5. K.R. Popper: 'The Propensity interpretation of the Calculus of probability, and the Quantum theory'. In: Observation and Interpretation, Proceedings of the Ninth Colston Research Society, (University of Bristol, Bristol 1957) pp. 65–70,88–9.
6. K.R. Popper: 'The Propensity Interpretation of Probability, British Journal for the Philosophy of Science. **10**, 25–42 (1959)
7. K.R. Popper: The Quantum theory and the Schism in Physics (Hutchinson, London 1982)
8. K.R. Popper: Realism and the Aim of Science (Hutchinson, London 1983)
9. K.R. Popper: A World of Propensities (Thoemmes press, Bristol 1990)
10. D. Miller: Critical Rationalism: A Restatement and Defence (Open Court, Chicago and Lasalle 1994)
11. D. Miller: 'Propensities and Indeterminism'. In: Karl Popper: Philosophy and Problems, ed. by A. O'Hear (Cambridge University Press, Cambridge 1996) pp. 121–47
12. M. Redhead: 'Popper and the Quantum theory' In: Karl Popper: Philosophy and Problems, ed. by A. O'Hear (Cambridge University Press, Cambridge 1996) pp. 163–176
13. R. Von Mises: Probability Statistics and Truth, 2nd English edn. (George and Alan Unwin, London 1961)
14. D. Gillies: 'Varieties of Propensity' British Journal for the Philosophy of Science. **52** (2000)
15. A.N. Kolmogorov: Foundations of the Theory of Probability, 2nd English edn. (Chelsea, New York 1956)
16. K.R. Popper: The Open Universe: An argument for Indeterminism (Hutchinson, London 1982)
17. A.N. Kolmogorov: 'Über die analytischen Methoden in der Wahrscheinlichkeitsrechnung' Mathematische Annalen **104**, 415–458 (1931)

18. W.C. Salmon: Causality and Explanation (Oxford University Press, Oxford and New York 1998)
19. R.C. Montague: 'Deterministic Theories'. In: Formal Philosophy, Selected Papers ed. by R.H. Thomason (Yale University press, New Haven 1974), pp. 303–359
20. J. Earman: A Primer on Determinism (D. Reidel, Dordrecht 1986)
21. W.C. Salmon: 'Propensities a Discussion Review' Erkenntnis. **14**, 183–216 (1979)
22. P. Humphreys: 'Why Propensities Cannot Be Probabilities' Philosophical Review. **94**, 557–70 (1985)
23. D. Miller: 'Single-Case Probabilities' Foundations of Physics. **21**, 1501–16 (1991)
24. C.S.I. McCurdy: 'Humphreys Paradox' Synthese. **108**, 105–125 (1996)
25. P. Milne: 'Can there be a Realist single Case Interpretation of Probability?' Erkenntnis. **25**, 129–32 (1986)
26. D. Gillies: 'Popper's Contribution to the Theory of Probability' In: Karl Popper: Philosophy and Problems, ed. by A. O'Hear (Cambridge University Press, Cambridge 1996) pp. 103–120.
27. P. Milne: 'A Note on Popper, Propensities and the Two Slit experiment' British Journal for the Philosophy of Science. **36**, 66–70 (1987)
28. L. Boltzmann: 'Bemerkungen über einige Probleme der mechanischen Wärmetheorie' Wien.Ber. **75**, 62 (1877)

Interpreting Probabilities: What's Interference Got to Do with It?

Tim Maudlin

Rutgers University, New Brunswick, NJ 08901, USA

There are several different questions which must be addressed when investigating the meaning and origin of probabilities in a physical theory. Among these questions are:

a. What are the probabilities probabilities *for*?
b. How are the probabilities to be calculated?
c. Where do the probabilities come into the account?
d. How are experimental outcomes analyzed in terms of the events to which the probabilities are attributed?
e. How are the probabilities to be understood?

These questions are, no doubt, obscure in several ways, but a useful way to get a handle on them is to see how some different theories might answer them.

In the various versions (both continuous and discrete) of the spontaneous localization theory, 1) the probabilities are for the wavefunction to evolve in a certain way (or to have a specified form after a period of evolution from a given initial state); 2) the probabilities are calculated using the collapse dynamics (giving results which approximate, but do not exactly match, Born's rule); 3) the probabilities are introduced directly into the dynamics; 4) the form of the wavefunction is interpreted as representing the results of experiments by a somewhat vague rule associating near-eigenfunctions of operators with physical states which have properties which are associated with the operators (this tortured explication is due to the so-called tails problem); 5) the probabilities are irreducible dynamical transition chances.

In Bohmian Mechanics, 1) the probabilities are for particles to have a certain configuration; 2) they are calculated by using the deterministic dynamics and conditionalizing on the effective wave-function and the initial probability distribution for particles in the universe. Assuming that the initial probability distribution is typical (relative to the measure given by the square of the wave-function), this recovers Born's rule for particle locations; 3) the probabilities for later times are inherited via the dynamics from the initial probability measure, conditionalizing on known outcomes of experiments; 4) results of experiments are analyzed in terms of particle configurations; 5) the probabilities are not fundamentally dynamical transition chances, and the understanding of typicality of initial conditions is a matter of some dispute, the same dispute which arises in classical statistical mechanics.

Perhaps some of these questions are ill-conceived, or inapplicable to certain physical theories. Still, one would like to have, for any theory, either a clear answer to the question or a clear explanation of why the question is a bad one. There is no canonical order in which these questions must be addressed, but it does seem most natural to answer the first one first. In any case, without a tolerably clear idea of how the first question is to be answered, it hard to know how even to approach several of the others.

Philosophers doubtless tend to focus on the first and last questions, i.e., on the physical ontology of the theory and on the nature of probability itself. From a mathematical point of view, the second question offers the most direct technical challenge. If one starts first with the problem of finding a way to get a measure which satisfies the axioms of the probability calculus, then one might hope to *discover* an answer to the first question by reflection on the measure: the measure itself will help determine what exactly it is a probability *of*. There is nothing wrong with this strategy in principle, but one does have to be sure that, in the end, the various other questions, beside the second, are clearly addressed.

Unlike Bohm's theory and the spontaneous reduction theories, the consistent histories approach seems to have followed this route. A particularly clear presentation along these lines is provided by Murray Gell-Mann in *The Quark and the Jaguar*. Gell-Mann identifies the failure to obey the axioms of probability theory as the *only* impediment to interpreting a particular function defined over pairs of histories as a probability:

> When the two histories in the pair are the same, D is a number between zero and one, like a probability. In fact, it can, under certain conditions, be interpreted as the probability of the history. To see what these conditions are, let us examine the relationship among the following quantities:
>
> $$D(A \text{ or } B, A \text{ or } B)$$
> $$D(A, A)$$
> $$D(B, B)$$
> $$D(A, B) + D(B, A).$$
>
> The first three quantities are numbers between zero and one and thus resemble probabilities. The last quantity can be positive or negative or zero, and is not a probability. The rule for calculating D in quantum mechanics is such that the first quantity is always the sum of the other three. But, if the last quantity is always zero when A and B are different, then $D(A \text{ or } B, A \text{ or } B)$ is just equal to $D(A, A) + D(B, B)$. In other words, if D is always zero when two histories are different, then D of a history and that same history always possess the additive property [i.e. the probability of the disjunction of two mutually exclusive events is the sum of their probabilities] and can therefore be interpreted as the probability of that history.
>
> The fourth quantity on the list is called the interference term between the histories A and B. If it is *not* zero for every pair of different

> histories in the set, then those histories cannot be assigned probabilities in quantum mechanics. They "interfere" with each other. [4], 142-2

Gell-Mann's idea is that the interference terms present the main difficulty facing the interpretation of the quantum formalism as a probabilistic theory. Histories which do not interfere can be assigned probabilities, and those probabilities can be understood "classically". It is a short step from here to a second conclusion: since the only empirical reason for demanding interference *terms* is the observation of interference *effects* (such as interference bands), in the absence of such effects there is no impediment to a classical understanding of the numbers calculated by his recipe as probabilities.

This reading make sense of what would otherwise be an extremely puzzling passage which follows. Gell-Mann sketches the usual Schrödinger cat scenario, then comments:

> The usual discussion of Schrödinger's cat goes on to describe alleged quantum interference between the live and dead cat scenarios. However, the live cat has considerable interaction with the rest of the world, through breathing, and even the dead cat interacts with the air to some extent... The live and dead cat scenarios decohere; there is no interference between them... Suppose the quantum event that determines the fate of the cat has already occurred, but we don't know what happened until we open a box containing the cat. Since the two outcomes decohere, this situation is no different from the classical one where we open a box inside of which the poor animal, arriving after a long airplane voyage, may be either dead or alive, with some probability for each. Yet reams of paper have been wasted on the supposedly weird quantum-mechanical state of the cat, both dead and alive at the same time. ibid. 152-3

The comment is curious, of course, because the existence of interference terms was certainly not *Schrödinger's* concern. He was worried in the first place about how to understand the physical state of a particle, since the wavefunction could not attribute it both a determinate position and momentum. If the wavefunction is a complete physical description of the particle, then the indeterminacy must be ontic: in certain circumstances, a particle simply fails to have a determinate position at all. The cat argument shows that if the wavefunction always obeys Schrödinger evolution, the indeterminacy will infect the macroscopic realm: it cannot be confined to the microscopic. Neither interference *terms* (in the mathematics) nor interference *effects* (as observable phenomena) enter into the argument at all.

If the wavefunction is complete, then reduction has several effects. It does, of course, eliminate (or greatly reduce) the off-diagonal terms in the wavefunction. This in turn eliminates certain sorts of physical effects. But the *primary* role of reduction (if the wavefunction is complete) is to answer the *first* question: to provide something (i.e. the form or development of the wavefunction itself) that the probabilities can be probabilities *for*. If the wavefunction is complete, then the probabilities (which reflect some sort of ignorance) must reflect ignorance about the wavefunction itself. Reduction is the chancy process about which we are ignorant: we don't know exactly how the wavefunction will collapse. *This* role of reduction cannot be derived merely from the disappearance of interference terms, or from the replacement of a pure state with a mixture. As John Bell put it:

> Perhaps it is useful to recall here just how the infamous postulate is formulated by von Neumann. If we look back we find that what von Neumann actually *postulates* is that "measurement"... an external intervention by R [the rest of the world] on S [the system]... causes the state $\phi = \sum c_n \phi_n$ to jump, with various probabilities, into ϕ_1 *or* ϕ_2 *or*... From the "*or*" here, replacing the "*and*", as a result of external intervention, von Neumann infers that the resulting density matrix, averaged over the several possibilities, has no interference terms between states of the *system* which correspond to different measurement results. I would emphasize several points here:
>
> a. Von Neumann presents the disappearance of coherence in the density matrix, not as a postulate, but as the *consequence* of a postulate. The *postulate* is made at the wavefunction level, and is just that already made by Dirac for example.
>
> b. I can not imagine von Neumann arguing in the opposite direction, that the lack of interference in the density matrix implies, without further ado, "or" replacing "and" at the wavefunction level. A special postulate to that effect would be required...
>
> [1], 25-6

If Gell-Mann's account of consistent histories, and in particular of the role of decoherence, is to hang together, then in that account the wavefunction must not be complete, and the probabilities are not probabilities for the quantum state itself to have a particular form. Decoherence is of use because it provides conditions under which the quantum state can be used to derive a probability measure (i.e. when the interference between histories disappears), not because, like real reduction, it directly provides events for the probabilities to be *of*.

What, then, are the probabilities in consistent histories *for*? One is tempted to say that they are for *histories* and leave it at that, but this is not sufficient. In a formal sense, of course, a history is a time-ordered sequence of projection operators. But in that sense, all histories exist as purely mathematical objects: none is more real than any other. Furthermore, in that sense the histories all exist irrespective of the appearance of interference terms when calculating D. Histories in the formal sense can be thought of as a sort of *description*, and the probability ascribed to one must be the probability that it *correctly* describes something. The something must be the physical universe itself, a universe whose career is *not* completely described by the wavefunction. That is, Gell-Mann's theory must be some sort of "hidden variables" theory, with the physical state of the universe, whatever it is, a) not determined by the complete form of the wavefunction and b) describable, in principle, by sequences of projection operators.

At this point, we are in danger of falling into a morass of confusion due to an ambiguity in the meaning of the term "history". On the one hand, a history is just a *story*, an *account of something*, which may be either true or false, accurate or inaccurate, complete or incomplete. In this sense, humans living 100,000 years ago were "pre-historic", i.e. they pre-dated written documents. On the other hand, we say that we want to *discover* or *find out* about that period of human history, or find out about the early history of the universe, meaning we want to find out *what actually happened* at a certain time. The existence of this second sort of history, i.e. what actually occurred, is obviously completely independent of the existence of the first kind of history. To avoid this confusion, let's call the first kind of history a "history" and the second kind "History". The object of writing histories is to produce an accurate account of History (i.e. of what actually happened in the past).

Using this terminology, then, Gell-Mann's theory appears to be that sequences of projection operators can be used as histories, the wavefunction can (in the appropriate circumstance) be used to attribute probabilities to these histories, *and the probability is to be construed as the probability that a given history accurately describes History.* The "appropriate circumstance" is simply the disappearance (or near disappearance) of the interference terms in the wavefunction (and consequent nonexistence of interference effects).

We are now is a position to ask several questions. First, is there any single history which *completely* describes History? If so, then the obvious question is: to which family of consistent histories does this master history belong? This way leads to the position Fay Dowker and Adrian Kent call the Unknown Set interpretation [2,3]. Or can histories from *incompatible* families each correctly describe History? This possibility is not ruled out simply because the families are incompatible with each other: that just shows that one cannot, in this way, assign joint probabilities to pairs of histories chosen one from each family. Finally, is the fact that a family is consistent (i.e. that interference terms between members disappear) *sufficient* for D to be a probability? If so, then exactly one history from each family correctly describes History. This leads to several rather obvious problems, given how peculiar and non-classical most consistent sets of histories are, but this is not the place to pursue those problems. It is enough to have seen how wide the gap between answering question 2 (to which Gell-Mann directly attends) and answering question 1 is. We have also seen why decoherence (i.e. the disappearance of off-diagonal terms) is so important in the consistent histories approach. It plays no such central role in either Bohm's theory or in the spontaneous localization theory. Indeed, in Bohm's theory one can not only assign probabilities for, e.g., the particle in the 2-slit experiment to go through either slit, one can also determine (*ex post facto*) which slit it went through, even when "interference bands" form.

There is one last observation to be made about Gell-Mann's theory. He characterizes it as a refinement of the "modern view", which he ascribes to Hugh Everett. He does, though, criticize Everett for his choice of vocabulary, and the criticism is telling:

> For example, his interpretation is often described in terms of "many worlds," whereas we believe that "many alternative histories of the universe" is what is really meant. Furthermore, the many worlds are all described as being "equally real," whereas we believe it is less confusing to speak of "many histories, all treated alike by the theory except for their different probabilities." To use the language we recommend is to address the familiar notion that a given system can have different possible histories, each with its own probability; it is not necessary to become queasy trying to conceive of many "parallel universes," all equally real. ibid., 138

It is clear from this passage that Gell-Mann's ontology is less confusing, and markedly less unusual, than Everett's. The notion of multiple possible histories (i.e. multiple histories, each of which might, for all we know, correctly describe the History of a system), each assigned its own probability, is indeed familiar from classical physics. These classical histories (e.g. possible trajectories through phase space) are not "equally real" in the simple sense that at most one of them correctly describes what happens to the system. What is absent from Gell-Mann's improved vocabulary, though, is any trace of the characteristic features of Everett's interpretation! In particular, Everett makes no use of the families of consistent histories, and appears to regard the uncollapsed wavefunction as complete. Furthermore, Everett's central notion, that of the "relative state" of a system, is nowhere to be found in Gell-Mann's story. As far as I can tell, there is little to recommend any association at all between Gell-Mann's and Everett's theories. For to the central interpretative question we have pursued, viz. "What are the probabilities probabilities *for*?", the two interpretations appear to give quite different answers.

This leaves us with a large question: how are *any* of our five questions to be answered in the relative state theory? I do not mean to venture any opinions on this question save the observation that the consistent histories account does not appear to offer us the resources to answer it.

References

1. Bell, J. S. 1990, "Against Measurement" in Miller, A. I. (ed.) (1990), Sixty-Two Years of Uncertainty, New York, Plenum Press, pp. 17-31
2. Dowker, F. and A. Kent (1995), "Properties of Consistent Histories," *Physical Review Letters* 75, 3038-3041
3. Dowker, F. and A. Kent (1996), "On the Consistent Histories Approach to Quantum Mechanics," *Journal of Statistical Physics* **82**, 1575-1646
4. Gell-Mann, M. (1994), *The Quark and the Jaguar*, New York, W. H. Freeman and Co.

Lecture Notes in Physics

For information about Vols. 1–538
please contact your bookseller or Springer-Verlag

Vol. 539: A. Borowiec, W. Cegła, B. Jancewicz, W. Karwowski (Eds.), Theoretical Physics. Fin de Siècle. Proceedings, 1998. XX, 319 pages. 2000.

Vol. 540: B. G. Schmidt (Ed.), Einstein's Field Equations and Their Physical Implications. Selected Essays. 1999. XIII, 429 pages. 2000.

Vol. 541: J. Kowalski-Glikman (Ed.), Towards Quantum Gravity. Proceedings, 1999. XII, 376 pages. 2000.

Vol. 542: P. L. Christiansen, M. P. Sørensen, A. C. Scott (Eds.), Nonlinear Science at the Dawn of the 21st Century. Proceedings, 1998. XXVI, 458 pages. 2000.

Vol. 543: H. Gausterer, H. Grosse, L. Pittner (Eds.), Geometry and Quantum Physics. Proceedings, 1999. VIII, 408 pages. 2000.

Vol. 544: T. Brandes (Ed.), Low-Dimensional Systems. Interactions and Transport Properties. Proceedings, 1999. VIII, 219 pages. 2000

Vol. 545: J. Klamut, B. W. Veal, B. M. Dabrowski, P. W. Klamut, M. Kazimierski (Eds.), New Developments in High-Temperature Superconductivity. Proceedings, 1998. VIII, 275 pages. 2000.

Vol. 546: G. Grindhammer, B. A. Kniehl, G. Kramer (Eds.), New Trends in HERA Physics 1999. Proceedings, 1999. XIV, 460 pages. 2000.

Vol. 547: D. Reguera, G. Platero, L. L. Bonilla, J. M. Rubí(Eds.), Statistical and Dynamical Aspects of Mesoscopic Systems. Proceedings, 1999. XII, 357 pages. 2000.

Vol. 548: D. Lemke, M. Stickel, K. Wilke (Eds.), ISO Surveys of a Dusty Universe. Proceedings, 1999. XIV, 432 pages. 2000.

Vol. 549: C. Egbers, G. Pfister (Eds.), Physics of Rotating Fluids. Selected Topics, 1999. XVIII, 437 pages. 2000.

Vol. 550: M. Planat (Ed.), Noise, Oscillators and Algebraic Randomness. Proceedings, 1999. VIII, 417 pages. 2000.

Vol. 551: B. Brogliato (Ed.), Impacts in Mechanical Systems. Analysis and Modelling. Lectures, 1999. IX, 273 pages. 2000.

Vol. 552: Z. Chen, R. E. Ewing, Z.-C. Shi (Eds.), Numerical Treatment of Multiphase Flows in Porous Media. Proceedings, 1999. XXI, 445 pages. 2000.

Vol. 553: J.-P. Rozelot, L. Klein, J.-C. Vial Eds.), Transport of Energy Conversion in the Heliosphere. Proceedings, 1998. IX, 214 pages. 2000.

Vol. 554: K. R. Mecke, D. Stoyan (Eds.), Statistical Physics and Spatial Statistics. The Art of Analyzing and Modeling Spatial Structures and Pattern Formation. Proceedings, 1999. XII, 415 pages. 2000.

Vol. 555: A. Maurel, P. Petitjeans (Eds.), Vortex Structure and Dynamics. Proceedings, 1999. XII, 319 pages. 2000.

Vol. 556: D. Page, J. G. Hirsch (Eds.), From the Sun to the Great Attractor. X, 330 pages. 2000.

Vol. 557: J. A. Freund, T. Pöschel (Eds.), Stochastic Processes in Physics, Chemistry, and Biology. X, 330 pages. 2000.

Vol. 558: P. Breitenlohner, D. Maison (Eds.), Quantum Field Theory. Proceedings, 1998. VIII, 323 pages. 2000

Vol. 559: H.-P. Breuer, F. Petruccione (Eds.), Relativistic Quantum Measurement and Decoherence. Proceedings, 1999. X, 140 pages. 2000.

Vol. 560: S. Abe, Y. Okamoto (Eds.), Nonextensive Statistical Mechanics and Its Applications. IX, 272 pages. 2001.

Vol. 561: H. J. Carmichael, R. J. Glauber, M. O. Scully (Eds.), Directions in Quantum Optics. XVII, 369 pages. 2001.

Vol. 562: C. Lämmerzahl, C. W. F. Everitt, F. W. Hehl (Eds.), Gyros, Clocks, Interferometers...: Testing Relativistic Gravity in Space. XVII,507 pages. 2001.

Vol. 563: F. C. Lázaro, M. J. Arévalo (Eds.), Binary Stars. Selected Topics on Observations and Physical Processes. 1999.IX, 327 pages. 2001.

Vol. 564: T. Pöschel, S. Luding (Eds.), Granular Gases. VIII, 457 pages. 2001.

Vol. 565: E. Beaurepaire, F. Scheurer, G. Krill, J.-P. Kappler (Eds.), Magnetism and Synchrotron Radiation. XIV, 396 pages. 2001.

Vol. 566: J. L. Lumley (Ed.), Fluid Mechanics and the Environment: Dynamical Approaches. VIII, 412 pages. 2001.

Vol. 567: D. Reguera, L. L. Bonilla, J. M. Rubí (Eds.), Coherent Structures in Complex Systems. IX, 465 pages. 2001.

Vol. 568: P. A. Vermeer, S. Diebels, W. Ehlers, H. J. Herrmann, S. Luding, E. Ramm (Eds.), Continuous and Discontinuous Modelling of Cohesive-Frictional Materials. XIV, 307 pages. 2001.

Vol. 569: M. Ziese, M. J. Thornton (Eds.), Spin Electronics. XVII, 493 pages. 2001.

Vol. 570: S. G. Karshenboim, F. S. Pavone, F. Bassani, M. Inguscio, T. W. Hänsch (Eds.), The Hydrogen Atom: Precision Physics of Simple Atomic Systems. XXIII, 293 pages. 2001.

Vol. 571: C. F. Barenghi, R. J. Donnelly, W. F. Vinen (Eds.), Quantized Vortex Dynamics and Superfluid Turbulence. XXII, 458 pages. 2001.

Vol. 572: H. Latal, W. Schweiger (Eds.), Methods of Quantization. XI, 224 pages. 2001.

Vol. 574: J. Bricmont, D. Dürr, M. C. Galavotti, G. Ghirardi, F. Petruccione, N. Zanghi (Eds.), Chance in Physics. XIII, 276 pages, 2001.

Vol. 575: M. Orszag, J. C. Retamal (Eds.), Modern Challenges in Quantum Optics. XXIII, 405 pages. 2001.

Vol. 579: R. Haug, H. Schoeller (Eds.), Interacting Electrons in Nanostructures. X, 227 pages. 2001.

Monographs
For information about Vols. 1–24
please contact your bookseller or Springer-Verlag

Vol. m 25: A. V. Bogdanov, G. V. Dubrovskiy, M. P. Krutikov, D. V. Kulginov, V. M. Strelchenya, Interaction of Gases with Surfaces. XIV, 132 pages. 1995.

Vol. m 26: M. Dineykhan, G. V. Efimov, G. Ganbold, S. N. Nedelko, Oscillator Representation in Quantum Physics. IX, 279 pages. 1995.

Vol. m 27: J. T. Ottesen, Infinite Dimensional Groups and Algebras in Quantum Physics. IX, 218 pages. 1995.

Vol. m 28: O. Piguet, S. P. Sorella, Algebraic Renormalization. IX, 134 pages. 1995.

Vol. m 29: C. Bendjaballah, Introduction to Photon Communication. VII, 193 pages. 1995.

Vol. m 30: A. J. Greer, W. J. Kossler, Low Magnetic Fields in Anisotropic Superconductors. VII, 161 pages. 1995.

Vol. m 31 (Corr. Second Printing): P. Busch, M. Grabowski, P.J. Lahti, Operational Quantum Physics. XII, 230 pages. 1997.

Vol. m 32: L. de Broglie, Diverses questions de mécanique et de thermodynamique classiques et relativistes. XII, 198 pages. 1995.

Vol. m 33: R. Alkofer, H. Reinhardt, Chiral Quark Dynamics. VIII, 115 pages. 1995.

Vol. m 34: R. Jost, Das Märchen vom Elfenbeinernen Turm. VIII, 286 pages. 1995.

Vol. m 35: E. Elizalde, Ten Physical Applications of Spectral Zeta Functions. XIV, 224 pages. 1995.

Vol. m 36: G. Dunne, Self-Dual Chern-Simons Theories. X, 217 pages. 1995.

Vol. m 37: S. Childress, A.D. Gilbert, Stretch, Twist, Fold: The Fast Dynamo. XI, 406 pages. 1995.

Vol. m 38: J. González, M. A. Martín-Delgado, G. Sierra, A. H. Vozmediano, Quantum Electron Liquids and High-Tc Superconductivity. X, 299 pages. 1995.

Vol. m 39: L. Pittner, Algebraic Foundations of Non-Com-mutative Differential Geometry and Quantum Groups. XII, 469 pages. 1996.

Vol. m 40: H.-J. Borchers, Translation Group and Particle Representations in Quantum Field Theory. VII, 131 pages. 1996.

Vol. m 41: B. K. Chakrabarti, A. Dutta, P. Sen, Quantum Ising Phases and Transitions in Transverse Ising Models. X, 204 pages. 1996.

Vol. m 42: P. Bouwknegt, J. McCarthy, K. Pilch, The W3 Algebra. Modules, Semi-infinite Cohomology and BV Algebras. XI, 204 pages. 1996.

Vol. m 43: M. Schottenloher, A Mathematical Introduction to Conformal Field Theory. VIII, 142 pages. 1997.

Vol. m 44: A. Bach, Indistinguishable Classical Particles. VIII, 157 pages. 1997.

Vol. m 45: M. Ferrari, V. T. Granik, A. Imam, J. C. Nadeau (Eds.), Advances in Doublet Mechanics. XVI, 214 pages. 1997.

Vol. m 46: M. Camenzind, Les noyaux actifs de galaxies. XVIII, 218 pages. 1997.

Vol. m 47: L. M. Zubov, Nonlinear Theory of Dislocations and Disclinations in Elastic Body. VI, 205 pages. 1997.

Vol. m 48: P. Kopietz, Bosonization of Interacting Fermions in Arbitrary Dimensions. XII, 259 pages. 1997.

Vol. m 49: M. Zak, J. B. Zbilut, R. E. Meyers, From Instability to Intelligence. Complexity and Predictability in Nonlinear Dynamics. XIV, 552 pages. 1997.

Vol. m 50: J. Ambjørn, M. Carfora, A. Marzuoli, The Geometry of Dynamical Triangulations. VI, 197 pages. 1997.

Vol. m 51: G. Landi, An Introduction to Noncommutative Spaces and Their Geometries. XI, 200 pages. 1997.

Vol. m 52: M. Hénon, Generating Families in the Restricted Three-Body Problem. XI, 278 pages. 1997.

Vol. m 53: M. Gad-el-Hak, A. Pollard, J.-P. Bonnet (Eds.), Flow Control. Fundamentals and Practices. XII, 527 pages. 1998.

Vol. m 54: Y. Suzuki, K. Varga, Stochastic Variational Approach to Quantum-Mechanical Few-Body Problems. XIV, 324 pages. 1998.

Vol. m 55: F. Busse, S. C. Müller, Evolution of Spontaneous Structures in Dissipative Continuous Systems. X, 559 pages. 1998.

Vol. m 56: R. Haussmann, Self-consistent Quantum Field Theory and Bosonization for Strongly Correlated Electron Systems. VIII, 173 pages. 1999.

Vol. m 57: G. Cicogna, G. Gaeta, Symmetry and Perturbation Theory in Nonlinear Dynamics. XI, 208 pages. 1999.

Vol. m 58: J. Daillant, A. Gibaud (Eds.), X-Ray and Neutron Reflectivity: Principles and Applications. XVIII, 331 pages. 1999.

Vol. m 59: M. Kriele, Spacetime. Foundations of General Relativity and Differential Geometry. XV, 432 pages. 1999.

Vol. m 60: J. T. Londergan, J. P. Carini, D. P. Murdock, Binding and Scattering in Two-Dimensional Systems. Applications to Quantum Wires, Waveguides and Photonic Crystals. X, 222 pages. 1999.

Vol. m 61: V. Perlick, Ray Optics, Fermat's Principle, and Applications to General Relativity. X, 220 pages. 2000.

Vol. m 62: J. Berger, J. Rubinstein, Connectivity and Superconductivity. XI, 246 pages. 2000.

Vol. m 63: R. J. Szabo, Ray Optics, Equivariant Cohomology and Localization of Path Integrals. XII, 315 pages. 2000.

Vol. m 64: I. G. Avramidi, Heat Kernel and Quantum Gravity. X, 143 pages. 2000.

Vol. m 65: M. Hénon, Generating Families in the Restricted Three-Body Problem. Quantitative Study of Bifurcations. XII, 301 pages. 2001.

Vol. m 66: F. Calogero, Classical Many-Body Problems Amenable to Exact Treatments. XIX, 749 pages. 2001.